Acoustic Sensing and Monitoring in Urban and Natural Environments

Acoustic Sensing and Monitoring in Urban and Natural Environments

Editor

Hector Eduardo Roman

MDPI

Basel • Beijing • Wuhan • Barcelona • Belgrade • Novi Sad • Cluj • Manchester

Editor
Hector Eduardo Roman
University of Milano-Bicocca
Milano
Italy

Editorial Office
MDPI AG
Grosspeteranlage 5
4052 Basel, Switzerland

This is a reprint of articles from the Special Issue published online in the open access journal *Sensors* (ISSN 1424-8220) (available at: https://www.mdpi.com/journal/sensors/special_issues/M94P5H0071).

For citation purposes, cite each article independently as indicated on the article page online and as indicated below:

Lastname, A.A.; Lastname, B.B. Article Title. *Journal Name* **Year**, *Volume Number*, Page Range.

ISBN 978-3-7258-2277-5 (Hbk)
ISBN 978-3-7258-2278-2 (PDF)
doi.org/10.3390/books978-3-7258-2278-2

Cover image courtesy of Dr. Hector Eduardo Roman

Contents

About the Editor vii

Preface ix

Hector Eduardo Roman
Editorial to the Special Issue "Acoustic Sensing and Monitoring in Urban and Natural Environments"
Reprinted from: *Sensors* **2024**, *24*, 6295, doi:10.3390/s24196295 1

Gerard Bota, Robert Manzano-Rubio, Lidia Catalán, Julia Gómez-Catasús and Cristian Pérez-Granados
Hearing to the Unseen: AudioMoth and BirdNET as a Cheap and Easy Method for Monitoring Cryptic Bird Species
Reprinted from: *Sensors* **2023**, *23*, 7176, doi:10.3390/s23167176 6

Isabelle L. Barnes and John E. Quinn
Passive Acoustic Sampling Enhances Traditional Herpetofauna Sampling Techniques in Urban Environments
Reprinted from: *Sensors* **2023**, *23*, 9322, doi:10.3390/s23239322 17

Roberto Benocci, Andrea Afify, Andrea Potenza, H. Eduardo Roman and Giovanni Zambon
Toward the Definition of a Soundscape Ranking Index (SRI) in an Urban Park Using Machine Learning Techniques
Reprinted from: *Sensors* **2023**, *23*, 4797, doi:10.3390/s23104797 28

Roberto Benocci, Andrea Afify, Andrea Potenza, H. Eduardo Roman and Giovanni Zambon
Self-Consistent Soundscape Ranking Index: The Case of an Urban Park
Reprinted from: *Sensors* **2021**, *23*, 3401, doi:10.3390/s23073401 50

Timothy Van Renterghem, Valentin Le Bescond, Luc Dekoninck and Dick Botteldooren
Advanced Noise Indicator Mapping Relying on a City Microphone Network
Reprinted from: *Sensors* **2023**, *23*, 5865, doi:10.3390/s23135865 65

Daniel Bonet-Solà, Pau Bergadà, Enric Dorca, Carme Martínez-Suquía and Rosa Ma Alsina-Pagès
Sons al Balcó: A Comparative Analysis of WASN-Based L_{Aeq} Measured Values with Perceptual Questionnaires in Barcelona during the COVID-19 Lockdown
Reprinted from: *Sensors* **2024**, *24*, 1650, doi:10.3390/s24051650 84

Ayoub Boumchich, Judicaël Picaut, Pierre Aumond, Arnaud Can and Erwan Bocher
Blind Calibration of Environmental Acoustics Measurements Using Smartphones
Reprinted from: *Sensors* **2024**, *24*, 1255, doi:10.3390/s24041255 108

Domenico Rossi, Antonio Pascale, Aurora Mascolo and Claudio Guarnaccia
Coupling Different Road Traffic Noise Models with a Multilinear Regressive Model: A Measurements-Independent Technique for Urban Road Traffic Noise Prediction
Reprinted from: *Sensors* **2024**, *24*, 2275, doi:10.3390/s24072275 132

Lei Gao, Chenzhi Cai, Chao Li and Cheuk Ming Mak
Numerical Analysis of the Mitigation Performance of a Buried PT-WIB on Environmental Vibration
Reprinted from: *Sensors* **2023**, *23*, 7666, doi:10.3390/s23187666 151

Andrés Eduardo Castro-Ospina, Miguel Angel Solarte-Sanchez, Laura Stella Vega-Escobar, Claudia Isaza and Juan David Martínez-Vargas
Graph-Based Audio Classification Using Pre-Trained Models and Graph Neural Networks
Reprinted from: *Sensors* **2024**, *24*, 2106, doi:10.3390/s24072106 . . . . **165**

Sam Lapp, Nickolus Stahlman and Justin Kitzes
A Quantitative Evaluation of the Performance of the Low-Cost AudioMoth Acoustic Recording Unit
Reprinted from: *Sensors* **2023**, *23*, 5254, doi:10.3390/s23115254 . . . . **177**

Drew Priebe, Burooj Ghani and Dan Stowell
Efficient Speech Detection in Environmental Audio Using Acoustic Recognition and Knowledge Distillation
Reprinted from: *Sensors* **2024**, *24*, 2046, doi:10.3390/s24072046 . . . . **189**

Justin Ivancic, Gamani Karunasiri and Fabio Alves
Directional Resonant MEMS Acoustic Sensor and Associated Acoustic Vector Sensor
Reprinted from: *Sensors* **2023**, *23*, 8217, doi:10.3390/s23198217 . . . . **200**

Justin Ivancic and Fabio Alves
Directional Multi-Resonant Micro-Electromechanical System Acoustic Sensor for Low Frequency Detection
Reprinted from: *Sensors* **2024**, *24*, 2908, doi:10.3390/s24092908 . . . . **224**

About the Editor

Hector Eduardo Roman

Hector Eduardo Roman is currently working as scientific collaborator at the Department of Physics and Department of Earth and Environmental Sciences of the University of Milano-Bicocca. He has been awarded a visiting professorship at the Pohang University of Science and Technology (POSTECH) working at the Division of IT Convergence Engineering (Pohang, Korea). He has been a Guest Scientist at the Max Planck Institute for the Physics of Complex Systems (MPIPKS, Germany) and Research Fellow at the Department of Physics of the University of Milan. He has been nominated Research Assistant (Privatdozent) at the University of Hamburg and the University of Giessen (Germany) and has been awarded a Fedeor Lynen Fellowship from the Alexander von Humboldt Foundation (Germany). His interests cover different areas of complex systems such as sensor networks, acoustic sensing and monitoring in natural/urban environments, atomic and soft matter physics, fractals, biomolecules, and finance. He is the co-editor/co-author of 8 books and has over 200 publications in international journals.

Preface

This reprint contains a collection of feature papers, written by internationally recognized experts, dealing with the issue of soundscape characterization in urban and natural environments, with the scope of expanding our knowledge of the way anthropogenic effects can affect the quality of life in populated areas and alter wildlife conservation in natural habitats. One of the aims is the development of new local quality-of-life indices based on environmental noise recording data for the purpose of suggesting specific strategies to mitigate anthropogenic noise pollution in general domains. The reprint should be useful to both experts in the field of soundscape analysis and researchers interested in pursuing new ideas in this rapidly evolving field of work.

The Editor has benefitted from an ongoing collaboration with Prof. Roberto Benocci and Prof. Giovanni Zambon from the Department of Earth and Environmental Sciences of the University of Milano-Bicocca.

Hector Eduardo Roman
Editor

sensors

MDPI

Editorial

Editorial to the Special Issue "Acoustic Sensing and Monitoring in Urban and Natural Environments"

Hector Eduardo Roman

Department of Physics, University of Milano-Bicocca, Piazza della Scienza 3, 20126 Milano, Italy; hector.roman@unimib.it

During the last decades, the great advances achieved in sensor technology and monitoring strategies have been instrumental to accurately quantify anthropogenic noise pollution in both urban and natural environments. Indeed, when lacking of human influence, a natural habitat soundscape constitutes a benchmark which allows us to estimate any negative impact that anthropogenic activity can have in a particular surrounding. The study of natural soundscapes can therefore be very useful for the development of new ideas and methodologies aimed at improving the quality of life in highly populated urban areas which may suffer from self-produced noise pollution. The latter is one of the greatest environmental threats to people's health, which has become an issue affecting millions of people worldwide.

Historically, human society develops more efficiently in urban areas which facilitate its great variety of activities. The emergence of large populated zones, however, is not the result of a long-term planning and therefore they are not optimized to yield high living standards. As a result, people living in large conglomerates must cohabit with the often deleterious actions of different agents, among which anthropogenic noise plays a central role. Only recently these issues have been considered seriously yielding the development of new branches of study and research. The latter are essential to guide and support efficient policy-making decisions aimed at finding the right intervention measures.

The aim of this Special Issue is to gather experts actively working in different fields of acoustic phenomena, such as sensing and monitoring techniques in either urban or natural environments, to help expanding our knowledge on soundscape monitoring and analysis contributions 1–7, road traffic and soundscape modeling contributions 8–10, and the development of more efficient sensors contributions 11–14 in support of such different endeavours.

In contribution 1, 'Hearing to the Unseen', Bota et al. present a work which can significantly improve on standard passive acoustic sensing (PAS) and environmental monitoring, the latter known to be challenging when huge amount of data needs to be collected. Specifically, they develop a new technique able to recognize species-specific sounds more efficiently, denoted as BirdNET, based on a novel machine learning tool for automated recognition and acoustic data processing. They apply BirdNET for detecting two cryptic forest bird species: Coal Tit (*Peripatus ater*) and the Short-toed Treecreeper (*Certhia brachydactyla*), illustrating the achievement of highly accurate recognition rates, typical of BirdNET performance. In addition, the software has been made freely available, encouraging researchers and managers to utilize it.

Understanding the impact of urbanization on the surrounding biodiversity and wildlife conservation is a main ecological issue. Often, the actual impact of a rapid urbanization is difficult to assess using traditional PAS methods, as discussed by Barnes and Quinn in contribution 2. They present a multimethod analysis of biodiversity in the rapidly urbanizing county Greenville (South Carolina, USA), based on several audio recodings in 25 locations along a predertermined trail, supplemented with visual assessments, an online database and the use of refugia tubes. The local species identification along the trail was employed to identify relationships between herpetofauna and acoustic indices (as

Citation: Roman, H.E. Editorial to the Special Issue "Acoustic Sensing and Monitoring in Urban and Natural Environments". *Sensors* **2024**, *24*, 6295. https://doi.org/10.3390/s24196295

Received: 12 May 2024
Accepted: 26 September 2024
Published: 29 September 2024

proxies for biodiversity), suggesting that the use of different sampling methods is crucial for achieving a more comprehensive and realistic evaluation of wildlife occupancy. This study should be of help to establish better conservation policies of biodiversity and appropriate urbanization guidelines in natural forested ecosystems.

Soundscape indices have been introduced with the aim of evaluating different contributions of the environmental sound components, providing accurate assessments of the "acoustic quality" within a complex habitat. In contribution 3, Benocci et al. apply machine learning (ML) algorithms (decision tree, random forest, adaptive boosting, and support vector machine) to optimize the (four) parameters determining the behavior of the soundscape ranking index (SRI). The latter was previously introduced by the authors to study sound recordings taken at 16 sites, distributed over a regular grid covering an area of about 22 hectares, within an urban park in the city of Milan surrounded by a variety of anthropophonic sources. The recordings span 3.5 h each over a period of four consecutive days. The authors find that two ML algorithms (decision tree and adaptive boosting) yield a set of parameters displaying a rather good classification performance (F1-scores: 0.70 and 0.71, respectively). The method is expected to prove useful when considering large amount of sound data, allowing for an efficient classification of the associated indices.

The above results are in quantitative agreement with a self-consistent estimation of the mean SRI values discussed in contribution 4. In the latter, a careful aural survey from a single day was performed in order to identify the presence of 19 predefined sound categories, within one minute of recording, for a total of 70 one-minute intervals. The resulting one minute histograms of the SRI values were used to define a dissimilarity function for each pair of sites. Dissimilarity increases significantly with the inter-site distance in real space, and optimal values of the 4 parameters were obtained by minimizing the standard deviation of the data, requiring consistency with a fifth parameter describing the power-law variation of dissimilarity with distance. This study can be useful to assess the quality of a soundscape in more general situations.

Predicting traffic noise over city-wide scales is challenging due to the relatively large amount of input information required to reach a sufficient accuracy. In contribution 5, the authors present a methodology by just relying on street categorization and a city microphone network. A simplified dynamic traffic model is employed to predict statistical and dynamical noise indicators, and to estimate the number of noise events. A standard sound propagation module is then employed to determine the noise levels. Finally, an ML technique elaborates the deterministic predictions of the different traffic parameter scenarios, choosing the one yielding the best accord with the indicators measured by the microphone network. The method is illustrated using data from the city of Barcelona, yielding results within a 2–3 dB precision, and number of events within a 30% accuracy. The current methodology allows considering a wide variety of noise indicators thus improving environmental noise assessment over city-wide scales.

A significant change in urban soundscapes occurred during the COVID-19 pandemic as a result of the imposed mobility and activity restrictions in the affected zones, yielding a dramatic reduction in noise pollution levels. The latter was measured in Barcelona, both before and during the lockdown, by means of a deployed acoustic sensor network consisting of 70 sound recording units, as discussed in contribution 6. Different noise indices were compared, together with a perceptual test conducted within the project Sons al Balcó during the lockdown. The analysis was based on a clustering procedure, separating objective from subjective data according to the predominant type of noise sources present in each sensor area. The areas were then classified as heavy, moderate and low-traffic areas. A reduction in noise indices values was found to be significantly correlated with an improved acoustic satisfaction and type of noise sources. These results suggest that objective calibrated data can be useful to estimate the subjective perception of urban soundscapes in cases lacking of input information.

To efficiently deal with major health and social issues in complex noisy environments, dynamical noise maps are generally needed to keep track of noise behavior in real time.

An alternative approach to existing real time recording methods has been proposed in contribution 7 based on the direct use of smartphones carried by mobile individuals as measurement tools of local noise levels. To successfully implement this methodology, an accurate calibration of the smartphones is required. The authors discuss the so called "blind calibration" procedure, in which measured noise levels from several smartphones, located within a limited area, are sent to a data center for a consistent evaluation of their values, allowing for their calibration. The article proposes to set up a blind calibration method by using data from the NoiseCapture (NC) smartphone application, and tests using NC datasets are discussed in support of the suggested model.

The accurate assessment of traffic noise levels in a given urban area is a primary requisite to complay with present road traffic noise (RTN) regulations. In contribution 8, Rossi et al. propose a new methodology based on an easy application of RTN models, without the need of relying on measured data for calibration in the first place. Equivalent continuous sound pressure levels were obtained using different emission models coupled to a sound propagation algorithm, by randomly generating traffic flows, speeds, and source–receiver distances. Finally, a multilinear regressive technique was developed to deal with the data in a managable way for applications. The procedure was validated using a set of long-term traffic and noise data, recorded using several sensors (sound level meters), car counters, and speed detectors, in Saint-Berthevin (France). The estimations provided by the multilinear regressions are in very good agreement with the field measurements, displaying mean absolute errors in the range (1.60–2.64) dB(A), suggesting that RTN models can be successfully integrated into a network of traffic sensors to forecast noise levels accurately.

Railway transit plays a dominant role in exerting vibration pollution in urban areas. A common technique employed to dissipate the associated waves consists in the infilling of periodic narrow excavation dugs in the soil located along the expected wave propagation paths. However, the presently used infilled ditches do not provide sufficient attenuation at low and medium frequencies. In contribution 9, Gao et al. suggest a novel solution to cope with the most disturbing lower frequencies. The new method is based on the use of acoustic metamaterials consisting of a periodic infilled trench combined with a wave-impeding block, typically embedded at a certain depth in the ground. The authors develop a 3D finite elements model to quantify the expected isolation performance of the suggested wave barriers. The results are very promising, suggesting that the combination of periodic infilled trench structures with a wave impedance block barrier can effectively attenuate the most disturbing low and medium frequencies, helping to mitigate environmental vibrations pollution.

Environmental sound studies gain considerable insight if an accurate classification of the different type of sound sources can be accomplished. In contribution 10, Castro-Ospina et al. consider the interesting idea of representing audio data using the language of graph theory, and apply it to the problem of sound classification. The new methodology is based on the use of pre-trained audio models to extract hidden features from audio files, information which is then represented as nodes of connected graphs. To solve multi-class audio classification problems, the graphs are trained using three graph neural networks (GNNs): graph convolutional networks, graph attention networks (GATs) and GraphSAGE. The GAT model emerges as the best performer yielding an accurate classification of environmental sounds, and an excellent identification of the type of land cover such as forest, savanna, or pasture. The study embodies a promising scenario to the use of learning techniques in the context of GNNs methodology for analyzing audio data related to environmental issues.

Autonomous recording units are widely used to record vocalizing species localized either in an indoor or outdoor context, of which AudioMoth is one of the most popular ones. Despite its extensive use, few quantitative tests of AudioMoth performance have been reported. Clearly, further information on the device is needed for the design of efficient field surveys, and for processing the recording data more accurately. In contribution 11, two types of tests on AudioMoth recorder performance characteristics are reported. In the first one, settings regarding different orientations and mounting conditions, using several devices, yielded little variations in acoustic performances. In particular, protecting plastic bags used in outdoor setups have little effect on sensitivity for frequencies below 10 kHz It is found that AudioMoth has an almost constant on-axis response, displaying some attenuation effects from sources located behind it. The latter can be of relevance when the device is positioned on a tree. In the second test, a variety of recording frequencies, gain settings, environmental temperatures, and battery types were considered. In particular, the lifespan of different types of batteries were measured, showing that lithium batteries work very well lasting for about 190 h at room temperature and sampling rate of 32 kHz, while at freezing temperatures they keep collecting data for a period twice as long as compared to their standard alkaline counterparts. This information should be of help to researchers performing lasting measurements under very different environmental conditions.

The work in contribution 12 deals with learning transfers from a 'teacher model' to a smaller 'student model'. Such distillation techniques are used to design efficient, lightweight student models for speech detection in bioacoustics. Taken the EcoVAD voice detection architecture as the teacher model, a comparative analysis is perfomed on the MobileNetV3-Small-Pi model aimed at working as a compact student architecture. Various configurations of the student models were analysed to identify the optimal performance, and different distillation techniques were studied to find the most effective method of model selection. The obtained distilled models exhibit comparable performances to the EcoVAD one, suggesting a possible approach to overcome present computational barriers for real-time ecological monitoring using compact devices.

In contributions 13 and 14, Ivancic et al. report recent developments of a small, lightweight, portable sensor well adapted to resolve quiet or distant acoustic sources and their location, in addition to underwater operations. The new acoustic device is based on a micro-electromechanical system (MEMS), and a novel design of a related acoustic vector sensor array (AVS) is discussed in contribution 13. In contribution 14, extensions of the MEMS acoustic sensor are presented to broaden the operating bandwidth by keeping a high signal-to-noise ratio, in particular at low frequencies. The new approach represents a significant improvement in sensor performance compared to standard MEMS sensor ones, in which the multi-resonant design plays a fundamental role to overcome the limitations of the standard devices which increase sensitivity at the expense of bandwidth.

Conflicts of Interest: The author declares no conflict of interest.

List of Contributions

1. Bota, G.; Manzano-Rubio, R.; Catalán, L.; Gómez-Catasús, J.; Pérez-Granados, C. Hearing to the unseen: AudioMoth and BirdNET as a cheap and easy method for monitoring cryptic bird species. *Sensors* **2023**, *23*, 7176. https://doi.org/10.3390/s23167176.
2. Barnes, I.L.; Quinn, J.E. Passive Acoustic Sampling Enhances Traditional Herpetofauna Sampling Techniques in Urban Environments. *Sensors* **2023**, *23*, 9322. https://doi.org/10.3390/s23239322.
3. Benocci, R.; Afify, A.; Potenza, A.; Roman, H.E.; Zambon, G. Toward the Definition of a Soundscape Ranking Index (SRI) in an Urban Park Using Machine Learning Techniques. *Sensors* **2023**, *23*, 4797. https://doi.org/10.3390/s23104797.
4. Benocci, R.; Afify, A.; Potenza, A.; Roman, H.E.; Zambon, G. Self-consistent soundscape ranking index: The case of an urban park. *Sensors* **2023**, *23*, 3401. https://doi.org/10.3390/s23073401.
5. Van Renterghem, T.; Le Bescond, V.; Dekoninck, L.; Botteldooren, D. Advanced Noise Indicator Mapping Relying on a City Microphone Network. *Sensors* **2023**, *23*, 5865. https://doi.org/10.3390/s23135865.
6. Bonet-Solà, D.; Bergadà, P.; Dorca, E.; Martínez-Suquía, C.; Alsina-Pagès, R.M. Sons al Balcó: A Comparative Analysis of WASN-Based LAeq Measured Values with Perceptual Questionnaires in Barcelona during the COVID-19 Lockdown. *Sensors* **2024**, *24*, 1650. https://doi.org/10.3390/s24051650.
7. Boumchich, A.; Picaut, J.; Aumond, P.; Can, A.; Bocher, E. Blind Calibration of Environmental Acoustics Measurements Using Smartphones. *Sensors* **2024**, *24*, 1255. https://doi.org/10.3390/s24041255.

8. Rossi, D.; Pascale, A.; Mascolo, A.; Guarnaccia, C. Coupling different road traffic noise models with a multilinear regressive model: A measurements-independent technique for urban road traffic noise prediction. *Sensors* **2024**, *24*, 2275. https://doi.org/10.3390/s24072275.
9. Gao, L.; Cai, C.; Li, C.; Mak, C.M. Numerical Analysis of the Mitigation Performance of a Buried PT-WIB on Environmental Vibration. *Sensors* **2023**, *23*, 7666. https://doi.org/10.3390/s23187666.
10. Castro-Ospina, A.E.; Solarte-Sanchez, M.A.; Vega-Escobar, L.S.; Isaza, C.; Martínez-Vargas, J.D. Graph-Based Audio Classification Using Pre-Trained Models and Graph Neural Networks. *Sensors* **2024**, *24*, 2106. https://doi.org/10.3390/s24072106.
11. Lapp, S.; Stahlman, N.; Kitzes, J. A quantitative evaluation of the performance of the low-cost AudioMoth acoustic recording unit. *Sensors* **2023**, *23*, 5254. https://doi.org/10.3390/s23115254.
12. Priebe, D.; Ghani, B.; Stowell, D. Efficient speech detection in environmental audio using acoustic recognition and knowledge distillation. *Sensors* **2024**, *24*, 2046. https://doi.org/10.3390/s24072046.
13. Ivancic, J.; Karunasiri, G.; Alves, F. Directional resonant MEMS acoustic sensor and associated acoustic vector sensor. *Sensors* **2023**, *23*, 8217. https://doi.org/10.3390/s23198217.
14. Ivancic, J.; Alves, F. Directional Multi-Resonant MEMS Acoustic Sensor for Low Frequency Detection. *Sensors* **2024**, *24*, 2908. https://doi.org/10.3390/s24092908.

sensors

MDPI

Article

Hearing to the Unseen: AudioMoth and BirdNET as a Cheap and Easy Method for Monitoring Cryptic Bird Species

Gerard Bota [1], Robert Manzano-Rubio [1], Lidia Catalán [2], Julia Gómez-Catasús [3,4] and Cristian Pérez-Granados [1,5,*]

1 Conservation Biology Group, Landscape Dynamics and Biodiversity Programme, Forest Science and Technology Center of Catalonia (CTFC), 25280 Solsona, Spain; gerard.bota@ctfc.cat (G.B.); robert.manzano@ctfc.cat (R.M.-R.)
2 Independent Researcher, 44002 Teruel, Spain; lidiacrispi@gmail.com
3 Terrestrial Ecology Group (TEG-UAM), Department of Ecology, Autonomous University of Madrid, 28049 Madrid, Spain; julia.gomez@uam.es
4 Research Centre in Biodiversity and Global Change (CIBC-UAM), Autonomous University of Madrid, 28049 Madrid, Spain
5 Ecology Department, Alicante University, 03080 Alicante, Spain
* Correspondence: cristian.perez@ua.es

Abstract: The efficient analyses of sound recordings obtained through passive acoustic monitoring (PAM) might be challenging owing to the vast amount of data collected using such technique. The development of species-specific acoustic recognizers (e.g., through deep learning) may alleviate the time required for sound recordings but are often difficult to create. Here, we evaluate the effectiveness of BirdNET, a new machine learning tool freely available for automated recognition and acoustic data processing, for correctly identifying and detecting two cryptic forest bird species. BirdNET precision was high for both the Coal Tit (*Peripatus ater*) and the Short-toed Treecreeper (*Certhia brachydactyla*), with mean values of 92.6% and 87.8%, respectively. Using the default values, BirdNET successfully detected the Coal Tit and the Short-toed Treecreeper in 90.5% and 98.4% of the annotated recordings, respectively. We also tested the impact of variable confidence scores on BirdNET performance and estimated the optimal confidence score for each species. Vocal activity patterns of both species, obtained using PAM and BirdNET, reached their peak during the first two hours after sunrise. We hope that our study may encourage researchers and managers to utilize this user-friendly and ready-to-use software, thus contributing to advancements in acoustic sensing and environmental monitoring.

Keywords: acoustic sensor; audio recognition; automated recognition software; autonomous recording unit; machine learning; Paridae; *Periparus ater*; passive acoustic monitoring; wildlife monitoring

Citation: Bota, G.; Manzano-Rubio, R.; Catalán, L.; Gómez-Catasús, J.; Pérez-Granados, C. Hearing to the Unseen: AudioMoth and BirdNET as a Cheap and Easy Method for Monitoring Cryptic Bird Species. *Sensors* **2023**, *23*, 7176. https://doi.org/10.3390/s23167176

Academic Editor: Hector Eduardo Roman

Received: 24 July 2023
Revised: 11 August 2023
Accepted: 13 August 2023
Published: 15 August 2023

1. Introduction

Nowadays, there exists an increasing demand for automated, efficient, and scalable ecological monitoring methodologies that possess the capability to address the ongoing decline in biodiversity [1]. Conventional field surveys, which rely on human presence in the field, may suffer from limitations and biases stemming from human expertise, while also being time-consuming and costly [1,2]. Fortunately, advancements in sensing technologies and computational capabilities now enable the execution of automated ecological surveys on a large scale, both spatially and temporally. Innovative biomonitoring techniques (see [3]) decrease the need for human presence in the field while reducing potential biases associated with human-based surveys. Although new technologies provide important improvements over traditional monitoring methods, their application is not exempt from considerations and limitations. For example, the technology considered should be selected taking into account the goals of the biodiversity monitoring scheme, the indicators to be measured, the accuracy for target taxa, as well as the available capacity of and budget for equipment [4,5].

Many animal species, ranging from small insects to whales, emit acoustic signals, so acoustic communication is widespread in the animal world, and very often species communicate using a sequence of distinct acoustic elements [6]. One of the emerging noninvasive and automated techniques for ecological monitoring is passive acoustic monitoring (PAM). This method involves the utilization of Autonomous Recording Units (ARUs) equipped with acoustic sensors (referred to as microphones hereafter) that are deployed in the field to obtain recorded acoustic data using a specified recording schedule. ARUs are sound recorders that can be programmed with specific time schedules to be unattended while operating in the field, with great battery autonomy and storage capacity and built to operate in outdoor conditions [7]. The subsequent analysis of these recorded sounds enables the detection and monitoring of individuals or ecosystems without disrupting their natural behavior [2].

PAM is a trending technique whose use to monitor species and ecosystems is increasingly gaining more and more attention (see reviews in [2,8]). This rise in popularity can be partially attributed to several factors. Firstly, the availability of affordable and efficient ARUs, such as AudioMoth [9,10], has facilitated the widespread adoption of PAM. Secondly, recent innovations in acoustic data processing techniques [11,12] have enhanced the analysis and interpretation of acoustic data obtained through PAM. Lastly, the development of low-cost yet high-quality microphones (e.g., [13]) has further contributed to the advancement of PAM. Passive acoustic surveys generate a significant amount of data, posing challenges for visual or acoustical verification of the recordings (but, see [14]). To address this issue, a wide range of audio signal recognition tools have been developed over the past decade to assist with audio processing and enable fast and accurate interpretation of the extensive acoustic data obtained from passive acoustic surveys. These tools encompass a spectrum of approaches, ranging from basic detectors that employ template matching methods (e.g., [15]) to advanced techniques, like deep learning and convolutional neural networks, which represent the current state-of-the-art in the field [12].

Birds are the most commonly monitored group of animals using PAM [2] and, consequently, the majority of advances in audio signal recognition have been focused on birds (see [11,16,17]). State-of-the-art techniques have demonstrated their ability to develop highly accurate bird recognition models. However, these sophisticated methods may pose challenges for implementation by managers, scientists, and the general public due to the significant level of informatics experience required [17]. Fortunately, a recently updated machine learning tool called BirdNET provides a free and user-friendly solution for automated audio recognition [11]. This ready-to-use tool has been widely adopted by the general public (over 1.1 million participants used the BirdNET APP during 2020, [18]) and by scientists (see review of applications in [19]), enabling an easy access and use of machine learning for automated wildlife recognition. BirdNET employs a deep neural network for automated detection and classification of wildlife vocalizations [11]. The tool divides the original sound recordings into 3-s segments and provides identification for over 6000 species of wildlife for each segment [11,18]. BirdNET can identify multiple species within the same segment, and each detection is accompanied by a quantitative confidence score, automatically provided by BirdNET, ranging from 0 to 1. This score reflects the probability of accurately identifying the species, with a score of 1 indicating a perfect match. The confidence score can be adjusted by the user as a threshold value, enabling the filtering of BirdNET output at a desired confidence level. Selecting a higher confidence score increases the percentage of correctly classified detections but may result in a lower number of detections overall. However, our current knowledge on how confidence score impacts BirdNET species detection accuracy is very limited (reviewed in [19]). Additionally, BirdNET allows the users to adjust the overlap of prediction segments, modify the sensitivity parameter, and to apply filters to classify sounds based on recording location, time period, or target species [11,19].

BirdNET, as a user-friendly tool, can be easily accessed through various friendly interfaces. It can be used in a smartphone (BirdNET App, [18]), enabling users to directly record bird sounds in the field. Alternatively, a web-based platform called BirdNET-API allows users to upload their recordings for analysis [20]. BirdNET functionality is also integrated into Raven Pro, an audio software developed by the Cornell Lab of Ornithology, and can be run on Windows or Python through the BirdNET-Analyzer, which is openly accessible on GitHub (https://github.com/kahst/BirdNET-Analyzer, accessed on 11 August 2023). While BirdNET was initially designed for bird species recognition, the most recent updates have expanded its capabilities to include a limited number of other species, such as frogs and primates [21,22]. However, the extent of BirdNET's ability in correctly identifying bird species vocalizations from sound recordings collected using omnidirectional microphones, the ones typically used in PAM, remains largely confined to a few case studies (as reviewed by [19]).

In this study, our objective is to conduct a comprehensive evaluation of the usefulness of using low-cost recorders (AudioMoth) and BirdNET for monitoring two cryptic forest bird species. For each species, we have set out the following aims: (1) Assessing the precision of BirdNET in correctly identifying bird vocalizations; (2) Determining the optimal confidence score threshold of each species, which might be useful to establish a reliable criterion for accepting BirdNET detections with a high level of confidence; (3) Estimating the percentage of presences automatically identified by BirdNET compared to human visual inspection of sonograms with the default values and using the optimal confidence score threshold; (4) Apply the method on a large field acoustic dataset aiming to describe the diel vocal behavior of the studied species, which we recorded during the daily period over the course of one month in two distinct habitat types. While our assessment is restricted to two bird species, we hope that our approach may be a valuable guidance to improve the overall quality of passive acoustic surveys using widely spread low-cost ARUs and free user-friendly automated audio processing software.

2. Materials and Methods

2.1. Study Species

The Coal Tit (*Periparus ater*) and the Short-toed Treecreeper (*Certhia brachydactyla*) were selected as the forest study bird species owing to their cryptic behavior and challenges for monitoring using visual cues. The Coal Tit is a small passerine that is resident in Europe, North Africa, and parts of Asia [23]. The distribution of the Short-toed Treecreeper is limited to Europe and North Africa [24]. The challenges associated with monitoring the Coal Tit are primarily related to its canopy habits, as they often inhabit the upper and outer parts of the forest canopy [25], making them potentially difficult to observe depending on forest structure. The Short-toed Treecreeper is a small passerine specialized in feeding on insects found in bark crevices. While its preference for feeding in trunks may make them more detectable, the genus *Certhia*, to which the Short-toed Treecreeper belongs, is characterized for its effective camouflage against tree trunks, making it challenging for humans to visually detect them in trunks [26]. Fortunately, both species exhibit high vocal activity. Male Coal Tits produce songs primarily during the breeding season, characterized by a common pattern of two or three ascending or descending elements that are repeated several times [23,27]. The song of the Short-toed Treecreeper is relatively simple but distinctive. It consists of a low-pitched vocalization that is repeated several times [28]. Given the high vocal activity of both species and the challenges associated with their visual detection, passive acoustic monitoring might be a reliable tool for monitoring their presence and providing valuable insights into their behavior.

2.2. Study Area

The study was carried out in two forest areas located in the Comunidad Foral de Navarra (Northern Spain). Both areas were located at about 600–800 m a.s.l. and separated around 30 km. The first forest was predominantly dominated by Sessile Oak (*Quercus petraea*), which was located in the municipality of Etxarri (42°58′10.96″ N, 1°53′6.33″ W). This forest is recognized as one of the well-preserved oak forests in Navarra. The second forested area, situated in the municipality of Aitzarotz (43°1′27.83″ N, 1°45′55.72″ W), was dominated by European Beech (*Fagus sylvatica*). The Navarra region experiences a humid and temperate climate, with average annual temperatures ranging between 8.5 and 14.5 °C and a typical annual rainfall regime between 1100 and 2500 mm.

2.3. *Acoustic Monitoring Protocol*

We deployed one AudioMoth recorder [9] in each monitored forest, which operated from 24 February to 23 March 2022. This period corresponds to the pre-breeding period for both species in temperate forests (first laying date around mid April, [29,30]). The recorders were securely housed in AudioMoth IPX7 cases (Open Acoustic Devices) and mounted on trees at a height of approximately 1.5 m above the ground. The recorders were configured to record audio at a sampling rate of 48 kHz, gain Med–High, and 16 bits per sample. The recording schedule spanned from 8 a.m. to 7 p.m., with a single 15-min recording captured at the beginning of each hour. This schedule was designed to encompass most of the daily hours, considering local sunrise occurring at 7:30 a.m. and sunset at 6:30 p.m. (sunrise and sunset estimates for 10 March in the study area). Following this protocol, we obtained a total of 11 15-min recordings per day and study location. This gave a total of 336 15-min recordings per site (87 h of recording per site) during the whole study period.

2.4. *Recording Analyses*

Acoustic recordings were analyzed using BirdNET (version 2.2.0, [11]). BirdNET was run using the default parameter values, including a sensitivity parameter of 1.0, a confidence score threshold of 0.1, and no overlap of prediction segments (0). We configured BirdNET to report detections exclusively for the Coal Tit and the Short-toed Treecreeper, therefore, avoiding the detection of nontarget species [31].

2.5. *Recognizer Performance*

We estimated the precision of BirdNET, a commonly used acoustic metric for assessing recognizer performance [32]. BirdNET precision was assessed without applying any filtering on the confidence score threshold. Precision was estimated as the proportion of BirdNET detections correctly classified by the total verified BirdNET detections [32]. To estimate precision, we randomly selected 309 BirdNET detections from the Coal Tit BirdNET output (28.5% of total detections) and 311 detections within the BirdNET output reported for the Short-toed Treecreeper (10.2% of total detections). We included a larger number of detections with lower confidence scores because there is a higher probability of mislabeling detections with lower confidence scores but also because there were a larger number of detections with low confidence scores in BirdNET's output. For each selected detection, the observer listened to and inspected the spectrogram in Raven Pro 1.6 [33] at the timestamp of the 3-s segment and reported whether the target species was present or absent. This process allowed us to determine the proportion of BirdNET detections that were correctly classified among the total verified BirdNET detections [32].

We also used the validation dataset described in the paragraph above to identify the confidence score threshold with a 95% probability of correct identification for each species. Following the approach outlined in [22], we back-transformed BirdNET's confidence scores into its original logit scale using the following equation:

$$\text{Logit score} = \ln(1/(1 - \text{confidence score}))$$

Next, for each species, we fitted a logistic regression to establish a relationship between the correct or incorrect classification of the validated detections as a response variable and the BirdNET logit-scale prediction score as an independent variable. The logistic regressions provide an equation that enables us to convert BirdNET scores into the probability of a given prediction being correct. For each species, the equations considering a probability of correct identification of 95% were as follows:

$$\text{logit(P)} = \ln(\text{intercept}) + 0.95 \times \ln(\text{logit-score})$$

The identified optimal score was used as a confidence score threshold to consider only BirdNET detections with a high probability of correct identification when describing the diel pattern of vocal activity. The diel pattern of vocal activity was described showing the percentage of BirdNET detections made per recording hour and location for each of the monitored species.

Finally, we also assessed the detection and identification performance of BirdNET for each bird species, by establishing a validation dataset consisting of referenced recordings. This involved a manual review of 100 recordings, with 50 recordings from each location among those recorded between 8 a.m. and 9 a.m. For each recording, a human observer assessed whether the Coal Tit and/or the Short-toed Treecreeper were detected by visually inspecting spectrograms in Raven Pro 1.6 [33]. Recordings were reviewed blindly without knowledge of the site location, date, or time of recording. To assess the effectiveness of BirdNET in detecting the Coal Tit and the Short-toed Treecreeper, we determined the percentage of presences detected by BirdNET compared to the total number of recordings with known presence in the validation dataset. For this evaluation, we examined every audio recording that BirdNET annotated as containing one or both species. An expert observer verified, by listening to or inspecting the spectrogram, whether the species was truly present or absent at the timestamp of the 3-s segment annotated by BirdNET. If the species was absent at the selected timestamp, additional detections were verified until the presence of the species was confirmed or until the last detection was checked. Those recordings with no BirdNET annotations were annotated as absences according to BirdNET output. We estimated the percentage of occurrences detected by BirdNET in comparison to a human verification in two scenarios: (1) Without applying any filtering to BirdNET's output (i.e., default values, all detections with confidence scores above 0.1 were included); (2) By filtering BirdNET's output by the optimal confidence score threshold for each species (see Section 3).

3. Results

3.1. Recognizer Performance

BirdNET precision was high for both the Coal Tit and the Short-toed Treecreeper, with mean values of 92.6% and 87.8%, respectively (Table 1). As expected, the confidence score threshold had a significant impact on the accuracy of bird vocalization to be correctly identified (Table 1, Figure 1). For the Coal Tit, the highest confidence score of a mislabeled detection was 0.267, indicating that all detections with a confidence score higher than this value were correctly classified. Similarly, for the Short-toed Treecreeper, the mislabeled detection with the highest confidence score had a value of 0.471. A summary table showing the overall BirdNET precision for both species across three confidence score categories can be seen in Table 1.

According to the logistic regressions the equations considering a 95% probability of correct identification for each species were:

$$\text{logit(P Coal Tit)} = \ln(0.154) + 0.95 \times \ln(9.300),\ p < 0.001,$$

$$\text{logit(P Short-toed Treecreeper)} = \ln(0.290) + 0.95 \times \ln(5.237),\ p < 0.001.$$

Therefore, the minimum confidence score to consider only detections with a 95% probability of correct identification was 0.247 for the Coal Tit and 0.335 for the Short-toed Treecreeper (Figure 1).

Table 1. Number of BirdNET detections (annotated by the software) and verified detections (correctly classified after human verification) for the Coal Tit and the Short-toed Treecreeper across three confidence score categories. The overall precision of BirdNET (in percentage, %) for each category and species is shown between brackets.

Confidence Score Category	Coal Tit		Short-Toed Treecreeper	
	Detections	Verified	Detections	Verified
(0.1–0.29)	157	134 (85.3%)	153	119 (77.8%)
(0.3–0.49)	77	77 (100%)	77	73 (94.8%)
>0.5	75	75 (100%)	81	81 (100%)
TOTAL	309	286 (92.6%)	311	273 (87.8%)

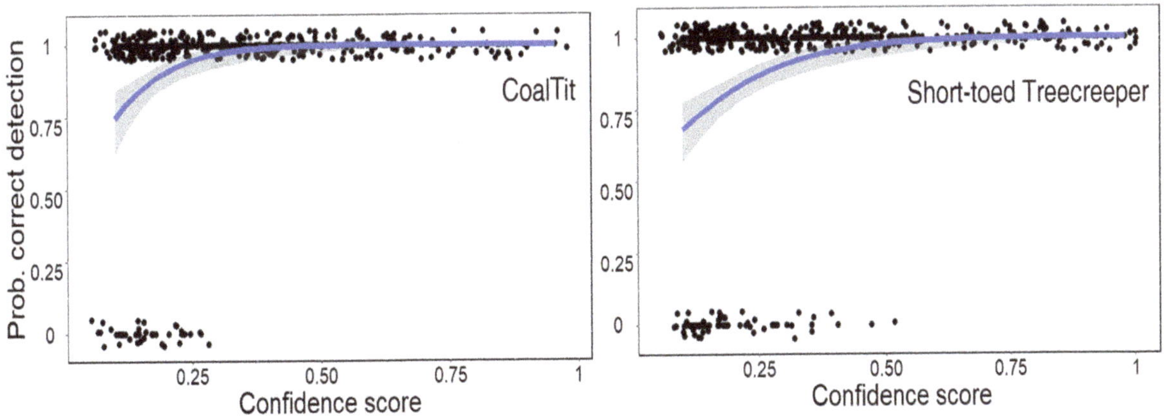

Figure 1. Results of the logistic regression showing the relationship between the probability of a correct BirdNET prediction and the confidence score of a given prediction for the (**left**) Coal Tit and the (**right**) Short-toed Treecreeper. Statistical analyses were performed using the BirdNET logit-scale of the prediction score (see Section 2) as an independent variable, but we represent the original confidence score of BirdNET for graphical purposes.

According to the validation dataset (100 15-min recordings), a human detected the presence of the Coal Tit and the Short-toed Treecreeper in 42 and 64 recordings, respectively (Table 2). When using BirdNET with the default confidence score the Coal Tit was detected in 38 recordings (90.5% of the recordings with known presence) while it was detected in 23 recordings (54.8%) using the optimal confidence score (Table 2). For the Short-toed Treecreeper, BirdNET correctly detected the species in 63 recordings (98.4% of the recordings with known presence) using the default confidence score and in 52 recordings (81.3%) using the optimal confidence score (Table 2). In both cases, there was a higher agreement between BirdNET and the human observer in terms of correctly predicting the presence or absence of each species when using the default confidence scores (95 and 92 recordings for the Coal Tit and Short-toed Treecreeper, respectively) compared to the optimal confidence score (81 and 86 recordings for the Coal Tit and Short-toed Treecreeper, respectively; see Table 2). However, the number of mislabeled recordings (i.e., where BirdNET annotated the presence of the species, but it was not confirmed) decreased from 1 to 0 when using the optimal confidence score instead of the default values for the Coal Tit and from 7 to 2 in the case of the Short-toed Treecreeper (Table 2).

Table 2. Confusion matrix of the ability of BirdNET to correctly detect the presence of the Coal Tit and the Short-toed Treecreeper in sound recordings. The validation dataset was composed of 100 15-min recordings manually reviewed for each species, in which the Coal Tit and the Short-toed Treecreeper were known to be present in 42 and 64 recordings, respectively.

	Coal Tit				Short-Toed Treecreeper			
	Default Values (>0.1)		Optimal Score (>0.247)		Default Values (>0.1)		Optimal Score (>0.335)	
	Detected	Not-Detected	Detected	Not-Detected	Detected	Not-Detected	Detected	Not-Detected
Presence	38	4	23	19	63	1	52	12
Absence	1	57	0	58	7	29	2	34

3.2. Diel Vocal Activity Pattern

The vocal activity of both species is described using those detections with a 95% probability of correct identification (thresholds of confidence score of 0.247 and 0.335 for the Coal Tit and the Short-toed Treecreeper, respectively). The diel pattern of vocal activity differed between species. The vocal activity pattern of the Coal Tit occurred primarily during the first two hours of the day, with 96.1% of all BirdNET detections (566 out of 589) occurring at 8 a.m. and 9 a.m. (Figure 2). The peak vocal activity of the Short-toed Treecreeper, in both sites, also occurred at 9 a.m., accounting for 23.0% of the total predictions (Figure 2). However, unlike the Coal Tit, the Short-toed Treecreeper sustained a high level of vocal activity throughout the morning, gradually decreasing until reaching minimal levels by the end of the day (Figure 2).

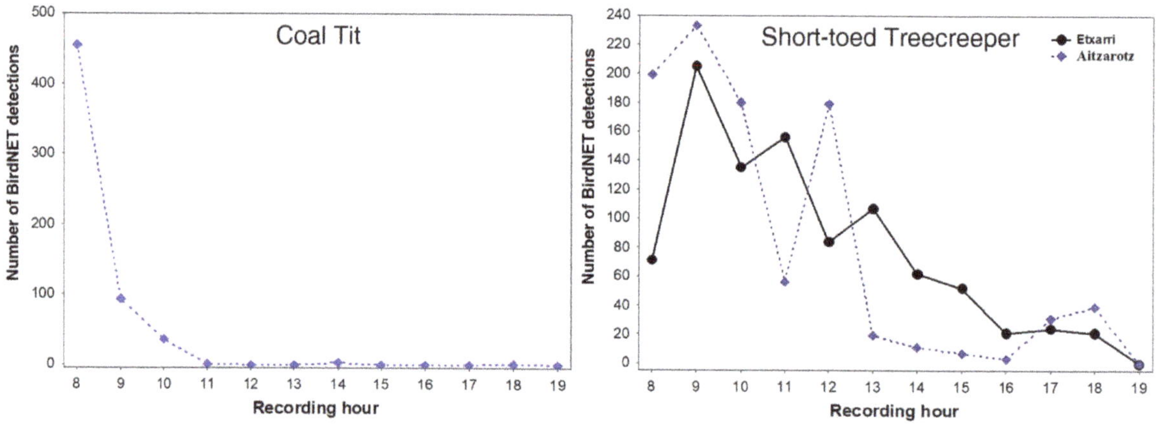

Figure 2. Diel pattern of vocal activity of the (**left**) Coal Tit and the (**right**) Short-toed Treecreeper in Northern Spain. Vocal activity was monitored using passive acoustic monitoring between 24 February and 23 March 2022 in two forested areas from 8:00 to 19:15. The first 15 min of each hour were recorded. The diel pattern is expressed as the total number of BirdNET detections with a 95% probability of correct identification per recording hour. The data shown is based on 568 BirdNET detections of the Coal Tit in Aitzarotz (data from Etxarri not shown in the graph since there were only 21 detections) and 1905 BirdNET detections (948 in Etxarri and 957 in Aitzarotz) of the Short-toed Treecreeper.

4. Discussion

In this article, we have demonstrated the effectiveness of using low-cost open-source acoustic sensors in combination with BirdNET, a readily available machine learning tool, for efficient monitoring of two cryptic bird species in forest environments: the Coal Tit and the Short-toed Treecreeper. While the use of acoustic sensors mounted in ARUs is well-established for monitoring wildlife and ecosystems (see, e.g., [1,2,34]), the analyses of acoustic data has posed challenges in terms of automation and scalability. However, recent advancements have aimed to address these challenges, with BirdNET being a notable contribution in this field [11,19]. BirdNET is a convolutional neural-network-based tool designed for processing acoustic data [11]. Although BirdNET has rarely been used in scientific studies, the existing evaluations have consistently reported a high accuracy in identifying bird species (reviewed by [19]) but also anurans and primates [21,22].

We have demonstrated that BirdNET achieved a high level of precision in correctly identifying the Coal Tit and the Short-toed Treecreeper, with a mean precision of 93% and 88%, respectively, when using the default confidence score threshold. We also observed that the precision of BirdNET was highly varied with the selection of the confidence score threshold, with no mislabeled identifications when using a confidence score threshold of 0.247 for the Coal Tit and 0.335 for the Short-toed Treecreeper. Our findings are in agreement with [35], who also reported improved precision in BirdNET when using a higher confidence score threshold for three North American bird species. Although the impact of the confidence score on BirdNET output may vary among species, the general pattern is consistent, with larger precision values obtained when using a high confidence score, but it lowers the proportion of predictions made and, therefore, the proportion of calls and presences detected [19]. However, our current knowledge of the specific impact of different confidence scores on BirdNET's ability to accurately detect species' presence in sound recordings is very limited (but, see [19,35] and next paragraph).

In our study, we determined the optimal confidence score for each monitored species, which was defined as the minimum confidence score required to consider only detections with a 95% probability of correct identification. The defined values might be used for future studies aiming to monitor the Coal Tit or the Short-toed Treecreeper. We used these optimal values as thresholds to assess the impact of confidence score BirdNET's ability to detect the presence of both species in sound recordings and to describe their diel pattern of vocal activity in a large acoustic dataset. As expected, when using the optimal confidence score, the number of detected presences decreased compared to using the default values. The percentage of presences detected by BirdNET using the optimal confidence score, instead of the default values, decreased more for the Coal Tit (from 90.5% to 54.8%) than for the Short-toed Treecreeper (from 98.4% to 81.3%). It is a surprising result since the optimal confidence score of the Short-toed Treecreeper was higher (0.335) than the one of the Coal Tit (0.247); therefore, a higher decrease in presences detected for the Short-toed Treecreeper would have been expected. One possible explanation for this surprising result might be related to a high vocal activity of the Short-toed Treecreeper, which may compensate for the potential decrease in the number of vocalizations detected when using a higher confidence score. Our findings might be also affected by different foraging strategies or territorial behavior of the target species and, therefore, by birds moving more often outside the detection range of the recorder. However, we were unable to include this factor in our analyses. In our study we focused on the ability of BirdNET to detect the species' presence in sound recordings, but further research could expand on this by assessing the ability of BirdNET to detect vocalizations, an acoustic metric known as recall rate, which is not frequently evaluated in BirdNET surveys [19].

The selection of an appropriate confidence score in BirdNET may depend on the priorities of the user (e.g., ability and time to verify more or less false positives) and research goals. A recent review on BirdNET suggested starting with a minimum confidence score of around 0.5 to assess BirdNET performance [19], while other authors have recommended that confidence scores of 0.7–0.8 should be in the appropriate range for most studies [36]. However, our study highlights the importance of conducting species-specific assessments (e.g., by creating independent validation datasets for each species) for choosing an appropriate confidence score and how this selection can vary depending on research goals. According to our data, if our objective is to detect the presence of both species in sound recordings it would be necessary to use low confidence scores (e.g., 0.1, default values). Otherwise, a significant number of presences will be undetected. On the other hand, if our goal is to study ecological processes, which usually require low-error estimates, such as describing the vocal behavior of a bird species, selecting a higher confidence value may be more appropriate. Although this selection may decrease the number of vocalizations detected, it would provide a more reliable description of the behavior (see [21]).

Finally, we linked BirdNET detections to ecological processes using a large acoustic dataset as an example of how this tool may help researchers and managers to improve acoustic monitoring programs and contribute to a better understanding of the ecological processes. The vocal activity of the Coal Tit was primarily concentrated in the first two hours after sunrise, with minimal vocal activity throughout the day. This pattern aligns with the typical vocal activity pattern observed in most passerines [37,38], and with the pattern described for two close-related species, the Blue Tit (*Cyanistes caeruleus*) and the Great Tit (*Parus major*) [39]. The vocal activity of the Short-toed Treecreeper, similar to the Coal Tit, peaked during the first hours after sunrise. However, the Short-toed Treecreeper showed sustained vocal activity throughout the morning, and the species even vocalized during the afternoon. This prolonged vocalization behavior might be related to the species strong vocal response towards other species, particularly to the closely related Common Treecreeper (*Certhia familiaris*, [40,41]), which coexists in the study area and may have stimulated the vocal activity of the Short-toed Treecreeper.

5. Conclusions

Our study has demonstrated the effectiveness of BirdNET, a new tool for processing acoustic data, in accurately identifying two cryptic bird species and detecting their presence in sound recordings. We have also highlighted the importance of carefully selecting the confidence score, as it has a significant impact on the output of BirdNET, and may potentially lead to poorly informed conclusions. In both species a higher confidence score reduces false positives but also results in fewer detections of species' presences. We hope that our assessment and the methods we employed, including calculating the optimal confidence score (see [22]), will encourage researchers and managers to make use of this freely available software (accessible on GitHub) that is user-friendly (e.g., can be run as a GUI from Windows) and ready-to-use (>6500 species already included). Additionally, the continuous development of BirdNET (last update in June 2023) will contribute to further improvements in acoustic sensing and monitoring in both urban and natural environments, including the ability to detect multiple species from various taxa simultaneously [19–22].

Author Contributions: Conceptualization, G.B. and C.P.-G.; methodology, G.B., L.C. and C.P.-G.; software, L.C., J.G.-C. and C.P.-G.; validation, L.C. and C.P.-G.; formal analysis, J.G.-C. and C.P.-G.; investigation, G.B., R.M.-R. and C.P.-G.; resources, G.B., R.M.-R. and C.P.-G.; data curation, G.B., L.C. and C.P.-G.; writing—original draft preparation, G.B. and C.P.-G.; writing—review and editing, all authors; visualization, G.B., R.M.-R. and C.P.-G.; supervision, G.B. and C.P.-G.; project administration, G.B., R.M.-R. and C.P.-G.; funding acquisition, G.B. and C.P.-G. All authors have read and agreed to the published version of the manuscript.

Funding: This research was partially funded by Red Eléctrica Española with the support of the Navarra Regional Government.

Institutional Review Board Statement: Not applicable.

Informed Consent Statement: Not applicable.

Data Availability Statement: Raw databases employed for BirdNET validation can be downloaded at: https://figshare.com/s/888467a40c77f46d463a, with https://doi.org/10.6084/m9.figshare.23736570, accessed on 11 August 2023.

Acknowledgments: CPG acknowledges the support of the Ministerio of Educación y Formación Profesional through the Beatriz Galindo Fellowship (Beatriz Galindo—Convocatoria 2020). JGC is funded by a Margarita Salas postdoctoral fellowship (CA4/RSUE/2022-00205) funded by the Spanish Ministry of Universities, the Recovery, Transformation, and Resilience Plan (key instrument for the development of Next Generation EU recovery funds) and Universidad Autónoma de Madrid.

Conflicts of Interest: The authors declare no conflict of interest.

References

1. Gibb, R.; Browning, E.; Glover-Kapfer, P.; Jones, K.E. Emerging opportunities and challenges for passive acoustics in ecological assessment and monitoring. *Methods Ecol. Evol.* **2019**, *10*, 169–185. [CrossRef]
2. Sugai, L.S.M.; Silva, T.S.F.; Ribeiro, J.W., Jr.; Llusia, D. Terrestrial passive acoustic monitoring: Review and perspectives. *BioScience* **2019**, *69*, 15–25. [CrossRef]
3. Lahoz-Monfort, J.J.; Magrath, M.J. A comprehensive overview of technologies for species and habitat monitoring and conservation. *BioScience* **2021**, *71*, 1038–1062. [CrossRef] [PubMed]
4. Schmeller, D.S.; Böhm, M.; Arvanitidis, C.; Barber-Meyer, S.; Brummitt, N.; Chandler, M.; Chatzinikolaou, E.; Costello, M.J.; Ding, H.; García-Moreno, J.; et al. Building capacity in biodiversity monitoring at the global scale. *Biodivers. Conserv.* **2017**, *26*, 2765–2790. [CrossRef]
5. Stephenson, P.J. Technological advances in biodiversity monitoring: Applicability, opportunities and challenges. *Curr. Opin. Environ. Sustain.* **2020**, *45*, 36–41. [CrossRef]
6. Kershenbaum, A. Entropy rate as a measure of animal vocal complexity. *Bioacoustics* **2014**, *23*, 195–208. [CrossRef]
7. Shonfield, J.; Bayne, E.M. Autonomous recording units in avian ecological research: Current use and future applications. *Avian Conserv. Ecol.* **2017**, *12*, 14. [CrossRef]
8. Desjonquères, C.; Gifford, T.; Linke, S. Passive acoustic monitoring as a potential tool to survey animal and ecosystem processes in freshwater environments. *Freshw. Biol.* **2020**, *65*, 7–19. [CrossRef]
9. Hill, A.P.; Prince, P.; Piña Covarrubias, E.; Doncaster, C.P.; Snaddon, J.L.; Rogers, A. AudioMoth: Evaluation of a smart open acoustic device for monitoring biodiversity and the environment. *Methods Ecol. Evol.* **2018**, *9*, 1199–1211. [CrossRef]
10. Lapp, S.; Stahlman, N.; Kitzes, J. A Quantitative Evaluation of the Performance of the Low-Cost AudioMoth Acoustic Recording Unit. *Sensors* **2023**, *23*, 5254. [CrossRef]
11. Kahl, S.; Wood, C.M.; Eibl, M.; Klinck, H. BirdNET: A deep learning solution for avian diversity monitoring. *Ecol. Inform.* **2021**, *61*, 101236. [CrossRef]
12. Stowell, D. Computational bioacoustics with deep learning: A review and roadmap. *PeerJ* **2022**, *10*, e13152. [CrossRef] [PubMed]
13. Darras, K.F.; Deppe, F.; Fabian, Y.; Kartono, A.P.; Angulo, A.; Kolbrek, B.; Mulyani, Y.A.; Prawiradilaga, D.M. High microphone signal-to-noise ratio enhances acoustic sampling of wildlife. *PeerJ* **2020**, *8*, e9955. [CrossRef]
14. Cameron, J.; Crosby, A.; Paszkowski, C.; Bayne, E. Visual spectrogram scanning paired with an observation–confirmation occupancy model improves the efficiency and accuracy of bioacoustic anuran data. *Can. J. Zool.* **2020**, *98*, 733–742. [CrossRef]
15. Nadimpalli, U.D.; Price, R.R.; Hall, S.G.; Bomma, P. A comparison of image processing techniques for bird recognition. *Biotechnol. Prog.* **2006**, *22*, 9–13. [CrossRef]
16. Priyadarshani, N.; Marsland, S.; Castro, I. Automated birdsong recognition in complex acoustic environments: A review. *J. Avian Biol.* **2018**, *49*, jav-01447. [CrossRef]
17. Xie, J.; Zhong, Y.; Zhang, J.; Liu, S.; Ding, C.; Triantafyllopoulos, A. A review of automatic recognition technology for bird vocalizations in the deep learning era. *Ecol. Inform.* **2022**, *73*, 101927. [CrossRef]
18. Wood, C.M.; Kahl, S.; Rahaman, A.; Klinck, H. The machine learning–powered BirdNET App reduces barriers to global bird research by enabling citizen science participation. *PLoS Biol.* **2022**, *20*, e3001670. [CrossRef]
19. Pérez-Granados, C. BirdNET: Applications, performance, pitfalls and future opportunities. *IBIS* **2023**, *165*, 1068–1075. [CrossRef]
20. Pérez-Granados, C. A first assessment of BirdNET performance at varying distances: A playback experiment. *Ardeola* **2023**, *70*, 257–269. [CrossRef]
21. Wood, C.M.; Kahl, S.; Barnes, S.; Van Horne, R.; Brown, C. Passive acoustic surveys and the BirdNET algorithm reveal detailed spatiotemporal variation in the vocal activity of two anurans. *Bioacoustics* **2023**, 1–12. [CrossRef]
22. Wood, C.M.; Barceinas Cruz, A.; Kahl, S. Pairing a user-friendly machine-learning animal sound detector with passive acoustic surveys for occupancy modeling of an endangered primate. *Am. J. Primatol.* **2023**, *85*, e23507. [CrossRef] [PubMed]
23. Tietze, D.T.; Martens, J.; Sun, Y.H.; Liu Severinghaus, L.; Päckert, M. Song evolution in the coal tit Parus ater. *J. Avian Biol.* **2011**, *42*, 214–230. [CrossRef]
24. Harrap, S. Short-toed Treecreeper (Certhia brachydactyla), version 1.0. In *Birds of the World*; del Hoyo, J., Elliott, A., Sargatal, J., Christie, D.A., de Juana, E., Eds.; Cornell Lab of Ornithology: Ithaca, NY, USA, 2020.
25. Brotons, L. Changes in foraging behaviour of the coal tit Parus ater due to snow cover. *Ardea* **1997**, *85*, 249–258.
26. Nokelainen, O.; Helle, H.; Hartikka, J.; Jolkkonen, J.; Valkonen, J.K. The Eurasian Treecreeper (Certhia familiaris) has an effective camouflage against mammalian but not avian vision in boreal forests. *IBIS* **2022**, *164*, 679–691. [CrossRef]
27. Goller, F. Der Gesang der Tannenmeise (Parus ater): Beschreibung und kommunikative Funktion. *J. Ornithol.* **1987**, *128*, 291–310. [CrossRef]
28. Osiejuk, T.S.; Kuczynski, L. Song functions and territoriality in Eurasian Treecreeper Certhia familiaris and Short-toed Treecreeper Certhia brachydactyla. *Acta Ornithol.* **2000**, *35*, 109–116.
29. Sanj, J.J.; Moreno, J.; Pancorbo, M.M. The significance of double broods in the Coal Tit Parus ater breeding in a montane coniferous forest in central Spain. *Ardeola* **1993**, *40*, 155–161.
30. Frías, O.; Villar, S.; Potti, J. Datos sobre la reproducción del Agateador Común (*Certhia brachydactyla*) en la Sierra de Ayllón (Sistema Central). *Anuario Ornitológico de Madrid* **1999**, *1999*, 108–113.

31. Manzano-Rubio, R.; Bota, G.; Brotons, L.; Soto-Largo, E.; Pérez-Granados, C. Low-cost open-source recorders and ready-to-use machine learning approaches provide effective monitoring of threatened species. *Ecol. Inform.* **2022**, *72*, 101910. [CrossRef]
32. Knight, E.; Hannah, K.; Foley, G.; Scott, C.; Brigham, R.; Bayne, E. Recommendations for acoustic recognizer performance assessment with application to five common automated signal recognition programs. *Avian Conserv. Ecol.* **2017**, *12*, 14. [CrossRef]
33. Cornell Lab of Ornithology. *Raven Pro: Interactive Sound Analysis Software*; Version 1.6.4; Computer Software; The Cornell Lab of Ornithology: Ithaca, NY, USA, 2023.
34. Ross, S.R.J.; O'Connell, D.P.; Deichmann, J.L.; Desjonquères, C.; Gasc, A.; Phillips, J.N.; Sehti, S.S.; Wood, C.M.; Burivalova, Z. Passive acoustic monitoring provides a fresh perspective on fundamental ecological questions. *Funct. Ecol.* **2023**, *37*, 959–975. [CrossRef]
35. Malamut, E.J. Using Autonomous Recording Units and Image Processing to Investigate Patterns in Avian Singing Activity and Nesting Phenology. Ph.D. Thesis, University of California, Los Angeles, CA, USA, 2022.
36. Sethi, S.S.; Fossøy, F.; Cretois, B.; Rosten, C.M. Management Relevant Applications of Acoustic Monitoring for Norwegian Nature–The Sound of Norway. In *NINA Report 2064*; Norwegian Institute for Nature Research: Trondheim, Norway, 2021.
37. Catchpole, C.K.; Slater, P.J. *Bird Song: Biological Themes and Variations*, 2nd ed.; Cambridge University Press: Cambridge, UK, 2008.
38. Gil, D.; Llusia, D. The bird dawn chorus revisited. In *Coding Strategies in Vertebrate Acoustic Communication*; Aubin, T., Mathevon, N., Eds.; Animal Signals and Communication; Springer: Cham, Switzerland, 2020; Volume 7, pp. 45–90.
39. Amrhein, V.; Johannessen, L.E.; Kristiansen, L.; Slagsvold, T. Reproductive strategy and singing activity: Blue tit and great tit compared. *Behav. Ecol. Sociobiol.* **2008**, *62*, 1633–1641. [CrossRef]
40. Gil, D. Increased response of the Short-Toed Treecreeper Certhia brachydactyla in sympatry to the playback of the song of the Common Treecreeper *C. familiaris*. *Ethology* **1997**, *103*, 632–641. [CrossRef]
41. Clouet, M.; Gerard, J.F. Interactions between sibling species of treecreepers Certhia familiaris and C. brachydactyla in the Pyrenees and the mistaken identity hypothesis. *Bird Study* **2020**, *67*, 385–392. [CrossRef]

sensors MDPI

Article

Passive Acoustic Sampling Enhances Traditional Herpetofauna Sampling Techniques in Urban Environments

Isabelle L. Barnes * and John E. Quinn

Department of Biology, Furman University, Greenville, SC 29613, USA; john.quinn@furman.edu
* Correspondence: barnes.isabelle00@gmail.com

Abstract: Data are needed to assess the relationships between urbanization and biodiversity to establish conservation priorities. However, many of these relationships are difficult to fully assess using traditional research methods. To address this gap and evaluate new acoustic sensors and associated data, we conducted a multimethod analysis of biodiversity in a rapidly urbanizing county: Greenville, South Carolina, USA. We conducted audio recordings at 25 points along a development gradient. At the same locations, we used refugia tubes, visual assessments, and an online database. Analysis focused on species identification of both audio and visual data at each point along the trail to determine relationships between both herpetofauna and acoustic indices (as proxies for biodiversity) and environmental gradient of land use and land cover. Our analysis suggests the use of a multitude of different sampling methods to be conducive to the completion of a more comprehensive occupancy measure. Moving forward, this research protocol can potentially be useful in the establishment of more effective wildlife occupancy indices using acoustic sensors to move toward future conservation policies and efforts concerning urbanization, forest fragmentation, and biodiversity in natural, particularly forested, ecosystems.

Keywords: frogs; mixed methods; trail; Piedmont; unsupervised classification

Citation: Barnes, I.L.; Quinn, J.E. Passive Acoustic Sampling Enhances Traditional Herpetofauna Sampling Techniques in Urban Environments. *Sensors* **2023**, *23*, 9322. https://doi.org/10.3390/s23239322

Academic Editor: Hector Eduardo Roman

Received: 7 October 2023
Revised: 20 November 2023
Accepted: 20 November 2023
Published: 22 November 2023

1. Introduction

The planet and the life that inhabits it are currently undergoing a sixth mass extinction marked by extensive declines in global biodiversity. One of the taxa that has been experiencing drastic population decline is herpetofauna, namely frogs. This poses a major environmental issue as frog species are a key indicator of environmental health, and the loss of frog populations can lead to a complete food web and ecosystem collapse [1]. Thus, the need to protect and monitor frog species in the wild is becoming more and more paramount.

One of the main drivers of herpetofauna population loss is habitat loss. A clear portion of this habitat loss comes in the form of urbanization and anthropogenic expansion. Urbanization is an increasingly evident problem for biodiversity as the human population continues to grow and spread. This is becoming so much of a dilemma, in fact, that residential development is projected to increase in area by 51% between the years 2003 and 2030 [2]. This projected increase will result in a decrease in undeveloped lands in many parts of the world, leading to a decrease in biodiversity in these areas as well as a loss of the ecosystem services that nature provides [3].

Deforestation because of urbanization creates threats across taxa from vegetation to avian species, mammals, and herpetofauna. Studies by Howell et al. [4], found a significant decrease in population growth rates and stability of federally threatened Chiricahua Leopard Frogs after their natural habitat of wetlands and canyon streams was modified and removed for the implementation of development and agriculture. They found that these changes in population growth were influenced widely by habitat removal, new habitat characteristics, and the demographics that the processes of urbanization enforced upon the new patches [4].

Researchers have long used a diversity of techniques to sample herpetofauna [5,6]. These include a variety of active and passive sampling techniques such as transects, pitfall traps, cover boards, funnel traps, refugia tubes, and glue traps. These methods are used to reduce researcher bias and extend the length of the sampling season [7]. However, each method is limited in terms of which type of herpetofauna is being sampled (e.g., pitfall traps only work for ground-dwelling taxa). Likewise, they can be time-consuming and costly. Thus, decisions need to be made to optimize data collection [8]. In addition to these well-established methods, new sensor technologies have been developed that may provide additionality to current sampling and monitoring methods, in particular passive acoustic sampling.

Automated recording units (ARUs) are an increasingly common acoustic method to measure biodiversity in real-time and across dispersed landscapes [9]. The ability of these acoustic sensors to collect large quantities of spatial data makes them useful monitoring tools [10]. With simple programming and installation, ARUs require little time for the researcher to be in the field [10,11]. ARUs also provide researchers with the ability to cover a larger spatio-temporal scale in relation to more customary sampling practices [1]. In the last 10 years, ARUs have been widely used to monitor the occupancy, movement, and behavioral patterns of birds and bats (e.g., [12,13]) but less frequently for frogs, toads, and other herpetofauna (but see MacLaren et al. [14]).

ARUs may add additional value for detecting the distribution of herpetofauna because multiple locations can be intensively sampled concurrently. Through analysis of frog vocalizations, the relative abundance and species richness can be determined for a location, allowing an estimate of herpetofauna species diversity to be reached [1]. This can improve researchers' ability to detect those species that may have a low detection probability or that require frequent sampling for detection [15]. In addition, the volume of data collected using ARUs opens the opportunity to leverage new machine learning tools to analyze these data, though unlike other species, for example bats, less work has been done to build effective tools for easy application when sampling herpetofauna.

In addition to species-specific detections, the ARUs allow for a diversity of different acoustic measures of an ecosystem that can reflect the broader ecological community as compared to a single species and have been shown to correlate with other measures of biodiversity (e.g., [16,17]). These measures, frequently defined as acoustic or soundscape indices, reflect the multiple dimensions of sound in an environment, including biophony, geophony, and anthrophony. These indices can be measured as continuous variables over time, thus describing the acoustic environment a frog or other species may experience along an urban to rural gradient including other wildlife species and vehicle traffic or other sources of human noise.

In this project, we compared traditional methods for biodiversity sampling and reporting (visual observations, citizen science, and refugia tubes) with new methods (ARUs, soundscape indices). We expected to see different, but complementary, patterns in the response of traditional and new sampling techniques, such that research on the diversity of herpetofauna in urban areas could potentially focus on a narrower set of tools and indicators.

2. Materials and Methods

2.1. Study Area

We collected data along the Swamp Rabbit Trail in Greenville County, SC, USA (Figure 1A). The county is in the Piedmont ecoregion, at the base of the Appalachian mountains, extends from northern Virginia all the way into central Alabama, and includes the northwestern corner of South Carolina, known as the Upstate. This area has experienced rapid urban development in the last century. The area has made the transition from a forest biome to a heterogenous populated forest anthrome due to the effects of human development and population growth [18]. This change has forged a novel ecological environment which mixes forest biome and urban development and has created a demand for conservation efforts to prevent further deterioration of the area.

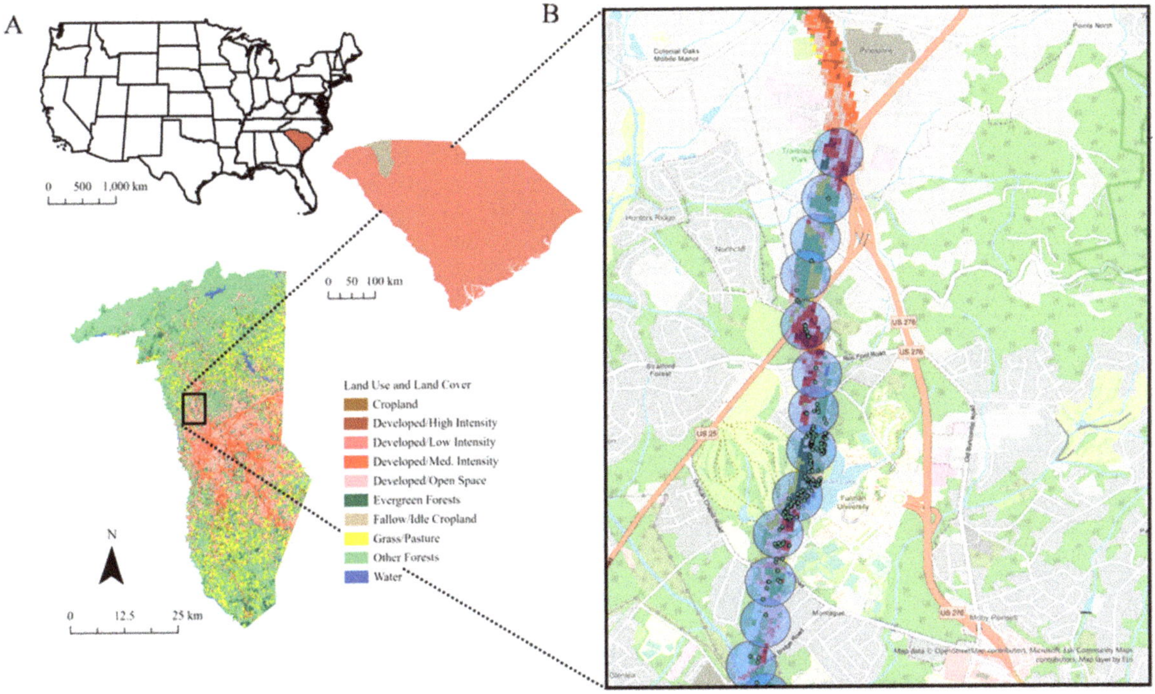

Figure 1. (**A**) The location of Greenville Co., SC in the southeastern United States including land use and land cover in the county as classified by the NLCD land cover types. (**B**) The same land use and land cover within 100 m of the trial and associated buffers for extracting iNaturalist data. Black circles indicate the 100 m buffer that was surveyed around the trail. The red, pink, green, and yellow pixels within the buffers indicate the land use and land cover from NLCD. The green points in the buffers indicate iNaturalist observations.

The Swamp Rabbit Trail (SRT, Figure 1B) is a 22-mile-long rails-to-trails development. The SRT is embedded in fragmented forested and developed areas in the South Carolina Piedmont ecoregion including urban and suburban development, subdivisions, and highway systems. The understory plant communities along the trail, not unlike many other disturbed sites in the region, are dominated by non-native and invasive vegetation. We specifically focused on six miles of the trail found adjacent to Furman University, spreading two miles northwest into the outskirts of Travelers Rest and four miles southwest towards downtown Greenville.

2.2. Data Collection

We used a variety of audio and visual collection methods to determine herpetofauna presence and abundance along the trail including refugia tubes, visual observation surveys, and iNaturalist. Our particular focus for this article is the adoption of automated recording units (ARU) which are sensors that are being used with increasing frequency for a diversity of environmental monitoring efforts (as discussed above).

We placed refugia tubes every 800 m (0.5 miles) along the trail in 2018 to enhance the attainment of visual observation data. We used polyvinyl chloride (PVC) pipe traps, following the methods in Boughton et al. [19], for the collection of treefrog species. We used opaque, white PVC pipe which was cut into 60 cm lengths with a capped bottom for the refugia tubes. The tubes were mounted in trees about 1.5 m high along the trail. In the base of each tube, we poured water to create a moist environment most like those in which anurans would be found naturally [19]. To keep our study period consistent, the refugia tubes were only routinely checked during the summer months of 2021 and 2022 to align with the ARU data.

Our observation data came from two sources. First, we conducted regular transect surveys along the trail to observe and hear individuals. Transects were 400 m along the trail and 50 m deep. Transect depth varied depending on the environment surrounding different sections of the trail. Observational surveys took place twice weekly. Second, we downloaded data from iNaturalist, filtered to Greenville, SC with only herpetofauna species selected. We included in this sample data from 2008 to the present to increase the number of observations included. As an added precaution to the quality range that can be obtained from citizen science databases, we also filtered the data to only include research-grade observations. We then used ArcGIS Pro to filter this dataset to only observations that were within 50 m of the trail.

We collected acoustic data using the SM2 from Wildlife Acoustics Inc. Recordings during the summer months (late May, June, and July) of 2021 and 2022 to ensure the optimum number of species that would be calling due to the ideal temperatures and weather [20]. Using procedures described in Sidie-Slettedahl et al. [21], we deployed the ARUs every 400 m (0.25 miles) along the trail. We hung recording devices between 1.5 and 2 m approximately 10 m off of the paved trail in a tree or shrub [22]. We programmed the ARUs at 48 dB gain (left and right) and to record for 10 min, every hour, on the hour, 24 h a day. An ARU was located at each sampling point for one week.

2.3. Data Processing

To identify the frequency of detections of each species in the audio recordings we used the Kaleidoscope Version 5.3.4 software from Wildlife Acoustics (Maynard, MA, USA). Specifically, we used Kaleidoscope as a form of unsupervised classification to calculate distinct sound clusters at each recording location. We set a minimum and maximum frequency range of 150 Hz and 5500 Hz. We set a minimum and maximum length of detection at 0.1 and 15 s, 3 maximum inter-syllable gap, and 1.0 as the max distance from the cluster. These values were chosen to avoid including the repetitive call of one individual multiple times in our dataset. Details on the clustering methods can be found at wildlifeacoustics.com. We then manually inspected each file in each cluster to identify each vocalizing amphibian.

To identify spatial and temporal patterns in acoustic indices we used the tuneR package [23] for the program R (R Core Team 2019) to read sound files. We then used the soundecology [24] and seewave [25] packages to obtain values of each index from channel 1. We measured anthrophony or technophony (human-derived sounds), which we defined following the literature (e.g., [26]) as sound occurring in the 1–2 kHz range, and biophony (ecologically derived sounds), which we defined as sound occurring in the 2–8 kHz frequency range (following [26]), for each sound file. We used the soundecology and seewave packages to obtain values of Normalized Difference Soundscape Index, Acoustic Complexity Index, Acoustic Diversity Index, Acoustic Evenness Index, total entropy, and Bioacoustic Index (abbreviated NDSI, ACI, ADI, AEI, H, and BAI, respectively) for each sound file. NDSI was an index of anthropogenic noise disturbance measuring the proportion of biophony to anthrophony [26]. ACI measured the variation in the intensity of sounds [27]. ADI and AEI both measure the distribution of sound power across frequency ranges [28]. ADI quantified this distribution using the Shannon diversity index, thus measuring sound diversity similarly to species diversity, while AEI used the Gini index of evenness, thus measuring sound evenness similarly to species evenness. H was a function of temporal energy dispersal and spectral energy dispersal [25]. BAI was a function of both power and frequency range of sound between 2000 and 11,000 Hz [29]. We made three passes with this modification at 80 Hz, 1000 Hz, and 2000 Hz, following Hyland et al. [30].

Because of the filters, Biophony and BAI are considered as shown with the 2000 Hz filter as both are only measured in this acoustic space. Anthrophony and NDSI are only considered at the 1000 Hz filter, because anthrophony is measured between 1000 and 2000 Hz and thus NDSI also does not have results for 80 Hz, and the 2000 Hz filter, because it is a ratio of biophony above 2000 Hz and anthrophony between 1000 Hz and 2000 Hz.

2.4. Data Analysis

We used the program R 4.0.2 (R Core Team 2019) to synthesize, visualize, and analyze the data with the ggplot and dplyr packages [31]. For individual species, count per unit effort (CPU) was calculated by dividing the total number of observations by the number of data inputs recorded at each location. To avoid pseudo-replication in the soundscape data, we averaged all values for each index for each sampling. Thus, a site was the unit of study for subsequent statistical analyses of ARU data. We tested for spatial relationships using linear regression with land use and land cover as explanatory variables for both CPU and each acoustic index. Significance was based on an alpha value of 0.05.

3. Results

In total, our different collection methods were able to detect 1419 herpetofauna observations (Figure 2; 1405 from the ARUs, 0 from refugia tubes, 11 from iNaturalist, and 3 visual observations). By including ARU data our number of detections increased by 15× and the number of species increased from 11 to 17. From the ARUs alone, we were able to accurately detect nine different anuran species: Cope's Gray Treefrog (*Dryophytes chrysoscelis*), American Bullfrog (*Lithobates catesbeianus*), American Toad (*Anaxyrus americanus americanus*), Fowler's Toad (*Anaxyrus fowleri*), Green Frog (*Lithobates clamitans*), Green Tree Frog (*Dryophytes cinereus*), Northern Cricket Frog (*Acris crepitans*), Pickeral Frog (*Lithobates palustris*), and Spring Peeper (*Pseudacris crucifer*). Cope's Gray Treefrog was the most frequently detected species at most locations. The greatest number of species was detected at mile marker 28.25 on the trail with nine of our species being detected by two different data collection methods. Three species were identified by multiple collection methods. For example, the American Toad was captured with all three methods. Meanwhile, Cope's Gray Treefrogs were found both audibly and visually through the ARU data as well as during visual transect monitoring. The Fowler's Toads were also identified by two of the collection methods: ARUs and iNaturalist.

However, with our other detection methods, we were also able to include other, less vocal herpetofauna groups including snakes, turtles, and skinks. We totaled eight non-anuran species identifications using visual and citizen science data collection: Broad-headed Skink (*Plestiodon laticeps*), Common Five-lined Skink (*Plestiodon fasciatus*), Deirochelyine Turtles, Eastern Box Turtle (*Terrapene carolina carolina*), Eastern Copperhead Snake (*Agkistrodon contortrix*), Eastern Garter Snake (*Thamnophis sirtalis sirtalis*), Eastern Kingsnake (*Lampropeltis getula*), and Slider Turtles (*Trachemys*). Also, by restricting the iNaturalist data to only our 2-year study period, we would have lost 81.1% of our dataset from iNaturalist.

These data also allow us to observe relationships between different species and land cover types. For example, the Green Frog displayed a significant, but different, relationship with the combination of both the developed and forested cover (Table 1, Figure 3). Though not significant, the Cope's Gray Treefrog, a species that was observed by every observation technique, showed slight associative responses between occupancy and the presence of water but lacked this response when compared to occupancy within forested environments (Table 1, Figure 3).

There were no statistically significant relationships ($p > 0.20$ for all indices) between the acoustic indices and measures of adjacent land use and land cover. While there were no statistical relationships between the indices and associated land use and land cover, we did find evidence of clear heterogeneity over space and time for many indices, allowing for better tracking of biodiversity over space and time. For example, both biophony and anthrophony varied between their respective minimum and maximum values between locations (Figure 4). In some locations the index value changes by nearly that same amount over a 24 h period. Perhaps of greater value are the clear outliers in the ACI and ADI and more so in the BAI that may suggest soundscapes with richer biodiversity or a greater impact of change.

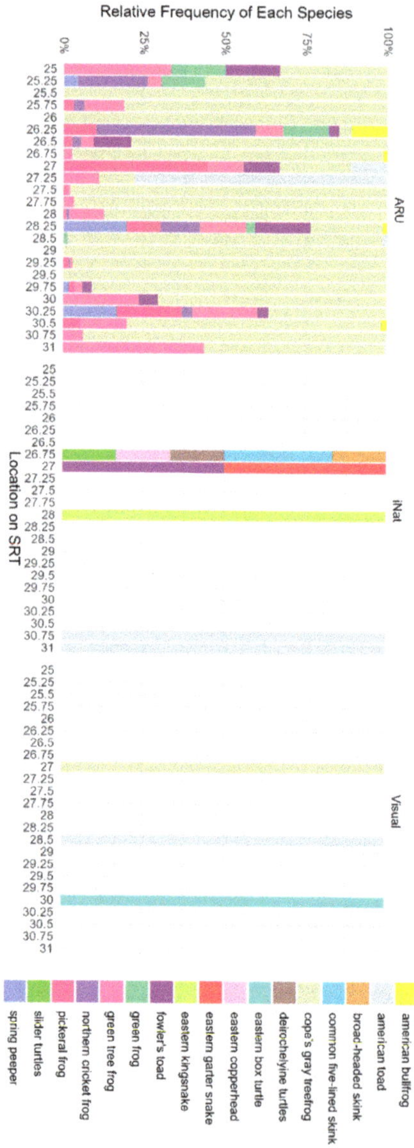

Figure 2. Species relative frequencies (colors) by location and observation technique. Data was collected along the Swamp Rabbit Trail, Greenville SC. ARU and visual observation data from Summer 2021 to 2022. iNaturalist data from 2016 to 2021.

Table 1. Regression estimates and standard error for the response of Cope's Gray Treefrog, Green Frog, and total species abundance, as a function of the percentage of water, forest, and development along the Swamp Rabbit Trail. Significant p values noted in bold and shown in Figure 3.

	Cope's Gray Treefrog			Green Frog			Total		
	Estimate	Std. Error	p Value	Estimate	Std. Error	p Value	Estimate	Std. Error	p Value
Intercept	0.243	0.120		2.599	0.139		0.054	0.035	
Water (%)	−0.029	0.024	0.250	−0.001	0.001	0.800	0.000	0.001	0.800
Forest (%)	0.000	0.002	0.920	**−0.026**	**0.001**	**0.0003**	−0.001	0.001	0.320
Developed (%)	0.001	0.003	0.810	**−0.026**	**0.001**	**0.0003**	0.001	0.001	0.255

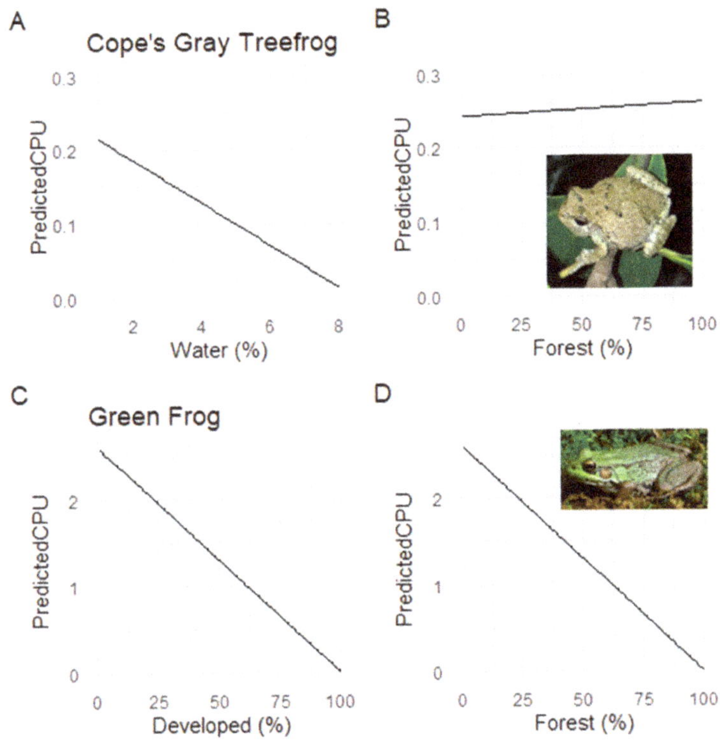

Figure 3. Count per unit effort (CPU) as a factor of habitat type for (**A**) Cope's Gray Treefrog and wetland habitat, (**B**) Cope's Gray Treefrog and forested habitat, (**C**) Green Frog and developed habitat, and (**D**) Green Frog and of forested habitat. Data collected along the Swamp Rabbit Trail, Greenville, SC, USA.

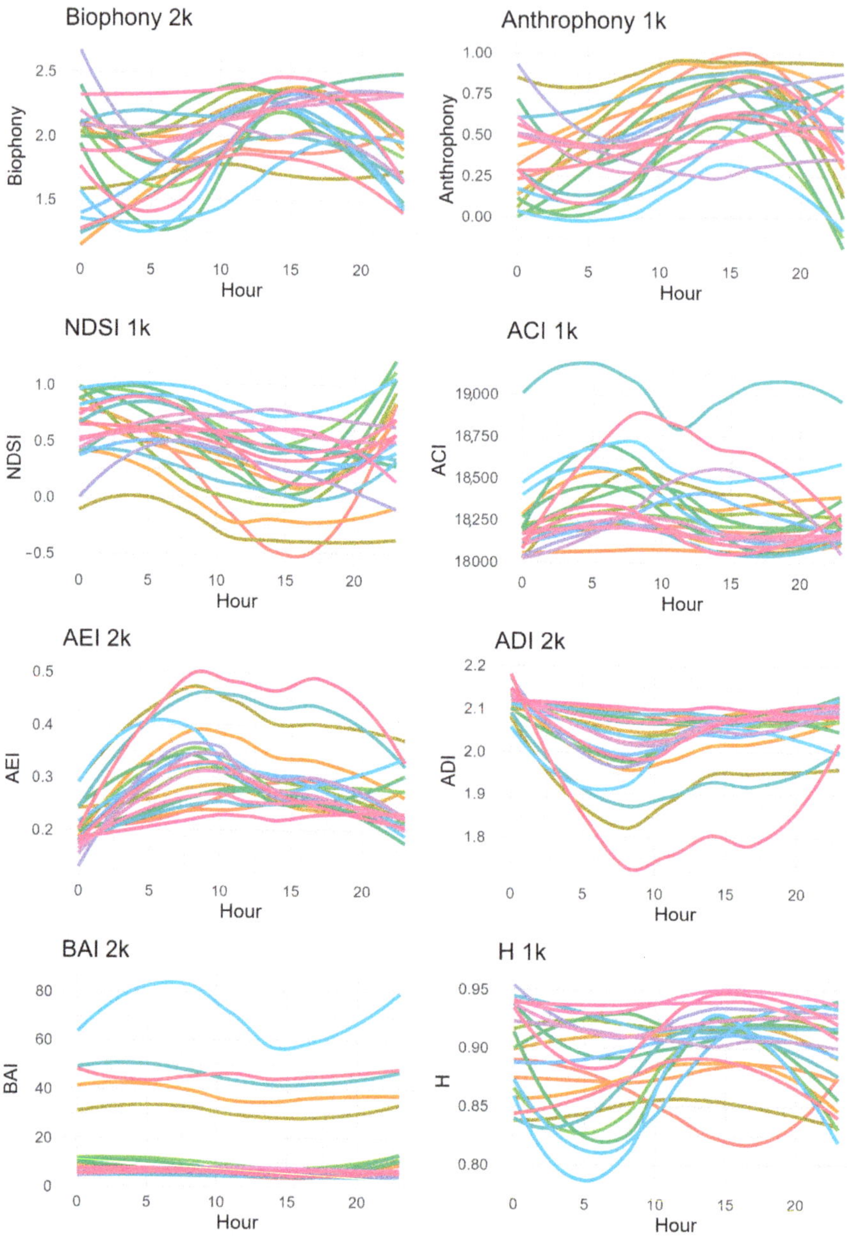

Figure 4. Variation in each filtered acoustic index across a 24 h time window at each location (line colors). Data collection using automated recording units (SM2 from Wildlife Acoustics) in spring and summer of 2021 and 2022. Acoustic indices calculated in R using the packages described in the methods. 1K and 2K represent the filter applied to each of the indices.

4. Discussion

A long-standing challenge in biodiversity research, and in particular herpetological research, is the inability to conduct a comprehensive survey of a location due to the variability of species over space and time. In this case study, our results clearly show variation in the detection of individual species identified through unsupervised classification and in communities identified through acoustic indices. Focusing on the former, by leveraging sensors, ARUs clearly contributed to the greatest number of detections, thus increasing estimates of herpetofauna diversity. However, they could not pick up everything, emphasizing the necessary variations in data collection methods to conduct a comprehensive analysis of biodiversity at a point or within a region. By using different detection and identification methods, we were able to create a more comprehensive index of the study site. There were, however, vocal species that we were unable to detect via audio data due to noise congestion or other factors. An example of this can be seen in Figure 1 where the American Toad (depicted by the color gray in the figure) was not detected by the ARUs at mile markers 30.75 and 31.0 on the Swamp Rabbit Trail, but we were able to visually identify them at that location using iNaturalist citizen science data. Likewise, the variability in the soundscape indices suggests there is variation in acoustics data that could be explained by the same variables as the occupancy data or, more likely, other measures of environmental change. While we did not separate what specific species were affecting each index, the five sites that stand out for the BAI warrant further investigation for conservation efforts for herpetofauna and other wildlife. Moreover, if there is evidence of a correlation between the BAI (or another index) and richness or occupancy of key herpetofauna, overall or at a given point in time, it may be that these indices can be used to track herpetofauna diversity until better machine learning tools become available to process these data.

As stated, each study method contains strengths and weaknesses. Some of these weaknesses were highlighted in data processing and analysis. Active sampling, through transect sampling and physical searching for taxa, allows for an increased probability of sample bias due to the likelihood of the researcher only looking in locations in which herpetofauna species are more likely to be found. Alternatively, passive sampling made possible with sensors embedded in ARUs reduce bias but also restrict the comprehensiveness of the study due to the specifications for a distinct type of taxa. This was seen through the lack of non-vocalizing species that were detected using ARUs, an expected result. Lastly, citizen science data has the potential to be an unreliable research source due to the likelihood of the public misidentifying an organism; however, it allows researchers to gain a larger understanding of species found both spatially and over a longer time scale. This potential for error and a less reliable dataset can be counteracted using filters, such as scientific-grade observations, within citizen science applications; however, this also greatly reduces the available dataset that the platform has to offer. Citizen science data is also restrictive as it may not include rare or elusive species in its observational skillset if a species had not been recorded prior.

Like ARUs, citizen science data collection has increased in frequency and usefulness. For ecology, this growth has been driven by the popular citizen science platform iNaturalist, which was launched in 2008 (inaturalist.org). Platforms such as iNaturalist allow researchers to surpass the confines of time, effort, and funding due to the immense volume of field observations that can be gathered by the larger public [32]. Through using data from iNaturalist, we were also able to include species that may have evaded our efforts in data collection despite their presence in the study site in our comprehensive data set. Many of these included species do not typically rely on sound and calls for communication, like skinks, salamanders, snakes, and turtles. Even though some of these groups may use acoustic communication, like a snake hissing or rattling, these vocalizations were not detected by our ARUs due to the greater amplitude of the surrounding forest noise. By using iNaturalist data, we were also able to account for species that may have been observed outside of our study timespan. Instead of only including data from the summer months of 2021 and 2022, we also were able to include observations from as far back as August 2016 to increase the comprehensiveness of our study.

One challenge with the ARU data that persists is the necessity of manual identification of species. We attempted to build an advanced classifier within Kaleidoscope. However, due to the lack of regularity of the individual species' vocalizations (a short chirp versus a call) and the inevitable variability of background noise overpowering the vocalizations themselves, we were unsuccessful. Future work could be oriented toward the formation of an efficient, advanced classifier for herpetofauna species similar to the ones that are currently seen for birds and bats. We also experienced shortcomings with our refugia tube collection method. Due to the lack of observations from these refugia tubes, we can only assume that the frogs had a bias to continue to choose natural refugia over an artificial one [33]. Thus, we concluded that refugia tubes were not a useful data collection method for our study site.

5. Conclusions

In conclusion, in this case study, the use of a variety of different collection methods led to a more comprehensive study of the herpetofauna along the study transect. This allows us to think further about how these relationships may vary seasonally and develop further occupancy patterns. Locally, this information can be used to further implement conservation strategies regarding the establishment of an effective buffer around the trail and the forested areas through which it runs. Future research should evaluate the added value of using multiple techniques, including ARUs, in other study regions. In addition researchers should continue to use this variety of collection methods to create a reliable index for other species.

Author Contributions: Conceptualization, I.L.B. and J.E.Q.; methodology, I.L.B. and J.E.Q.; software, I.L.B.; validation, I.L.B.; formal analysis, I.L.B.; investigation, I.L.B.; resources, I.L.B. and J.E.Q.; data curation, I.L.B.; writing—original draft preparation, I.L.B.; writing—review and editing, I.L.B. and J.E.Q.; visualization, I.L.B.; supervision, J.E.Q.; project administration, I.L.B. and J.E.Q.; funding acquisition, J.E.Q. All authors have read and agreed to the published version of the manuscript.

Funding: This research was funded by the Summer Fellowship Program within the Biology Department at Furman University.

Institutional Review Board Statement: Not applicable.

Informed Consent Statement: Not applicable.

Data Availability Statement: Data will be accessible within Harvard Dataverse following publication of the research and accompanying manuscripts.

Acknowledgments: We thank the Furman University Biology Department, Furman Undergraduate Research, and members of the CHESS lab (Ashley Razo, Caroline Vickery, Holly Brown, Sydney Anderson, Sydney McManus, Marchant Harris, Jake Gerardi, Emilia Hyland).

Conflicts of Interest: The authors declare no conflict of interest. The funders had no role in the design of the study; in the collection, analyses, or interpretation of data; in the writing of the manuscript; or in the decision to publish the results.

References

1. Xie, J.; Michael, T.; Zhang, J.; Roe, P. Detecting frog calling activity based on acoustic event detection and multi-label learning. *Procedia Comput. Sci.* **2016**, *80*, 627–638. [CrossRef]
2. Eakin, C.; Calhoun, A.J.K.; Hunter Jr., M. L. Indicators of wood frog (*Lithobates sylvaticus*) condition in a suburbanizing landscape. *Ecosphere* **2019**, *10*, e02789. [CrossRef]
3. Pyles, M.V.; Prado-Junior, J.A.; Magnago, L.F.; de Paula, A.; Meira-Neto, J.A. Loss of biodiversity and shifts in aboveground biomass drivers in tropical rainforests with different disturbance histories. *Biodivers. Conserv.* **2018**, *27*, 3215–3231. [CrossRef]
4. Howell, H.J.; Rothermel, B.B.; White, N.; Searcy, C.A. Gopher tortoise demographic responses to a novel disturbance regime. *J. Wildl. Manag.* **2020**, *84*, 56–65. [CrossRef]
5. Mengak, M.T.; Guynn, D.C. Pitfalls and Snap Traps for Sampling Small Mammals and Herpetofauna. *Am. Midl. Nat.* **1987**, *118*, 284. [CrossRef]
6. Ribeiro-Júnior, M.A.; Gardner, T.A.; Ávila-Pires, T.C.S. Evaluating the Effectiveness of Herpetofaunal Sampling Techniques across a Gradient of Habitat Change in a Tropical Forest Landscape. *J. Herpetol.* **2008**, *42*, 733. [CrossRef]
7. Trimble, M.J.; van Aarde, R.J. A note on polyvinyl chloride (PVC) pipe traps for sampling vegetation-dwelling frogs in South Africa. *Afr. J. Ecol.* **2013**. [CrossRef]
8. Field, S.A.; Tyre, A.J.; Thorn, K.H.; O'Connor, P.J.; Possingham, H.P. Improving the efficiency of wildlife monitoring by estimating detectability: A case study of foxes (*Vulpes vulpes*) on the Eyre Peninsula, South Australia. *Wildl. Res.* **2005**, *32*, 253. [CrossRef]

9. Sugai, L.S.M.; Silva, T.S.F.; Ribeiro, J.W., Jr.; Llusia, D. Terrestrial passive acoustic monitoring: Review and perspectives. *BioScience* **2019**, *69*, 15–25. [CrossRef]
10. Acevado, M.A.; Villanueva-Rivera, L.J. Using automated digital recording systems as effective tools for the monitoring of birds and amphibians. *Wildl. Soc. Bull.* **2006**, *34*, 211–214. [CrossRef]
11. Jorge, F.C.; Machado, C.G.; Siqueira da Cunha Nogueira, S.; Nogueira-Filho, S.L.G. The effectiveness of acoustic indices for forest monitoring in Atlantic rainforest fragments. *Ecol. Indic.* **2018**, *91*, 71–76. [CrossRef]
12. Schindler, A.R.; Gerber, J.E.; Quinn, J.E. An ecoacoustic approach to understand the effects of human sound on soundscapes and avian communication. *Biodiversity* **2020**, *21*, 15–27. [CrossRef]
13. Caldwell, K.L.; Carter, T.C.; Doll, J.C. A comparison of bat activsity in a managed central hardwood forest. *Am. Midl. Nat.* **2019**, *181*, 225–244. [CrossRef]
14. MacLaren, A.R.; Crump, P.S.; Forstner, M.R. Optimizing the power of human performed audio surveys for monitoring the endangered Houston toad using automated recording devices. *PeerJ* **2021**, *9*, 119–135. [CrossRef]
15. Quinn, J.E.; Brandle, J.R.; Johnson, R.J.; Tyre, A.J. Application of detectability in the use of indicator species: A case study with birds. *Ecol. Indic.* **2011**, *11*, 1413–1418. [CrossRef]
16. Quinn, J.E.; Schindler, A.E.; Blake, L.; Schaffer, S.K.; Hyland, E. Loss of winter wonderland: Proximity to different road types has variable effects on winter soundscapes. *Landsc. Ecol.* **2021**, *37*, 381–391. [CrossRef]
17. Allen-Ankins, S.; McKnight, D.T.; Nordberg, E.J.; Hoefer, S.; Roe, P.; Watson, D.M.; McDonald, P.G.; Fuller, R.A.; Schwarzkopf, L. Effectiveness of acoustic indices as indicators of vertebrate biodiversity. *Ecol. Indic.* **2023**, *147*, 109937. [CrossRef]
18. Brown, M.G.; Quinn, J.E. Zoning does not improve the availability of ecosystem services in urban watersheds. A case study from Upstate South Carolina, USA. *Ecosyst. Serv.* **2018**, *34*, 254–265. [CrossRef]
19. Boughton, R.G.; Staiger, J.; Franz, R. Use of PVC Pipe Refugia as a Sampling Technique for Hylid Treefrogs. *Am. Midl. Nat.* **2000**, *144*, 168–177. [CrossRef]
20. Dorcas, M.E.; Gibbons, W. *Frogs & Toads of the Southeast*; University of Georgia Press: Athens, GA, USA, 2008; Volume 264.
21. Sidie-Slettedahl, A.M.; Jensen, K.C.; Johnson, R.R.; Arnold, T.W.; Austin, J.E.; Stafford, J.D. Evaluation of autonomous recording units for detecting 3 species of secretive marsh birds. *Wildl. Soc. Bull.* **2015**, *39*, 626–634. [CrossRef]
22. Campos-Cerqueira, M.; Aide, T.M. Improving distribution data of threatened species by combining acoustic monitoring and occupancy modeling. *Methods Ecol. Evol.* **2016**, *7*, 1340–1348. [CrossRef]
23. Ligges, U.; Krey, S.; Mersmann, O.; Schnackenberg, S. tuneR: Analysis of Music and Speech. 2018. Available online: https://CRAN.R-project.org/package=tuneR (accessed on 6 October 2023).
24. Villanueva-Rivera, L.J.; Pijanowski, B.C. Soundecology: Soundscape Ecology. R Package Version 1.3.3. 2018. Available online: https://CRAN.R-project.org/package=soundecology (accessed on 6 October 2023).
25. Villanueva-Rivera, L.J.; Pijanowski, B.C.; Doucette, J.; Pekin, B. A primer of acoustic analysis for landscape ecologists. *Landsc. Ecol.* **2011**, *26*, 1233. [CrossRef]
26. Sueur, J.; Aubin, T.; Simonis, C. Seewave: A free modular tool for sound analysis and synthesis. *Bioacoustics* **2008**, *18*, 213–226. [CrossRef]
27. Kasten, E.P.; Gage, S.H.; Fox, J.; Joo, W. The remote environmental assessment laboratory's acoustic library: An archive for studying soundscape ecology. *Ecol. Inform.* **2012**, *12*, 50–67. [CrossRef]
28. Pieretti, N.; Farina, A.; Morri, D. A new methodology to infer the singing activity of an avian community: The Acoustic Complexity Index (ACI). *Ecol. Indic.* **2011**, *11*, 868–873. [CrossRef]
29. Boelman, N.T.; Asner, G.P.; Hart, P.J.; Martin, R.E. Multitrophic invasion resistance in Hawaii: Bioacoustics, field surveys, and airborne remote sensing. *Ecol. Appl.* **2007**, *17*, 2137–2144. [CrossRef]
30. Hyland, E.B.; Schulz, A.; Quinn, J.E. Quantifying the Soundscape: How filters change acoustic indices. *Ecol. Indic.* **2023**, *148*, 110061. [CrossRef]
31. Wickham, H.; Averick, M.; Bryan, J.; Chang, W.; McGowan, L.D.; Francois, R.; Grolemund, G.; Hayes, A.; Henry, L.; Hester, J.; et al. Welcome to the tidyverse. *J. Open Source Softw.* **2019**, *4*, 1–6. [CrossRef]
32. Beninde, J.; Delaney, T.W.; Gonzalez, G.H.; Shaffer, B. Harnessing iNaturalist to quantify hotspots of urban biodiversity: The Los Angeles case study. *Front. Ecol. Evol.* **2023**, *11*, 983371. [CrossRef]
33. Martin, F. *PVC Pipe Samplers for Hylid Frogs: A Cautionary Note*; Herpetological Natural History; US Department of Energy Environmental Services and Technology Department: Oak Ridge, TN, USA, 2004.

sensors

MDPI

Article

Toward the Definition of a Soundscape Ranking Index (SRI) in an Urban Park Using Machine Learning Techniques

Roberto Benocci [1,*], Andrea Afify [2,3], Andrea Potenza [1], H. Eduardo Roman [2,*] and Giovanni Zambon [1]

[1] Department of Earth and Environmental Sciences (DISAT), University of Milano-Bicocca, Piazza della Scienza 1, 20126 Milano, Italy; a.potenza@campus.unimib.it (A.P.); giovanni.zambon@unimib.it (G.Z.)
[2] Department of Physics, University of Milano-Bicocca, Piazza della Scienza 3, 20126 Milano, Italy; a.afify@campus.unimib.it or a.afify@nexid.it
[3] NEXiD Edge, NEXiD, Via Fabio Filzi 27, 20124 Milano, Italy
* Correspondence: roberto.benocci@unimib.it (R.B.); hector.roman@unimib.it (H.E.R.)

Abstract: The goal of estimating a soundscape index, aimed at evaluating the contribution of the environmental sound components, is to provide an accurate "acoustic quality" assessment of a complex habitat. Such an index can prove to be a powerful ecological tool associated with both rapid on-site and remote surveys. The soundscape ranking index (SRI), introduced by us recently, can empirically account for the contribution of different sound sources by assigning a positive weight to natural sounds (biophony) and a negative weight to anthropogenic ones. The optimization of such weights was performed by training four machine learning algorithms (decision tree, DT; random forest, RF; adaptive boosting, AdaBoost; support vector machine, SVM) over a relatively small fraction of a labeled sound recording dataset. The sound recordings were taken at 16 sites distributed over an area of approximately 22 hectares at Parco Nord (Northern Park) of the city Milan (Italy). From the audio recordings, we extracted four different spectral features: two based on ecoacoustic indices and the other two based on mel-frequency cepstral coefficients (MFCCs). The labeling was focused on the identification of sounds belonging to biophonies and anthropophonies. This preliminary approach revealed that two classification models, DT and AdaBoost, trained by using 84 extracted features from each recording, are able to provide a set of weights characterized by a rather good classification performance (F1-score = 0.70, 0.71). The present results are in quantitative agreement with a self-consistent estimation of the mean SRI values at each site that was recently obtained by us using a different statistical approach.

Keywords: soundscape; ecoacoustic indices; soundscape ranking index (SRI); urban parks; machine learning

Citation: Benocci, R.; Afify, A.; Potenza, A.; Roman, H.E.; Zambon, G. Toward the Definition of a Soundscape Ranking Index (SRI) in an Urban Park Using Machine Learning Techniques. *Sensors* **2023**, *23*, 4797. https://dx.doi.org/10.3390/s23104797

Academic Editor: Benoit Vozel

Received: 21 March 2023
Revised: 8 May 2023
Accepted: 12 May 2023
Published: 16 May 2023

1. Introduction

Among the elements used to evaluate the environmental status as a whole, one is strictly connected to the acoustic quality of a habitat, recognized as a vital dimension of wildlife conservation [1,2]. The induced modifications prompted by the encroaching urbanization with increasingly excessive human noise and a lack of gradients between natural and built environments can lead to direct deleterious effects on biodiversity as documented in recent works [3–6].

The diffusion of passive acoustic monitoring with a large memory capability, and the possibility of analyzing acoustic recordings by extracting specific spectral and level characteristics through ecoacoustic indices (see [7] for a review), allow us to retrieve important information about the unique assemblage of sounds across space and time. Such habitat characteristics are collectively referred to as a soundscape [8,9], the latter recognized as a distinct feature or ecological "signature" of a landscape [10,11].

Such characteristics can indeed be reflected in ecoacoustics indices calculated over predefined time intervals. Thus, they integrate the acoustic dynamics of an ecosystem, consisting of vocalizing species, anthropogenic noise, and natural phenomena [12], into a set of time series that can be proved to explain observed changes in habitat status [13], providing insights on species diversity and human impacts across a wide range of terrestrial [14–16] and aquatic environments [17,18]. The validation of ecoacoustic indices calculation is usually sound-truthed by specialized operators that classify hours of recordings according to predefined sound categories.

This identification procedure of sound sources is highly time consuming and requires specific knowledge of animal vocalizations. This necessarily limits its applicability to small datasets [19,20]. A cumulative approach that provides a qualitative description of the recorded sound (e.g., many/few vocalizing birds, many/few birds species, high/low traffic noise, etc.) partially improved the validation process, showing good matching between the "manual" identification of acoustic categories and ecoacoustic indices [21]. Thus, the need for employing unsupervised methods to process large amounts of information, independently of human intervention, is evident. Having access to such techniques can allow us to study huge datasets on very different time and spatial scales, prompting the disentangling of information hidden within the complex network of interest. To this end, we resort to machine learning (ML) techniques. The latter are currently widely used to train models using empirical data for a plethora of applications, such as translation, text classification, web searching, image recognition, and speech recognition. For instance, some of the first relevant applications were developed for the classification of handwritten digits [22] and the automatic composition of music [23,24].

It is widely recognized that ML techniques have the ability to learn generic spatial features [25,26], suitable in particular for image-related tasks. Recent applications of deep learning (DL) and ML computations for studying soundscape dynamics show promising results in terms of species identification [27,28], the separation of audio sources by using a set of techniques aimed at recovering individual sound sources when only mixtures are accessible [29], and also unsupervised classifications by means of convolutional neural networks (CNNs) [11]. In urban areas, ML models have been applied for predicting long-term acoustic patterns from short-term sound pressure level measurements [30] and for detecting anomalous noise sources prior to computing traffic noise maps [31]. CNNs have been applied to soundscape classification [32,33], species-specific recognition [34–36], and the identification of multiple and simultaneous acoustic sources using a two-stage classifier able to determine, in real time, simultaneous urban acoustic events taking advantage of physical redundancy from a wireless acoustic sensors network (WASN) in the city of Barcelona [37].

Several soundscape sources have been classified using two CNN classifiers to distinguish between biophony and anthropophony in the city of London by training CNN models on a limited quantity of labeled sound samples [28]. Their results exceed the analysis performed by multiple acoustic indices. Other attempts successfully provided sound sources identification at the price of a huge "manual" procedure in approximately 60,000 sound recordings [38]. The prediction of soundscape components, including quiet periods and microphone interference, was also performed by training a CNN with a huge dataset collected over four years across Sonoma County (California) by citizen scientists with high precision [39].

Other examples of ML applications to soundscape prediction can be found in [40–44]. In [40], the authors present a method for the automatic recognition of the soundscape quality of urban recordings by applying four different support vector machine (SVM) regressors to a combination of spectral features. In [41], a mixture of features (temporal, spectral, and perceptual) was used to classify urban sound events belonging to nine different categories. In [42], the detection and classification of acoustic events were obtained by using a modified Viterbi decoding process in combination with weighted finite-state transducers (WFSTs). In [43], acoustic indicators collected from the city of Barcelona were used to train several clustering algorithms, demonstrating the possibility of parceling the city based on the noise levels in the area. In [44], an unsupervised learning technique was applied to group the nodes of a WASN in clusters with the same behavior, thus recognizing complex patterns on this basis. Other studies make use of ML techniques together with signal processing to classify acoustic events at subsequent stages (layers) [45].

In [46], two types of supervised classifiers, namely artificial neural networks (ANNs) and a Gaussian mixture model (GMM), were compared to determine the primary noise source in the acoustic environment. In [47], the authors combined local features and short-term sound recording features with long-term descriptive statistics to create a deep convolutional neural network for classifying urban sound events. In [48], four well-known deep neural networks were fine-tuned for bird species classification.

As can be appreciated, the use of ML techniques have mostly found applications in soundscape studies by correlating different noise events to the perception of the population with the aim of automatically detecting potentially disturbing noise events (see also [49]). When using traditional ML algorithms, the choice of appropriate features of the audio file either in the form of frequency content, dynamic information, or both, always represents the first step in the analysis. In this regard, the most widely used feature is represented by the mel-frequency cepstral coefficient (MFCC).

Urban parks represent a unique area of study as retrieving source-specific information from geophony, biophony, and anthropophony remains a challenging task due to interference and confounding factors arising from the simultaneous presence of different sound sources. It should be emphasized that, in the existing literature, the question of defining an indicator that 'summarizes' the information about the acoustic environment unbound from human perceptual nature has not generally been considered. In order to fill this gap, we devised an index enabling us to quantify the quality of the local environment sound in a simple fashion. The index is referred to as the soundscape ranking index (SRI), and is assembled by weighting the different soundscape components (geophony, biophony, and anthropophony) present in a habitat/environment. The identification of the different soundscape components requires a time-intensive "manual" labeling or sound truthing of the recorded audio files, and is usually performed by a single expert.

In this work, we studied the possibility of predicting the SRI at an urban park in the city of Milan (Italy) from the extracted spectral features of audio recordings. This task was pursued by applying a set of ML classification models to our dataset collected over an extended area, where the resulting indexes were grouped, for simplicity and in accord with our previous works, into three main categories denoted as "poor", "intermediate", and "good" environment sound qualities. This classification was obtained according to the different contributions of the environment sound sources. These groups are influenced by the choice of the set of weights attributed to each soundscape component (typically the weight is given a positive value in the case of the presence of biophonies and a negative value in the presence of anthropophonies). Here, we used the classification capabilities of the chosen ML algorithms to fine-tune the soundscape weights, thus obtaining the optimal separation of the area of study in terms of local environmental sound qualities. The use of ML techniques to study the SRI will allow us to consider much larger datasets than those studied by means of supervised methods requiring human intervention. Work along this line is in progress.

The paper is organized as follows. Section 2 describes the area of study and the instrumentation used. The formulation of SRI in terms of weighting the different soundscape components is discussed together with the ML optimization procedure used. The classification models and spectral features representing the dataset are described in detail. The results are presented in Section 3, where the different models are also discussed on the basis of the validation procedures. In Section 4, we summarize the main achievements of the present work and outline possible future developments along the present lines.

2. Materials and Methods

In this section, we briefly describe the area of study, instrumentation, and recording pattern. We include the description of the SRI index, the scheme used for its prediction and optimization based on the features extracted from the spectral analysis of audio recordings in the form of ecoacoustic indices and mel-frequency cepstral coefficients (MFCCs), and the manual labeling. We also illustrate the classification models used to predict the manual labeling from the extracted features of the audio recordings.

2.1. Area of the Study

The Parco Nord (Northern Park) in the city of Milan extends over an area of approximately 790 hectares and is located within a highly urbanized area. Approximately 45% of its surface is dedicated to natural green spaces and vegetation, whereas the remaining surface is devoted to agricultural activities and infrastructures. The area of study is a tree-covered parcel of approximately 22 hectares encircled by agricultural fields, lawns, paths, and roads (see Figure 1). It has a semi-natural structure that is characterized by herbaceous layers of nemoral flora, shrub and arboreal layers, and dead wood. The area is crossed by numerous paths and is mainly used for recreational activities. It contains an artificial lake of approximately 300 m^2 located approximately 250 m from the edge of the bush. The main traffic noise sources are the A4 highway and the Padre Turoldo street, located to the north at around 100 m from the wooded parcel. There is also the presence of a small airport (Bresso airport) on the west side at around 500 m from the tree-line edge.

Figure 1. Area surrounding the grid of sensors indicated as numbered spots. Red spots indicate the active recording sites, and yellow spots indicate sites with recording disruption. In the figure, the A4 highway, Padre Turoldo Street, and Bresso airport runaway are indicated.

2.2. Audio Recorders

We used low-cost digital audio recorders produced by the SMT Security (Figure 2a). They were set to measure continuously with a sampling rate of 48 kHz and were equipped with a two-week lifetime powerbank. Before using low-cost devices, it was necessary to verify their possible different frequency responses in the frequency range of interest. Thus, initial tests were devoted to selecting those recorders with a frequency response within 5% of the average spectrum calculated over all recorders. The average spectrum was computed using a 512-point FFT analysis by applying a white noise as a sound source. Full details of the the frequency characterization of the recorders can be found in [50]. The results are reported in Figure 2b, where a reduction in sensitivity for frequencies higher than 10 kHz is observed.

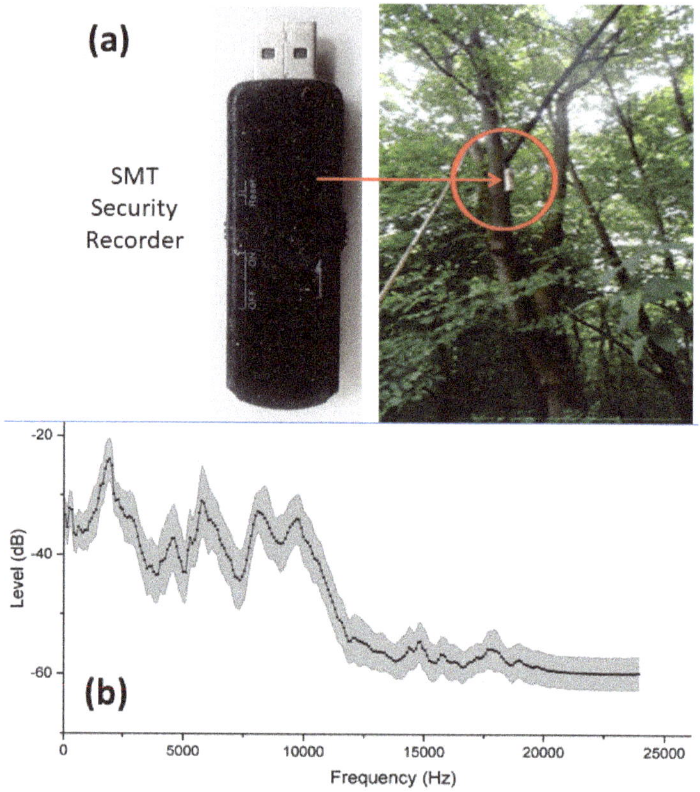

Figure 2. (**a**) SMT Security recorder and its practical location (red circle) on a tree. (**b**) Average spectrum (computed using a 512-point FFT) of sound level response of the recorders [dB] vs. frequency in the range (0–24) kHz. The gray band around the curve corresponds to one standard deviation of the response calculated over several SMT Security recorders. Sensitivity decreases for frequencies higher than 10 kHz.

2.3. Measurement Scheme

The 22 recorders were initially positioned on a regular grid, as shown in Figure 1, covering an area of approximately $270 \times 500\,\text{m}^2$, plus another grid with an area of $300 \times 250\,\text{m}^2$ for the southern part of the parcel. The recordings were scheduled for the period of greatest singing activity of the avifauna and repeated over four days, namely on 25–28 May 2015, from 06:30 a.m. to 10:00 a.m. (CET), corresponding to 3.5 h for each site and for each recording session. Unfortunately, six recorders did not work properly (see yellow spots in Figure 1) and thus the audio files analyzed in this study involved only 16 sites.

2.4. Aural Survey

In this section, we describe the scheme adopted for the aural analysis of audio files in order to quantify distinctive sound features. An aural survey was carried out to quantify the biophonies, anthropophonies, and geophonies. In particular, a single expert listened carefully to the recordings according to the following scheme: one minute listened to for every three minutes of continuous recording, for a total of seventy minutes of listening per site. The expert focused on quantifying the biophonic activity (mainly avian vocalizations) and the tecnophonic sources, evaluating the parameters reported in the scheme of Figure 3.

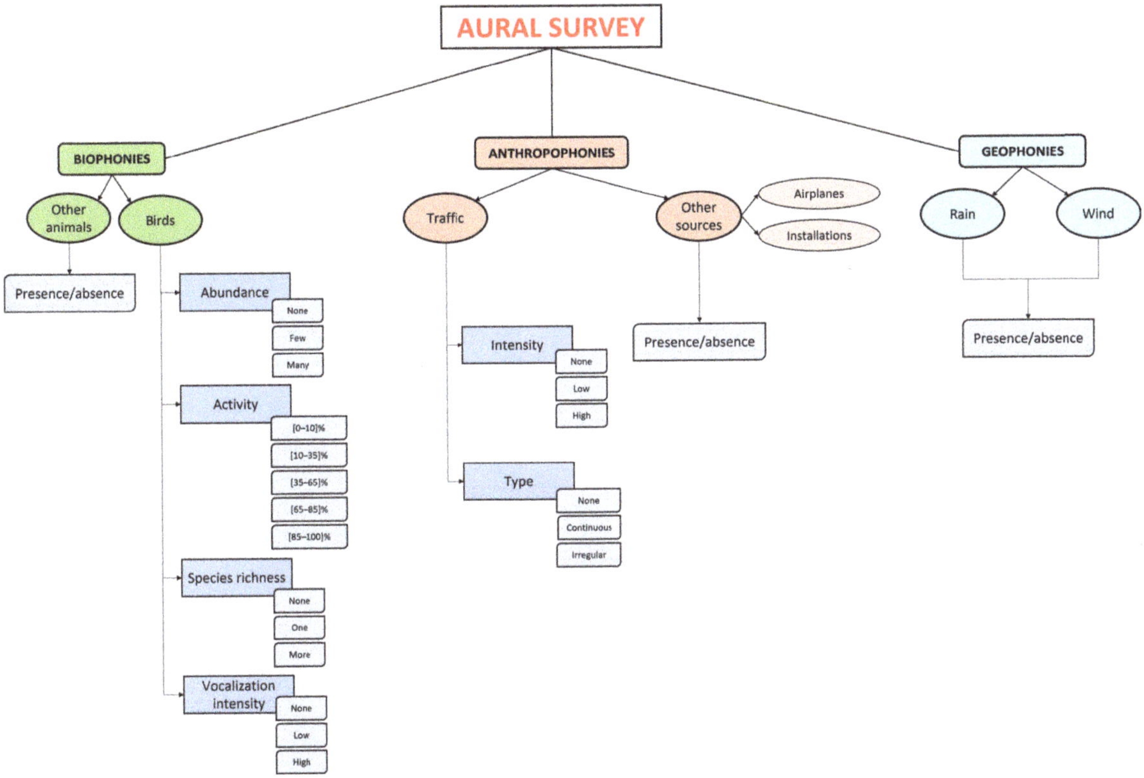

Figure 3. Classification of sound sources considered for the aural survey corresponding to six categories: birds, other animals, road traffic noise, other noise sources (airplanes, trains), rain, and wind. For each category, the following attributes were considered. Birds: (1) individual abundance, (2) perceived singing activity (%), (3) species richness, (4) vocalization intensity. Other biological sound sources: presence–absence. Anthropogenic noise: (1) noise intensity, (2) typology of traffic. Other anthropogenic sources: presence–absence. Geophonies were absent in the considered recordings.

Each soundscape component was analyzed to extract information about the sound source and its occurrence and intensity (see Figure 3). Following this criterion, the avian vocalizations were the most studied. For each minute listened, four parameters were evaluated: (1) individual abundance (no–few–many subjects), (2) perceived singing activity expressed as the percentage of time occupied by avian vocalizations (0–100%), (3) species richness (none–one–more than one species), and (4) vocalization intensity (no–low–high intensity). For other biological sound sources, such as other animals and people, just the presence–absence indicator was used.

The anthropogenic noise is mainly attributable to road traffic. For this source, two parameters were evaluated (see Figure 3): (1) noise intensity (no–low–high intensity) and (2) the typology of traffic (no–continuous–intermittent traffic). Other sources, such as installations and airplanes, were also studied using the presence–absence indicator. Finally, given their poor contribution to the soundscape of the area during the measurement campaign, geophonies such as rain and wind were not considered.

2.5. The Soundscape Ranking Index, SRI

We wish to quantify the quality of the local environment sound by means of a single index, SRI, as proposed recently [51]. In the following, we briefly recall the definition of the SRI, introduced to describe, on average, the local environment sound,

$$\mathrm{SRI}_{N_T} = \frac{1}{N_T} \sum_{r=1}^{N_T} \sum_{i=0}^{n_c} c_{N_i} N_{i,r}, \tag{1}$$

where N_T refers to the total number of recordings, $n_c + 1$ is the total number of identified categories (birds, other animals, road traffic noise, other noise sources, and rain and wind)—here, $n_c = 4$ and $N_{i,r} = 1$ if the ith sound category is present at the recording r; otherwise $N_{i,r} = 0$—and c_{N_i}s are coefficients chosen within the ranges displayed in Table 1 [51].

In the present study, a single audio recording, i.e., $N_T = 1$, was considered. The reason for this choice relies on the need to compare the present new results with those discussed by us in a previous work [52]. It can also be seen as providing a "snapshot" of the local soundscape. It should be emphasized that the present work is a first attempt to estimate the SRI for the Northern Park in Milan using ML techniques. Extensions of this analysis to more audio recordings is under consideration and the results will be considered elsewhere. Each set of calculated spectral features is expected to be correlated to a series of manually recognized sound categories within the single audio recording (we can provide the audio recording data upon request). In this specific case, Equation (1) becomes (see also [52])

$$\mathrm{SRI}_\ell = \sum_{i=0}^{n_c} c_{N_i} N_{i,\ell}, \tag{2}$$

where the subindex ℓ refers to the ℓth recording and the coefficients take on the following values: $c_{N_i} > 0$ ($c_{N_i} \to c_+, c_{++}$) when a sound category is associated with a natural sound where we have split the values into two subranges ($+$, $++$), and $c_{N_i} < 0$ ($c_{N_i} \to c_-, c_{--}$) for a potential disturbing event, also split into two subranges. The absence of bird vocalization is regarded as neutral, $c_{N_0} = 0$. In Table 1, we report the assumed ranges of variability for c_{N_i}. Note that in [52] we used the notation $P(1) = c_{++}$, $P(2) = c_+$, $P(3) = c_{N_0}$, $P(4) = c_-$, and $P(5) = c_{--}$.

Table 1. Range of variation in the coefficient c_{N_i} assigned to each sound category, $i = 0, \ldots, 4$, to be used in Equation (2). In this case, we aribitrarily chose $-5 \le c_{N_i} \le 5$ for convenience.

c_{N_i}	Range	P(i) [52]
c_{++}	$[2, 5]$	P(1)
c_+	$[0, 2]$	P(2)
c_{N_0}	0	P(3)
c_-	$[-2, 0]$	P(4)
c_{--}	$[-5, -2]$	P(5)

Thus, Equation (2) provides a number that is expected to be representative of the environmental sound quality. Following our previous works [51,52], we chose three intervals of the SRI to define the environmental sound quality, for a single recording denoted simply as ℓ, given by

$$\begin{aligned} \mathrm{SRI}_\ell &< 0 \quad \text{[poor quality]}, \\ 0 \le \mathrm{SRI}_\ell &\le 2 \quad \text{[medium quality]}, \\ \mathrm{SRI}_\ell &> 2 \quad \text{[good quality]}. \end{aligned} \tag{3}$$

It should be emphasized that the choice of these intervals for classifying the SRI is rather arbitrary. Nonetheless, they are based on rather generic features attributed to human perception, in the sense that the poor (good) quality interval reflects a prevalence of anthropogenic (biophonic) sound sources, whereas the intermediate quality interval reflects a balance/co-existence of both types of sound sources.

2.6. SRI *Optimization Procedure*

We first searched for the sound categories, reported in Table 2, by manually labeling the audio recording (see [52] for additional details). Each category i was assigned a weight, c_{N_i}, according to the attributes present in the audio recording (see column 'Attribute' in Table 2). The singing activity was assigned a weight that depends on the percentage of birds singing in each recording, e.g., for a singing activity in the interval (0, 25]%, we assigned a weight of $0.25 \times c_{++}$, whereas, for (25, 50]%, we assigned a weight of $0.50 \times c_{++}$, etc.

Then, we calculated the spectral features of the sound recording and implemented the optimization scheme illustrated in Figure 4. In order to achieve this, we assumed that the coefficients c_{N_i} can vary over the intervals reported in Table 1.

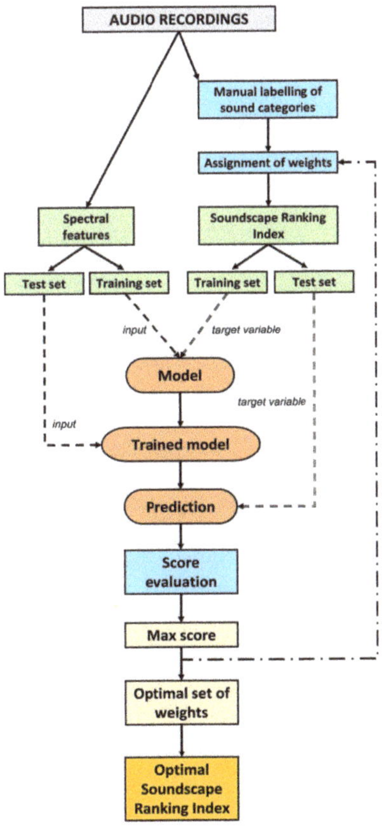

Figure 4. Scheme of the optimization procedure for the SRI according to the following steps: (1) assignment of weights to each sound category; (2) splitting of extracted spectral features and of the corresponding SRI into test and training sets; (3) running of classification models; (4) computation of classification score; (5) selection of optimal SRI according to the highest classification score.

Table 2. Coefficient c_{N_i} assigned to each sound category to be used in Equation (2).

Category	Attribute	c_{N_i}
Birds singing	no few many	c_{N_0} c_+ c_{++}
Birds species	no $\lesssim 2$ >2	c_{N_0} c_+ c_{++}
Singing activity (%)	0 (0, 25] (25, 50] (50, 75] (75, 100]	$0.00 \times c_{++}$ $0.25 \times c_{++}$ $0.50 \times c_{++}$ $0.75 \times c_{++}$ $1.00 \times c_{++}$
Traffic type	no traffic continuous intermittent	c_+ c_- c_{--}
Traffic intensity	zero low high	c_+ c_- c_{--}
Other sound sources	absent present	c_{N_0} c_{--}

In order to proceed, both the spectral features and SRIs need to be split into "training" and "test" sets. The training set is used as the input for each classification model, whereas the performance of the test set is quantified according to the metrics described in Section 2.9. Here, we implicitly assumed that the optimal classification outcome produces the optimal separation/distance among the sites. This consideration comes from the analysis performed in a previous work where sites were clustered on the basis of distances (dissimilarities) calculated over the extracted spectral features of the audio files recorded in the area of study [51,52]. In the classification process, SRIs represent the target variable to be predicted by each model. This process is repeated for each combination of weights, c_{N_i}, where $i > 0$, that is varied in the assigned interval. We considered a variation step $\Delta c = 0.1$ for each of the four intervals for c_{N_i}, where the total number of choices is given by the product of the number of possible values for each coefficient c_{N_i}: 20 values for c_+ and c_- each, and 30 for c_{++} and c_{--} each, yielding $20^2 \times 30^2 = 360{,}000$ combinations. The set of weights that define the optimal SRI was then obtained on the basis of a classification score as defined in Section 2.9.

2.7. Classification Models

In this section, we provide a brief description of the classification models used to predict the manual labeling sound categories, and thus the SRI index, from the ecoacoustic indices. In general, machine learning methods are better able to address multicollinearity issues and capture the potential non-linear relationships among variables. While binary classification is a more common application of them, they are also widely used for regressions when the target is continuous. In our case, we used classification algorithms for trinary (poor/medium/good) categories for determining the SRI.

The models taken into consideration in this study, which were implemented in Python programming language [53], are the following:

- Decision tree (DT);
- Random forest (RF);
- Support vector machine (SVM);
- Adaptive boosting (AdaBoost).

The supervised classification models implemented for the SRI optimization procedure were trained on the 80% of the data and tested on the remaining 20%. Data were split using a stratified procedure to keep the proportions between the classes of the corresponding target variable. Furthermore, class weights were used to take into account the class imbalance of the data set. In fact, training an algorithm with a skewed distribution of the classes can be achieved by giving different weights to both the majority and minority classes. The difference in weights will influence the classification of the classes during the training phase. The whole purpose is to penalize the misclassification made by the minority class by setting a higher class weight and, at the same time, reducing the weight for the majority class. The weight for the jth class of the target variable was chosen as follows:

$$W_j = \frac{1}{|j|}\frac{T}{N}, \tag{4}$$

where T is the total number of data items, N is the number of classes of the target variable, and $|j|$ is the number of items in the jth class.

2.7.1. Decision Trees

A decision tree (DT) is a non-parametric supervised learning method used to predict the value of a target variable by learning simple decision rules inferred from the data features [54,55]. DTs can be used to classify a set of data items using the inferred rules to recursively partition the feature space until each partition is pure or a stopping criterion is reached. Specifically, a DT learns a sequence of if-then statements, with each statement involving one feature and one split point. The topmost node in a DT is known as the root node and is constituted by the whole set of items. The root node is split starting from those variables that lead to the greatest degree of homogeneity.

Several measures were designed to evaluate the impurity of a partition, of which the Gini impurity (GI) is among the most popular ones [56]. Then, following the same criterion, each subsample, called a node, is split recursively into smaller nodes alongside single variables, and according to threshold values that identify two or more branches. Finally, when a node is no longer split into further nodes, either because a stopping criterion is reached or because it is pure, it becomes a leaf of the tree. An item is assigned to the class that has been associated with the leaf that it reaches.

2.7.2. Random Forest

A random forest is a meta estimator that fits a number of decision tree classifiers on various sub-samples of the dataset, and uses averaging to improve the predictive accuracy and control overfitting. Random forests (RFs), or random decision forests, are an ensemble of learning methods used for classification, regression, and other tasks that operates by constructing a multitude of decision trees during the training procedure. For classification tasks, the output of the random forest is the class that is selected by most trees. In other words, it fits a number of decision tree classifiers on various sub-samples of the dataset and uses averaging to improve the predictive accuracy and control overfitting [57]. For this reason, RF generally outperforms decision tree models.

2.7.3. Support Vector Machine

Support vector machines (SVMs) are supervised machine learning models that can be used for both classification and regression purposes [58,59]. They were initially devised as a binary classifier. SVMs map training data to points in space in order to maximize the width of the gap between the two categories. Thus, new data are mapped into that same space and predicted to belong to a category based on which side of the gap that they fall.

For multiclass classification, the same idea is employed by decomposing the multiclassification problem into multiple binary classification problems. This can be achieved by mapping data points to a high dimensional space to gain mutual linear separation between every two classes. This is called a *One-vs.-One* approach, which breaks down the multiclass problem into multiple binary classification problems using a binary classifier per each pair of classes. Another approach that can be used is the so-called *One-vs.-All* approach. In this case, a binary classifier per each class is used. The latter approach is used for the SR optimization procedure.

In general, a data point is viewed as a p-dimensional vector (a list of p numbers) and we want to know whether we can separate such points with a $(p-1)$-dimensional hyperplane. This is called a linear classifier. There are many hyperplanes that might classify the data, but the goal of SVMs is to find the best hyperplane that represents the largest separation, or margin, between the classes. It is defined so that it is as far as possible from the closest data points from each of the classes. SVMs are effective in high-dimensional spaces even if the number of dimensions is greater than the number of samples. SVMs can efficiently perform a non-linear classification using what is called the kernel trick, implicitly mapping their inputs into high-dimensional feature spaces [60].

For the sake of clarity, let us consider a simple separable classification method in multidimensional space. Given two classes of examples clustered in feature space, any reasonable classifier hyperplane should pass between the means of the classes. One possible hyperplane is the decision surface that assigns a new point to the class whose mean is closer to it. This decision surface is geometrically equivalent to computing the class of a new point by checking the angle between two vectors: the vector connecting the two cluster means and the vector connecting the mid-point on that line with the new point. This angle can be formulated in terms of a dot product operation between vectors. The decision surface is implicitly defined in terms of the similarity between any new point and the cluster mean—a kernel function. This simple classifier is linear in the feature space whereas, in the input domain, it is represented by a kernel expansion in terms of the training examples.

Radial basis function (RBF) kernels are the most generalized form of kernelization and are one of the most widely used kernels due to its similarity to Gaussian distribution [61]. The RBF kernel function for two points x and y computes the similarity or how close they are to each other. This kernel can be mathematically represented as follows:

$$K(x,y) = \exp\left(-\gamma\,||x-y||^2\right), \tag{5}$$

where γ is a hyperparameter that is inversely proportional to the standard deviation σ and $||x-y||$ is the Euclidean distance between two points x and y. The RBF kernel support vector machines are implemented using the scikit-learn library [62].

2.7.4. AdaBoost

Adaptive boosting has been a very successful technique for solving two-class classification problems. It was first introduced in [63] with the AdaBoost algorithm. In going from two-class to multi-class classification, most boosting algorithms have been restricted to reducing the multi-class classification problem to multiple two-class problems, e.g., [63–65]. The natural multi-class extension of the two-class AdaBoost was obtained with the algorithm stagewise additive modeling using a multi-class exponential loss function (SAMME) [66].

The core principle of AdaBoost is to fit a sequence of weak learners (i.e., models that are only slightly better than random guessing, such as small decision trees) on repeatedly modified versions of the data. The predictions from all of them are then combined through a weighted majority vote (or sum) to produce the final prediction. The data modifications at each so-called boosting iteration consist of applying weights, $(w_1, w_2, \ldots, w_N)$, to each of the training samples. Initially, the weights are set to $w_i = 1/N$ so that the first step simply trains a weak learner on the original data. For each successive iteration, the sample weights are individually modified and the learning algorithm is reapplied to the reweighted data. At a given step, those training examples that were incorrectly predicted by the boosted model induced at the previous step have their weights increased, whereas the weights are decreased for those that were predicted correctly. As iterations proceed, examples that are difficult to predict receive ever-increasing influence. Each subsequent weak learner is thereby forced to concentrate on the examples that are missed by the previous ones in the sequence [57].

2.8. Feature Extraction

In this section, we describe the features that were employed in the machine learning process. Feature extraction starts from the audio recordings and builds derived values (features) containing salient or summative information about the measured data. This process is intended to help the learning procedure by providing significant information about the content of the recordings. Here, we essentially used two types of features: those based on ecoacoustic indices and those based on mel-frequency cepstral coefficients (MFCCs) (see below).

2.8.1. Ecoacoustic Indices

Ecoacoustic indices (ECOs) are generally used to quantify the soundscape in both marine and terrestrial habitats, and are grouped into categories aiming at quantifying the sound amplitude and its level of complexity and weighting the importance of geophonies, biophonies, and technophonies (soundscape). In this work, we focused on the following set of ecoacoustic indices:

- The acoustic entropy index (H) highlights the evenness of a signal's amplitude over time and across the available range of frequencies [67].
- The acoustic complexity index (ACI) accounts for the modulation in intensity of a signal over changing frequencies [68].
- The normalized difference soundscape index (NDSI) accounts for the anthropogenic disturbance by computing the ratio between technophonies and biological acoustic signals [69].
- The bio-acoustic index (BI) is calculated as the area under the mean frequency spectrum above a threshold characteristic of the biophonic activity [15].
- The dynamic spectral centroid (DSC) indicates the center of mass of the spectrum [70].
- The acoustic diversity index (ADI) provides a measure of the local biodiversity at the community level without any species identification [70].
- The acoustic evenness index (AEI) provides reverse information of ADI with high values identifying recordings with the dominance of a narrow frequency band [70].

The ecoacoustic indices were calculated in the R environment (version 3.5.1 [36]). Specifically, the fast Fourier transform (FFT) was computed by the function *spectro* available in the R package "seewave" [71] in the frequency interval (0.1–12) kHz based on 1024 data points corresponding to a frequency resolution of $FR = 46.875$ Hz and, therefore, to a time resolution $TR = 1/FR = 0.0213$ s. The ecoacoustic indices were computed using the R package "soundecology" [72]. A dedicated script running in the "R" environment was written to calculate the DSC index. Two patterns of calculation were used:

- For each one-minute recording, we computed seven cumulative indices. Each recording is thus represented by seven features (seven indices).
- For each one-minute recording, we computed each index with a one-second time-step. Then, we calculated seven statistical descriptors (over 60 values): minimum, maximum, mean, median, skewness, kurtosis, and standard deviation. Each recording is thus represented by 49 features (seven indices times seven statistical descriptors).

Table 3 reports a summary of the extracted features employed in the classification process.

Table 3. Type of features extracted from the audio recording: characteristics and numerousness. Here we use the abbreviations: ecoacoustic indices (ECOs), mel-frequency cepstral coefficients (MFCCs) (see also Section 2.8.2).

Type of Feature	Characteristics	Number of Features
Seven ecoacoustic indices (7 ECOs)	one-minute integration time	7
Seven ecoacoustic indices and seven statistical descriptors (49 ECOs)	one-second integration time	49
Twelve MFCCs and seven statistical descriptors (84 MFCCs)	one-second time window	84
Twelve MFCCs (1428 MFCCs)	one-second time window	1428

2.8.2. Mel-Frequency Cepstral Coefficients (MFCCs)

The mel-frequency cepstrum (MFC) has become a convenient alternative for obtaining a reduced amount of data from each audio recording while keeping the core spectral information. The MFC is a representation of sound based on a linear discrete cosine transform (DCT) of a log-power spectrum on a non-linear MEL scale of frequency [73] The latter is a perceptual scale of pitches judged by listeners to be equally spaced from one another (logarithmically distributed human auditory bands). Thus, after getting the spectrum onto the MEL scale, by applying filter banks and the logarithm of energies in each filter bank, the last step is to calculate the MFCCs [74]. This is carried out by fitting the cosines to the calculated log-energies using the DCT. MFCCs are the coefficients that collectively describe the MFC; that is, the amplitudes of the resulting spectrum. In most applications, the number of coefficients is twelve. This number represents a trade-off between an accurate description of the spectrum and dimensionality reduction of our feature space.

The calculation of the MFCCs was performed in the R environment using the default number of MEL filter banks; that is, 40 logarithmically distributed bands over the whole spectrum. Another important issue is the selection of the most convenient time window size for extracting the features of the data set. In this regard, we have to keep in mind that the dataset is obtained by computing a fixed number of features from an audio recording, usually referred to as a "window".

A large time window size may capture relevant events but would result in a dataset with few instances. On the other hand, a small time window would result in a larger data set but may split the relevant events into several windows. For this reason, as we are trying to classify a summative description of the audio files (information described in Figure 3), we used a one-second time window as representative for distinguishing different sound characteristics. This number was selected to frame and window each audio file using a Hamming window with an overlap of 50%. In addition, in this case, we used two patterns of calculation (see Table 3):

- For each one-minute recording, we computed 12 MFCCs in a one-second time window. Then, we calculated seven statistical descriptors resulting from each audio recording: minimum, maximum, mean, median, skewness, kurtosis, and standard deviation. This corresponds to 84 features (12 MFCCs times 7 statistical descriptors).
- For each one-minute recording, we computed 12 MFCCs in a one-second time window. This corresponds to 1428 features (12 MFCCs times 119 time windows: 60 s window with 50% overlap).

2.9. Metrics

The performance of a model can be evaluated by the use of specific metrics that quantify the capability of the model to correctly predict one's target. In our case, the performance of a model was evaluated based on a selection of the optimal set of weights, c_{N_i}, reported in Equation (2) and described in Table 2, which contributes to the definition of the SRI.

A confusion matrix is generally the starting point for calculating each metric. A confusion matrix is a table used to describe the performance of a classification model on a set of (training and test) data for which the true values are known. A strong discrepancy between the results obtained between training and test data may be indicative of an overfitting issue. It generally contains the following information: true positives, TPs, and true negatives, TN, which are the observations that are correctly predicted, and false positives, FP, and false negatives, FN, which occur when the actual class contradicts the predicted class. The derived metrics are the following [75]:

Precision: This represents the ratio of correctly predicted positive observations to the total predicted positive observations. A high precision is related to a low false positive rate:

$$\text{Precision} = \frac{\text{TP}}{\text{TP} + \text{FP}}. \tag{6}$$

Recall (Sensitivity): This is the ratio of correctly predicted positive observations to all observations in the actual class. Thus, the recall tells us the proportion of correctly identified positives:

$$\text{Recall} = \frac{\text{TP}}{\text{TP} + \text{FN}}. \tag{7}$$

F1-score: This is defined as the harmonic mean of precision and recall [75]. Therefore, the F1-score takes both FPs and FNs into account:

$$\text{F1-score} = \frac{2 \cdot (\text{Recall} \cdot \text{Precision})}{\text{Recall} + \text{Precision}}. \tag{8}$$

This metric is useful in case both precision and recall are equally important. In our case, we decided to refer to the F1-score as the classification measure.

As a validation of the results, we used the k-fold cross-validation technique. It consists of an iterative procedure used to evaluate machine learning models. The procedure has a single parameter called k that refers to the number of folds that a given data sample has to be split into. This technique returns stratified folds; that is, folds obtained by preserving the percentage of samples for each class. At each kth iteration, the kth fold is used as the test set, whereas the other folds are used to train the model.

3. Results and Discussions

The results presented in this section refer to the audio files recorded on 25 May 2015, from 06:30 a.m. to 10:00 a.m. (CET). As described in Section 2.6, from the extracted features of all the audio recordings for which we had the corresponding labeling of sound categories, we ran four machine learning models to attempt a prediction of the soundscape ranking index calculated assigning a set of weights to each sound category. The range of variation in the above mentioned weights is reported in Table 1, and the best combination is calculated by the highest score provided by each classification model. The optimal set of weights are obtained by the highest classification measure, which, in our case, is the F1-score. Using the optimal set of weights, we calculated the SRI and derived a map of the environment sound quality of the area of study.

The machine learning algorithms selected usually work better on small data sets than deep neural networks. In fact, the latter require extremely large datasets to achieve high performances. Furthermore, a large dataset was not readily available and would be expensive and time-consuming to acquire. Another consideration when selecting classical machine learning algorithms concerns hyper-parameter tuning and the interpretability of these kinds of models. The underlying mechanisms of random forest, Adaboost, and SVMs are more straightforward than those of deep neural networks. Enhancing the interpretability also results in an easier tuning of hyper-parameters. However, for our preliminary study, we leaned on the default values as reported in [76], with the exception of the max depth parameter (used in DT and RF) used to control the size of the tree to prevent overfitting.

For each model, we split the entire dataset, consisting of 1220 audio files, into a training and test set with the following proportion: 80% of the dataset used for training and 20% of the dataset used for testing. As the reference measure to search for the optimal classification we used the F1-score, which is more suited to our case, i.e., an uneven class distribution due to the limited sample size numerousness. Table 4 reports the results of the four models for each of the four extracted features.

Table 4. Summary of results obtained for decision tree (DT), random forest (RF), support vector machine (SVM), adaptive boosting (AdaBoost) models, and four extracted features. Range of weights values and classification measures are reported. Precision, recall, and F1-score are provided with their standard deviations.

DT	c_{++}	c_{+}	c_{-}	c_{--}	Precision	Recall	F1-Score
(7 ECO)	[2.0, 2.1]	[1.9, 2.0]	[−1.2, −1.0]]	−3.6	0.64 ± 0.32	0.71 ± 0.10	0.62 ± 0.15
(49 ECO)	[2.3, 2.6]	[1.5, 1.7]	[−1.2, −1.0]	[−4.72, −4.0]	0.64 ± 0.26	0.67 ± 0.03	0.63 ± 0.13
(84 MFCC)	[2.0, 2.3]	[1.8, 2.0]	[−1.2, −1.0]	[−4.4, −3.9]	0.68 ± 0.24	0.73 ± 0.09	0.68 ± 0.10
(1428 MFCC)	[2.5, 2.6]	[1.4, 1.5]	−1.0	[−5.0, −4.8]	0.63 ± 0.22	0.64 ± 0.12	0.62 ± 0.12
RF							
(7 ECO)	2.0	2.0	[−1.5, −1.4]	[−2.6, −2.7]	0.63 ± 0.28	0.75 ± 0.16	0.63 ± 0.12
(49 ECO)	2.0	1.6	−0.3	−4.7	0.70 ± 0.28	0.78 ± 0.14	0.69 ± 0.12
(84 MFCC)	[2.0, 2.1]	[0.7, 0.8]	−0.1	[−4.3, −4.1]	0.71 ± 0.22	0.78 ± 0.15	0.71 ± 0.15
(1428 MFCC)	2.0	2.0	−1.7	−2.8	0.68 ± 0.15	0.75 ± 0.16	0.70 ± 0.03
SVM							
(7 ECO)	2.5	1.6	−1.0	−4.7	0.60 ± 0.18	0.60 ± 0.06	0.59 ± 0.10
(49 ECO)	[4.3, 5.0]	[1.9, 2.0]	[−1.2, −1.0]	[−2.3, −2.0]	0.33 ± 0.58	0.33 ± 0.56	0.33 ± 0.47
(84 MFCC)	[4.3, 5.0]	[1.9, 2.0]	[−1.2, −1.0]	[−2.3, −2.0]	0.33 ± 0.58	0.33 ± 0.56	0.33 ± 0.47
(1428 MFCC)	[4.3, 5.0]	[1.8, 2.0]	[−1.2, −1.0]	[−2.3, −2.0]	0.33 ± 0.58	0.33 ± 0.56	0.33 ± 0.47
AdaBoost							
(7 ECO)	[2.4, 2.6]	[1.3, 1.7]	[−1.8, −1.5]	[−2.2, −2.0]	0.60 ± 0.19	0.64 ± 0.18	0.62 ± 0.15
(49 ECO)	[2.3, 2.4]	2.0	[−1.8, −1.7]	[−3.4, −3.2]	0.68 ± 0.05	0.70 ± 0.05	0.69 ± 0.02
(84 MFCC)	2.9	2.0	−1.7	−2.5	0.65 ± 0.24	0.78 ± 0.15	0.70 ± 0.14
(1428 MFCC)	[2.0, 2.4]	[1.5, 1.8]	[−1.2, −1.0]	[−4.1, −3.3]	0.66 ± 0.09	0.68 ± 0.04	0.67 ± 0.04

In particular, the table contains the values of the weights and the corresponding classification measures. The weights can vary in an interval, meaning that the optimal classification measure (F1-score) can be obtained for a different combination of weights. The table also contains precision and recall as classification measures. All three measures are given in terms of the mean value ± standard deviation calculated over all classification classes defined by Equation (3).

In general, we can observe an increase in the classification performance as we provide more detailed information about the spectral content of each recording from 7 to 84 extracted features. For the 1428 features for each recording (1428 MFCCs), we observe a general drop (with the exception of the RF model) in the performance, more likely due to information redundancy contained in the time series. This redundancy is smoothed out by considering the statistical descriptors of the same time series (84 MFCCs).

A similar consideration can be carried out for the ecoacoustic indices. In this case, the 7 ECO features, derived by integrating the ecoacoustic indices over the whole length of the recording (1 minute), contain more condensed information; thus, it appears to not be enough to represent the complexity of the soundscape in a single summative index. On the other hand, the 49 ECO features provide a better representation of the spectral dynamics within each single audio recording. AdaBoost and RF models perform better. The AdaBoost model yields an F1-score of 0.70 (precision 0.65 and recall 0.78) with 84 MFCCs, and an F1-score of 0.69 (precision 0.68 and recall 0.70) with 49 ECO features. The RF model yields an F1-score of 0.71 (precision 0.71 and recall 0.78) with 84 MFCCs, and an F1-score of 0.70 (precision 0.68 and recall 0.75) with 1429 MFCCs. Hence, the RF model results in a slightly higher classification performance.

The highest metric ranking leads to the definition of nearly different sets of coefficients c_{N_i} assigned to each sound category to be used in Equation (2). The DT model provides similar coefficients to the RF model. On the other hand, the SVM model gives the worst classification performance, and the resulting coefficients c_{N_i} are completely discordant from the results of the other models.

Discussions

The possibility of deriving an overall soundscape index summarizing the contribution of all the sound components and being able to rank them in terms of sound "quality" can be one of the empowering ecological tools used to help rapid on-site field and remote surveys. Here, we tested four extracted features (see Table 3) from recordings taken on 25 May 2015, from 06:30 a.m. to 10:00 a.m. (CET), referring to a measurement campaign over an area of approximately 22 hectares located in the Parco Nord of Milan.

The idea of predicting an overall ranking index from a limited number of recordings, as initally defined in [51] and complemented in [52], was further developed in this paper, representing a first attempt to summarize the herd of ecoacoustic indices and spectral features for describing specific aspects of the audio-spectral content of a recording. This preliminary approach, based on ML techniques, revealed that two classification models, RF and AdaBoost, are able to provide rather good classification measures (F1-score = 0.70–0.71) using 84 extracted features from each recording. The two models are "evolutions" of DTs, of which AdaBoost required intensive time–machine calculation to complete the whole weight scan. In order to check for the possible overfitting of the models, we implemented a procedure called k-fold cross-validation, which refers to the number of groups, k, that a given data sample is to be split into. In this case, we tested the following values: k = (2, 5, 10), and repeated the operation 200, 100, and 100 times, respectively. Figure 5 illustrates the results in terms of the associated kernel density distributions.

We find that the RF model provides a more robust classification as its F1-score distribution presents maxima at higher values than for the AdaBoost model (see Table 5) for all the k-groups of split samples considered. As k increases, we also observe a spreading of the distribution due to a less numerous dataset.

Table 5. Mean F1-score ± standard deviation calculated for the distributions illustrated in Figure 5: random forest (RF) and adaptive boosting (AdaBoost) models, where k = (2, 5, 10).

k	RF	AdaBoost
2	0.62 ± 0.02	0.49 ± 0.04
5	0.62 ± 0.03	0.51 ± 0.05
10	0.62 ± 0.04	0.52 ± 0.08

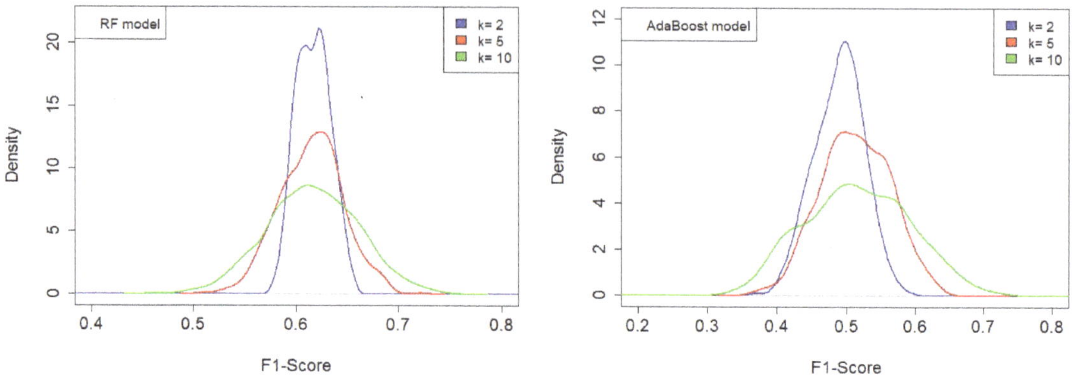

Figure 5. Kernel density distributions of F1-score for RF and AdaBoost, obtained using the k-fold splitting of the dataset for the following number of splits: k = (2, 5, 10), and repeating the operation 200, 100, and 100 times, respectively. (**Left panel**) The RF model with the following combination of weights: $c_{++} = 2.0$, $c_{+} = 1.6$, $c_{-} = -0.3$, $c_{--} = -4.7$. (**Right panel**) The AdaBoost model with the following combination of weights: $c_{++} = 2.9$, $c_{+} = 2.0$, $c_{-} = -1.7$, $c_{--} = -2.5$. RF model provides a more robust classification as its F1-score distribution presents maxima at higher values than for the AdaBoost model.

In order to further validate the obtained results, we computed an SRI map over the study area based on the weights obtained for the RF model, which is reported in Table 4. For each of the 16 sites, we considered the median value of the SRI computed over all the measurements corresponding to the labeled recordings. The results are shown in Figure 6.

As expected, lower SRI values (poor/medium soundscape quality) are found close to the traffic noise sources (Sites 1–4), where the presence of higher traffic noise and less bird singing activity contribute significantly to this outcome. On the other hand, sites belonging to the park interior are less influenced by traffic noise and host higher biodiversity (many birds of different species singing). This result is reflected by the higher SRI values (good soundscape quality). Sites (5–8) are at intermediate positions and show a sort of transient behavior of SRI values (medium/good soundscape quality).

In Figure 6 (left panel), we show the actual continuously changing SRI values, whereas, in the right panel, they are selected as in Equation (3) to obtain a simplified picture. Both maps are fully compatible with the results obtained in [50], where the statistical analysis based on the computed ecoacoustic indices revealed the presence of a two-cluster separation, and also with a more recent estimation of the SRI based on a self-consistent statistical analysis [52]. The latter gives the optimized parameter values of $c_{++} = 2.29$, $c_{+} = 0.766$, $c_{-} = -1.528$, and $c_{--} = -2.262$, which are consistent with those reported in Figures 5 and 6. Such a cluster separation is in agreement with the results of the aural survey (see Section 2.4) aimed at determining the sound components at the 16 sites (biophonies, technophonies, and geophonies).

Figure 6. SRI map obtained using the following combination of weights as for the RF model: $c_{++} = 2.0$, $c_{+} = 1.6$, $c_{-} = -0.3$, $c_{--} = -4.7$. (**Left panel**) Range of continuous SRI variability: small differences in SRI at different sites are highlighted. (**Right panel**) Range of variability of SRI as defined by Equation (3): two main clusters are depicted, confirming previous analysis (see [50,52]). The legends indicate the ranges of variability of the SRI, and the sensor numbers correspond to the active ones as described in Figure 1.

4. Conclusions

The study of the soundscape within urban parks represents an increasingly important issue as they represent the link between natural habitats and highly populated urban areas. The evaluation of the soundscape is usually carried out through the help of ecoacoustics analysis and thus the use of well-known ecoacoustic indices. In this work, we gathered spectral information in the form of ecoacoustic indices and MFCCs to train four machine learning models to predict a single index (the soundscape ranking index, SRI) carrying information of different sound sources and, in addition, providing a soundscape ranking among different locations within the urban park.

The SRI has the advantage of yielding a quick overview of an environment given a set of extracted spectral features. We found that the seven statistical descriptors calculated for the 12 MFCCs (for a total of 84 features) are able to determine the optimal combination of weights that leads to a quite high classification score. Values for the F1-score of approximately 0.70 and 0.71 were obtained for AdaBoost and RF models, respectively. However, the RF model proved to be more robust when tested using the k-fold cross-validation procedure. Indeed, the information carried by the SRI represents a summative representation of the soundscape quality, which is essentially driven by the prevalence of the sound sources acting locally. As such, the SRI can be used to rapidly provide maps of environment sound quality on the basis of few audio recordings. The splitting of the SRI into three main intervals may somehow be adjusted by considering its quantization into smaller bins. This will allow us to obtain finer shades of the environment sound quality.

Mapping the SRI yielded similar results to those recently obtained in [50] through a simpler statistical approach and using a self-consistent SRI computation able to visualize the internal structure of the soundscape in the same habitat [52]. For these reasons, we may conclude that the SRI can become a useful tool for helping policy makers follow up the soundscape evolution in "natural" habitats within urban zones. More specifically, it can be employed, once fully developed, to evaluate the impact of noise-mitigating measures on "pocket" parks, urban parks, and residential redevelopment areas, thus allowing one to follow up the soundscape evolution in "natural" habitats within urban zones.

As already stated in the introduction, the availability of a small labeled dataset can undermine the performance of ML models, which represents a limitation of the present study. However, the obtained performance can be considered satisfactory and can represent a benchmark for future developments. As a future development of the present work, we envisage the use of larger labeled datasets. This can be achieved by using additional recordings with the corresponding aural survey, and/or via data augmentation using Monte Carlo techniques. We also envisage the application of NN models to develop more efficient classification schemes. In many situations, there is also the need to use different sound recorders to map extended areas simultaneously. Indeed, this procedure can introduce a bias in the analysis owing to different frequency responses of each sound recorder. This issue also needs to be addressed in our future works.

Author Contributions: Conceptualization, R.B., A.A. and G.Z.; methodology, R.B.; investigation, A.A., R.B., A.P. and H.E.R.; software, A.A.; formal analysis, R.B. and A.A.; resources, G.Z.; data curation, A.P.; writing—original draft preparation, R.B.; writing—review and editing, R.B., H.E.R and A.A.; visualization, R.B., H.E.R. and A.P.; funding acquisition, G.Z. All authors have read and agreed to the published version of the manuscript.

Funding: This research received no external funding.

Institutional Review Board Statement: Not applicable.

Informed Consent Statement: Not applicable.

Data Availability Statement: Data are available upon request.

Conflicts of Interest: The authors declare no conflict of interest.

References

1. Dumyahn, S.L.; Pijanowski, B.C. Soundscape conservation. *Landsc. Ecol.* **2011**, *26*, 1327–1344. [CrossRef]
2. Schafer, R.M. *The Soundscape: Our Sonic Environment and the Tuning of the World*; Simon and Schuster: New York, NY, USA, 1993.
3. Barber, J.R.; Crooks, K.R.; Fristrup, K.M. The costs of chronic noise exposure for terrestrial organisms. *Trends Ecol. Evol.* **2010**, *25*, 180–189. [CrossRef] [PubMed]
4. Doser, J.W.; Hannam, K.M.; Finley, A.O. Characterizing functional relationships between technophony and biophony: A western New York soundscape case study. *Landsc. Ecol.* **2020**, *35*, 689–707. [CrossRef]
5. Francis, C.D.; Newman, P.; Taff, B.D.; White, C.; Monz, C.A.; Levenhagen, M.; Petrelli, A.R.; Abbott, L.C.; Newton, J.; Burson, S.; et al. Acoustic environments matter: Synergistic benefits to humans and ecological communities. *J. Environ. Manag.* **2017**, *203*, 245–254. [CrossRef] [PubMed]
6. Lawson, G.M. Networks cities and ecological habitats. In *Networks Cities*; Qun, F., Brearley, J., Eds.; China Architecture and Building Press: Beijing, China, 2011; pp. 250–253. Available online: https://eprints.qut.edu.au/40229/ (accessed on 1 May 2023).
7. Sueur, J.; Farina, A.; Gasc, A.; Pieretti, N.; Pavoine, S. Acoustic indices for biodiversity assessment and landscape investigation. *Acta Acust. United Acust.* **2014**, *100*, 772–781. [CrossRef]
8. Krause, B. The Loss of Natural Soundscapes. *Earth Isl. J.* **2002**, *17*, 27–29. Available online: www.earthisland.org/journal/index.php/magazine/archive (accessed on 1 May 2023).
9. Pijanowski, B.C.; Farina, A.; Gage, S.H.; Dumyahn, S.L.; Krause, B.L. What is soundscape ecology? An introduction and overview of an emerging new science. *Landsc. Ecol.* **2011**, *26*, 1213–1232. [CrossRef]
10. Pavan, G. Fundamentals of Soundscape Conservation. In *Ecoacoustics: The Ecological Role of Sounds*; Farina, A., Gage, S.H., Eds.; Wiley: New York, NY, USA, 2017; pp. 235–258. [CrossRef]
11. Sethi, S.S.; Jones, N.S.; Fulcher, B.D.; Picinali, L.; Clink, D.J.; Klinck, H.; Orme, C.D.L.; Wrege, P.H.; Ewers, R.M. Characterizing soundscapes across diverse ecosystems using a universal acoustic feature set. *Proc. Natl. Acad. Sci. USA* **2020**, *117*, 17049–17055. [CrossRef]
12. Lellouch, L.; Pavoine, S.; Jiguet, F.; Glotin, H.; Sueur, J. Monitoring temporal change of bird communities with dissimilarity acoustic indices. *Methods Ecol. Evol.* **2014**, *5*, 495–505. [CrossRef]
13. Kasten, E.P.; Gage, S.H.; Fox, J.; Joo, W. The remote environmental assessment laboratory's acoustic library: An archive for studying soundscape ecology. *Ecol. Inform.* **2012**, *12*, 50–67. [CrossRef]
14. Eldridge, A.; Guyot, P.; Moscoso, P.; Johnston, A.; Eyre-Walker, Y.; Peck, M. Sounding out ecoacoustic metrics: Avian species richness is predicted by acoustic indices in temperate but not tropical habitats. *Ecol. Indic.* **2018**, *95*, 939–952. [CrossRef]
15. Boelman, N.T.; Asner, G.P.; Hart, P.J.; Martin, R.E. Multitrophic invasion resistance in hawaii: Bioacoustics, field surveys, and airborne remote sensing. *Ecol. Appl.* **2007**, *17*, 2137–2144. [CrossRef] [PubMed]
16. Benocci, R.; Brambilla, G.; Bisceglie, A.; Zambon, G. Eco-Acoustic Indices to Evaluate Soundscape Degradation Due to Human Intrusion. *Sustainability* **2020**, *12*, 10455. [CrossRef]

17. Bertucci, F.; Parmentier, E.; Berten, L.; Brooker, R.M.; Lecchini, D. Temporal and spatial comparisons of underwater sound signatures of different reef habitats in Moorea Island, French Polynesia. *PLoS ONE* **2015**, *10*, e0135733. [CrossRef]
18. Harris, S.A.; Shears, N.T.; Radford, C.A. Ecoacoustic indices as proxies for biodiversity on temperate reefs. *Methods Ecol. Evol.* **2016**, *7*, 713–724. [CrossRef]
19. Pérez-Granados, C.; Traba, J. Estimating bird density using passive acoustic monitoring: A review of methods and suggestions for further research. *Ibis* **2021**, *163*, 765–783. [CrossRef]
20. Shonfield, J.; Bayne, E.M. Autonomous recording units in avian ecological research: Current use and future applications. *Avian Conserv. Ecol.* **2017**, *12*, 14. [CrossRef]
21. Benocci, R.; Roman, H.E.; Bisceglie, A.; Angelini, F.; Brambilla, G.; Zambon, G. Eco-acoustic assessment of an urban park by statistical analysis. *Sustainability* **2021**, *13*, 7857. [CrossRef]
22. LeCun, Y.; Boser, B.; Denker, J.S.; Henderson, D.; Howard, R.E.; Hubbard, W.; Jackel, L.D. Backpropagation applied to handwritten zip code recognition. *Neural Comput.* **1989**, *1*, 541–551. [CrossRef]
23. Lewis, J.P. Creation by refinement: A creativity paradigm for gradient descent learning networks. In Proceedings of the IEEE 1988 International Conference on Neural Networks, San Diego, CA, USA, 24–27 July 1988; pp. 229–233. [CrossRef]
24. Todd, P.M. A sequential neural network design for musical applications. In *1988 Connectionist Models Summer School*; Touretzky, D., Hinton, G., Sejnowski, T., Eds.; Morgan Kaufmann: San Mateo, CA, USA, 1988; pp. 76–84.
25. Cavallari, G.B.; Ribeiro, L.S.; Ponti, M.A. Unsupervised representation learning using convolutional and stacked auto-encoders: A domain and cross-domain feature space analysis. In Proceedings of the 31st SIBGRAPI Conference on Graphics, Patterns and Images (SIBGRAPI), Parana, Brazil, 29 October–1 November 2018; IEEE: Piscataway, NJ, USA, 2018; pp. 440–446. [CrossRef]
26. Ponti, M.A.; Ribeiro, L.S.F.; Nazare, T.S.; Bui, T.; Collomosse, J. Everything you wanted to know about deep learning for computer vision but were afraid to ask. In Proceedings of the 30th SIBGRAPI Conference on Graphics, Patterns and Images Tutorials (SIBGRAPI-T), Niteroi, Brazil, 17–18 October 2017; IEEE: Piscataway, NJ, USA, 2017; pp. 17–41. [CrossRef]
27. Christin, S.; Hervet, É.; Lecomte, N. Applications for deep learning in ecology. *Methods Ecol. Evol.* **2019**, *10*, 1632–1644. [CrossRef]
28. Fairbrass, A.J.; Firman, M.; Williams, C.; Brostow, G.J.; Titheridge, H.; Jones, K.E. CityNet–Deep learning tools for urban ecoacoustic assessment. *Methods Ecol. Evol.* **2019**, *10*, 186–197. [CrossRef]
29. Lin, T.H.; Tsao, Y. Source separation in ecoacoustics: A roadmap towards versatile soundscape information retrieval. *Remote Sens. Ecol. Conserv.* **2020**, *6*, 236–247. [CrossRef]
30. Navarro, J.M.; Pita, A. Machine Learning Prediction of the Long-Term Environmental Acoustic Pattern of a City Location Using Short-Term Sound Pressure Level Measurements. *Applied Sciences* **2023**, *13*, 1613. [CrossRef]
31. Orga, F.; Socoró, J.C.; Alías, F.; Alsina-Pagès, R.M.; Zambon, G.; Benocci, R.; Bisceglie, A. Anomalous Noise Events Considerations for the Computation of Road Traffic Noise Levels: The DYNAMAP's Milan Case Study. In Proceedings of the 24th International Congress on Sound and Vibration, ICSV 2017, London, UK, 23–27 July 2017. Available online: http://hdl.handle.net/2072/376268 (accessed on 1 May 2023).
32. Piczak, K.J. Environmental sound classification with convolutional neural networks. In Proceedings of the IEEE 25th International Workshop on Machine Learning for Signal Processing (MLSP), Boston, MA, USA, 17–20 September 2015; IEEE: Piscataway, NJ, USA, 2015; pp. 1–6. [CrossRef]
33. Salamon, J.; Bello, J.P. Deep convolutional neural networks and data augmentation for environmental sound classification. *IEEE Signal Process. Lett.* **2017**, *24*, 279–283. [CrossRef]
34. Ward, J.H. Hierarchical grouping to optimize an objective function. *J. Am. Stat. Assoc.* **1963**, *58*, 236–244. [CrossRef]
35. Ruff, Z.J.; Lesmeister, D.B.; Appel, C.L.; Sullivan, C.M. Workflow and convolutional neural network for automated identification of animal sounds. *Ecol. Indic.* **2021**, *124*, 107419. [CrossRef]
36. R Core Team. *R: A Language and Environment for Statistical Computing*; R Foundation for Statistical Computing: Vienna, Austria, 2018. Available online: Https://www.R-project.org/ (accessed on 28 April 2022).
37. Vidaña-Vila, E.; Navarro, J.; Stowell, D.; Alsina-Pagès, R.M. Multilabel Acoustic Event Classification Using Real-World Urban Data and Physical Redundancy of Sensors. *Sensors* **2021**, *21*, 7470. [CrossRef]
38. Mullet, T.C.; Gage, S.H.; Morton, J.M.; Huettmann, F. Temporal and spatial variation of a winter soundscape in south-central Alaska. *Landsc. Ecol.* **2016**, *31*, 1117–1137. [CrossRef]
39. Quinn, C.A.; Burns, P.; Gill, G.; Baligar, S.; Snyder, R.L.; Salas, L.; Goetz, S.J.; Clark, M.L. Soundscape classification with convolutional neural networks reveals temporal and geographic patterns in ecoacoustic data. *Ecol. Indic.* **2022**, *138*, 108831. [CrossRef]
40. Giannakopoulos, T.; Siantikos, G.; Perantonis, S.; Votsi, N.E.; Pantis, J. Automatic soundscape quality estimation using audio analysis. In Proceedings of the 8th ACM International Conference on Pervasive Technologies Related to Assistive Environments, Corfu, Greece, 1–3 July 2015; pp. 1–9. [CrossRef]
41. Tsalera, E.; Papadakis, A.; Samarakou, M. Monitoring, profiling and classification of urban environmental noise using sound characteristics and the KNN algorithm. *Energy Rep.* **2020**, *6*, 223–230. [CrossRef]
42. Lojka, M.; Pleva, M.; Kiktová, E.; Juhár, J.; Čižmár, A. Ear-tuke: The acoustic event detection system. In Proceedings of the Multimedia Communications, Services and Security: 7th International Conference, MCSS 2014, Krakow, Poland, 11–12 June 2014; Springer International Publishing: Berlin, Germany, 2014; pp. 137–148. [CrossRef]

43. Pita, A.; Rodriguez, F.J.; Navarro, J.M. Cluster analysis of urban acoustic environments on Barcelona sensor network data. *Int. J Environ. Res. Public Health* **2021**, *18*, 8271. [CrossRef] [PubMed]
44. Pita, A.; Rodriguez, F.J.; Navarro, J.M. Analysis and Evaluation of Clustering Techniques Applied to Wireless Acoustics Sensor Network Data. *Appl. Sci.* **2022**, *12*, 8550. [CrossRef]
45. Luitel, B.; Murthy, Y.S.; Koolagudi, S.G. Sound event detection in urban soundscape using two-level classification. In Proceedings of the 2016 IEEE Distributed Computing, VLSI, Electrical Circuits and Robotics (DISCOVER), Mangalore, India, 13–14 August 2016; pp. 259–263. [CrossRef]
46. Maijala, P.; Shuyang, Z.; Heittola, T.; Virtanen, T. Environmental noise monitoring using source classification in sensors. *Appl. Acoust.* **2018**, *129*, 258–267. [CrossRef]
47. Ye, J.; Kobayashi, T.; Murakawa, M. Urban sound event classification based on local and global features aggregation. *Appl. Acoust.* **2017**, *117*, 246–256. [CrossRef]
48. Gómez-Gómez, J.; Vidaña-Vila, E.; Sevillano, X. Western editerranean wetland birds dataset: A new annotated dataset for acoustic bird species classification. *Ecol. Inform.* **2023**, *75*, 102014. [CrossRef]
49. Brambilla, G.; Confalonieri, C.; Benocci, R. Application of the intermittency ratio metric for the classification of urban sites based on road traffic noise events. *Sensors* **2019**, *19*, 5136. [CrossRef] [PubMed]
50. Benocci, R.; Potenza, A.; Bisceglie, A.; Roman, H.E.; Zambon, G. Mapping of the Acoustic Environment at an Urban Park in the City Area of Milan, Italy, Using Very Low-Cost Sensors. *Sensors* **2022**, *22*, 3528. [CrossRef] [PubMed]
51. Benocci, R.; Roman, H.E.; Bisceglie, A.; Angelini, F.; Brambilla, G.; Zambon, G. Auto-correlations and long time memory of environment sound: The case of an Urban Park in the city of Milan (Italy). *Ecol. Indic.* **2022**, *134*, 108492. [CrossRef]
52. Benocci, R.; Afify, A.; Potenza, A.; Roman, H.E.; Zambon, G. Self-consistent Soundscape Ranking Index: The Case of an Urban Park. *Sensors* **2023**, *23*, 3401. [CrossRef]
53. Python. Available online: https://www.python.org/ (accessed on 12 June 2022).
54. Kamiński, B.; Jakubczyk, M.; Szufel, P. A framework for sensitivity analysis of decision trees. *Cent. Eur. J. Oper. Res.* **2018**, *26*, 135–159. [CrossRef]
55. Quinlan, J.R. Simplifying decision trees. *Int. J. Man-Mach. Stud.* **1987**, *27*, 221–234. [CrossRef]
56. Jost, L. Entropy and diversity. *Oikos* **2006**, *113*, 363–375. [CrossRef]
57. Hastie, T.; Friedman, J.H.; Tibshirani, R. *The Elements of Statistical Learning: Data Mining, Inference, and Prediction*; Springer: NewYork, NY, USA, 2009; Volume 2. [CrossRef]
58. Cortes, C.; Vapnik, V. Support-vector networks. *Mach. Learn.* **1995**, *20*, 273–297. [CrossRef]
59. Support Vector Machines. Available online: Https://scikit-learn.org/stable/modules/svm.html (accessed on 12 May 2022).
60. Aizerman, M.A.; Braverman, E.M.; Rozonoer, L.I. Theoretical foundations of the potential function method in pattern recognition learning. *Autom. Remote Control* **1964**, *25*, 821–837.
61. Radial Basis Function Kernel. Available online: Https://en.wikipedia.org/wiki/Radial_basis_function_kernel (accessed on 1 May 2023).
62. Scikit-Learn Implementation of SVM. Available online: https://scikit-learn.org/stable/auto_examples/svm/plot_rbf_parameters.html (accessed on 1 May 2023).
63. Freund, Y.; Schapire, R.E. A decision-theoretic generalization of on-line learning and an application to boosting. *J. Comput. Syst. Sci.* **1997**, *55*, 119–139. [CrossRef]
64. Schapire, R. Using output codes to boost multiclass learning problems. In Proceedings of the Fourteenth International Conference on Machine Learning, Nashville, TN, USA, 8–12 July 1997; Morgan Kauffman: Burlington, MA, USA, 1997; Volume 97, pp. 313–321. Available online: http://rob.schapire.net/papers/Schapire97.pdf (accessed on 1 May 2023).
65. Schapire, R.E.; Singer, Y. Improved boosting algorithms using confidence-rated predictions. In Proceedings of the Eleventh Annual Conference on Computational Learning Theory, Madison, WI, USA, 24–26 July 1998; pp. 80–91. [CrossRef]
66. Hastie, T.; Rosset, S.; Zhu, J.; Zou, H. Multi-class adaboost. *Stat. Its Interface* **2009**, *2*, 349–360. [CrossRef]
67. Sueur, J.; Pavoine, S.; Hamerlynck, O.; Duvail, S. Rapid acoustic survey for biodiversity appraisal. *PLoS ONE* **2008**, *3*, e4065. [CrossRef]
68. Pieretti, N.; Farina, A.; Morri, D. A new methodology to infer the singing activity of an avian community: The Acoustic Complexity Index (ACI). *Ecol. Indic.* **2011**, *11*, 868–873. [CrossRef]
69. Grey, J.M.; Gordon, J.W. Perceptual effects of spectral modifications on musical timbres. *J. Acoust. Soc. Am.* **1978**, *63*, 1493–1500. [CrossRef]
70. Yang, W.; Kang, J. Soundscape and sound preferences in urban squares: A case study in Sheffield. *J. Urban Des.* **2005**, *10*, 61–80. [CrossRef]
71. Seewave: Sound Analysis and Synthesis. Available online: https://cran.r-project.org/web/packages/seewave/index.html (accessed on 28 April 2022).
72. Soundecology: Soundscape Ecology. Available online: https://cran.r-project.org/web/packages/soundecology/index.html (accessed on 28 April 2022).
73. Davis, S.; Mermelstein, P. Comparison of parametric representations for monosyllabic word recognition in continuously spoken sentences. *IEEE Trans. Acoust. Speech Signal Process.* **1980**, *28*, 357–366. [CrossRef]

4. Sahidullah, M.; Saha, G. Design, analysis and experimental evaluation of block based transformation in MFCC computation for speaker recognition. *Speech Commun.* **2012**, *54*, 543–565. [CrossRef]
5. Precision-Recall. Available online: https://scikit-learn.org/stable/auto_examples/model_selection/plot_precision_recall.html (accessed on 13 December 2022).
6. Supervised Learning. Available online: https://scikit-learn.org/stable/supervised_learning.html#supervised-learning (accessed on 3 May 2023).

sensors

MDPI

Article

Self-Consistent Soundscape Ranking Index: The Case of an Urban Park

Roberto Benocci [1,*], Andrea Afify [2], Andrea Potenza [1], H. Eduardo Roman [2,*] and Giovanni Zambon [1]

1 Department of Earth and Environmental Sciences (DISAT), University of Milano-Bicocca, Piazza della Scienza 1, 20126 Milano, Italy

2 Department of Physics, University of Milano-Bicocca, Piazza della Scienza 3, 20126 Milano, Italy; a.afify@campus.unimib.it

* Correspondence: roberto.benocci@unimib.it (R.B.); hector.roman@unimib.it (H.E.R.)

Abstract: We have performed a detailed analysis of the soundscape inside an urban park (located in the city of Milan) based on simultaneous sound recordings at 16 locations within the park. The sound sensors were deployed over a regular grid covering an area of about 22 hectares, surrounded by a variety of anthropophonic sources. The recordings span 3.5 h each over a period of four consecutive days. We aimed at determining a soundscape ranking index (SRI) evaluated at each site in the grid by introducing 4 unknown parameters. To this end, a careful aural survey from a single day was performed in order to identify the presence of 19 predefined sound categories within a minute, every 3 minutes of recording. It is found that all SRI values fluctuate considerably within the 70 time intervals considered. The corresponding histograms were used to define a dissimilarity function for each pair of sites. Dissimilarity was found to increase significantly with the inter-site distance in space. Optimal values of the 4 parameters were obtained by minimizing the standard deviation of the data, consistent with a fifth parameter describing the variation of dissimilarity with distance. As a result, we classify the sites into three main categories: "poor", "medium" and "good" environmental sound quality. This study can be useful to assess the quality of a soundscape in general situations.

Keywords: soundscape; soundscape ranking index (SRI); urban parks; acoustic sensor networks

Citation: Benocci, R.; Afify, A.; Potenza, A.; Roman, H.E.; Zambon, G. Self-Consistent Soundscape Ranking Index: The Case of an Urban Park. *Sensors* **2021**, *23*, 3401. https://doi.org/10.3390/s23073401

Academic Editor: Antonio Guerrieri

Received: 28 February 2023
Revised: 15 March 2023
Accepted: 20 March 2023
Published: 23 March 2023

1. Introduction

The acoustic quality of a habitat is generally recognized as a primordial requirement of wildlife conservation [1,2]. It has become customary to evaluate the environmental status of particularly exposed areas, such as large populated urban zones, in addition to natural habitats. As a result of increasing urbanization all over the world, the associated, and in many cases excessive human technophonies, usually called "noise", together with a lack of planning to permit for a smooth transition between a built environment and a natural one, can yield several deleterious effects on biodiversity. An exhaustive picture on these issues has been documented in recents works [3–6].

The use of passive acoustic monitoring has become the main tool to study a particular habitat [7–10]. Its diffusion among researchers and technicians has evolved considerably together with its increasing memory capability. An important step towards its utilization on large scales came with the development of specific spectral and level characteristics of sound through the analysis of ecoacoustic indices (for a review, see [11]). This, in addition, allows the retrieval of important information about the way different sounds are assembled across both space and time. This new characteristic of the habitat sound is commonly referred to as the soundscape [12,13], and it has been recognized as a distinct feature or, more importantly, as an ecological "signature" of a landscape [14–16].

Urban parks are sources of natural sounds, but in many cases, they are mixed with anthropogenic noise, such as vehicle traffic noise, from surrounding built areas [17,18]. Many research works have shown that Brazilian and Italian parks exceed noise level thresholds due to the presence of traffic noise as the main source of disturbance [19–21]. These results also give evidence that the traffic noise perception in urban parks can undermine the potential beneficial effects of natural sounds [22].

The ecoacoustics indices merge the complex acoustic dynamics of an ecosystem, consisting of vocalizing species, anthropogenic noise, and natural phenomena [23], into sets of time series, allowing us to get a picture of the environmental changes at a given habitat [24]. Furthermore, they provide new insights on species diversity and human impacts across a wide range of terrestrial [25–29] and aquatic environments [30,31]. The validation of the calculation of ecoacoustic indices is usually sound-truthed by specialized operators who classify hours of recordings according to predefined sound categories.

The identification of sound sources by operators is highly time consuming and requires specific knowledge of animal vocalizations, thus limiting its applicability to small datasets [32,33]. However, a cumulative approach based on a qualitative description of the recorded sound (e.g., many/few vocalizing birds, many/few birds species, high/low traffic noise, etc.) was shown to improve the validation process, showing a good matching between the dentification of acoustic categories from the audio recording and the ecoacoustic indices [34].

In this work, we address the question of quantifying the quality of a local soundscape from a set of audio recordings at each site of a grid covering the area of interest. The intervention of an expert is required, who is expected to recognize the different sound categories determining the soundscape. The main goal is to associate a soundscape ranking index (SRI) with each site of the grid, in order to classify the quality of the local environmental sound into one of the following three categories: "poor", "medium" and "good". This is achieved by introducing a sound dissimilarity function between sites, found to be consistent with a power-law behavior with the inter-site distance. We conclude with a discussion of possible future work along these lines.

2. Materials and Methods

2.1. Area of the Study

The Parco Nord (Northern Park) of Milan (Italy) covers an area of approximately 790 hectares and is located within a highly urbanized area (see Figure 1). About 45% of its surface is dedicated to natural green spots and vegetation, while the remaining parts are devoted to agricultural activities and infrastructures. The area of study is a tree-covered parcel of approximately 22 hectares enclosed by agricultural fields, lawns, paths and roads. It has a semi-natural structure, which is characterized by herbaceous layers of nemoral flora, shrub and arboreal layers and dead woods.

The zone is crossed by numerous paths, and it is mainly used for recreational activities. It contains an artificial lake of about 300 m^2 surface area, located at approximately 250 m from the edge of the bush. The main traffic noise sources are the A4 highway and the Padre Turoldo street, both located north of the park at around 100 m from the wooded parcel. There is also the presence of a small airport (Bresso airport) on the west side at around 500 m from the tree-line edge.

2.2. Audio Recorders

Mapping the environmental sounds over large areas may require important financial commitments that could be alleviated by the availability of very low-cost recorders (VLCRs). Thus, we used SMT security digital audio recorders with a 48 kHz sampling rate and equipped with a two-week lifetime powerbank. The main disadvantage of using very low-cost recorders is the possibility that they present dissimilar microphone sensitivities. This may cause a different frequency and level response to a sound exposure. The response of each recorder was evaluated in terms of the following:

- Computation of the acoustic complexity index (ACI, see [27]) for a white noise source.
- Behavior of each VLCR at different frequencies.

The ACI computes the relative variation of recorded amplitudes of adjacent temporal steps in each frequency bin, as determined by the FFT analysis in the frequency range 0–24 kHz. Thus, its computation provides overall evidence of possible anomalies in the frequency response. Anomalies were identified when the recorder response strongly departed from the average ACI value. Based on these results, we selected audio recorders characterized by a response within 3% of the average ACI value. The full description of the procedure is reported in [35].

2.3. Measurement Scheme

The 22 recorders were initially positioned on two regular grids (see Figure 1), the first one (northen part) covering an area of approximately 500 $\times$ 270 m^2, and the second grid (southern part) covering an area of about 300 $\times$ 270 m^2. The recordings were scheduled for the period of greatest singing activity of the avifauna [10] and repeated over four days, namely on 25–28 May 2015, from 06:30 a.m. (UTC +2) to 10:00 a.m., corresponding to a 3.5 h long recording session for each day. Unfortunately, six recorders, numbered as 7, 9, 14, 15, 16 and 21 and indicated by the yellow spots in Figure 1, did not work properly, and thus, the audio files analyzed in this study reduced to only 16 sites.

Figure 1. The Parco Nord (Northern Park) area of study. The sensor network is indicated by the numbered dots (1–22), deployed inside the tree-covered zone of the park. The network consists of two regular grids, the northern and the southern ones, composed of 16 and 6 nodes, respectively. Those in red color are the actually used sensors (total number 16), while those in yellow are the six ones discarded due to malfunctioning. One can see the A4 highway, the Padre Turoldo street and the artificial lake to the northern part of the park, while to the west side of it, the small Bresso airport runaway is located.

2.4. Aural Survey

In order to quantify the biophonies [36], anthrophonies and geophonies [37], an aural survey was performed on one of the four days of recordings, of 3.5 h total duration for each of the 16 sites. Specifically, a single expert listened carefully to the 16 recordings according to the following scheme. For each site, the daily record duration of 210 min was grouped into 70 intervals of 3 min each, of which only the first 1-min interval was listened, while the following 2 min of recording were skipped. This operation spanned several weeks of careful work to accurately examine the whole set of recordings. The expert focused on quantifying the biophonic activity (mainly avian vocalizations) and technophonic sources (mainly traffic noise, including trains and airplanes, and other sources such as park maintenance activities), according to the scheme discussed in [38] and reported in detail in Table 1.

Table 1. Sound categories corresponding to the 19 identified sources. (First column) Sound sources. (Second column) Quantity/Duration/Level. (Third column) Parameter index n_i ($n_i \in \{1-5\}$ and $i = 1,\ 19$) associated with the ith category. (Fourth column) Identification feature.

Sound Source	Quantity	Parameters	Identification
Birds number	Many	1	Many birds
	Few	2	Few birds, no traffic
	None	3	No birds, no other sources
Birds species	>1	1	Many species
	1	2	One species, no traffic
	None	3	No birds, no other sources
Birds sound duration	100%	1	60 s
	75%	1	45 s
	50%	1	30 s
	25%	1	15 s
	0%	1	0 s
Traffic level	None	2	Few birds, no traffic
	Low	4	Continuous/low traffic
	High	5	Intermittent/high traffic
Traffic type	None	2	No traffic
	Continuous	4	Continuous/low traffic
	Intermittent	5	Intermittent/high traffic
Trains	Present	5	Trains
Airplanes	Present	5	Airplanes, other sources

For each single minute of listening, bird vocalizations and non-biophonic sources were searched. The former was subdivided into three main categories according to the quantity of birds (many, few, none), number of species (>1, 1, none) and sound durations, (100, 75, 50, 25, 0)%. The third column displays the index n of the parameter $P(n)$ ($n = 1, 5$), associated with each subcategory ($i = 1, 19$). The fourth column is aimed at providing a more specific characterization of the way the subcategory could be identified. Regarding the bird sound durations, the latter is expressed in terms of the percentage of bird singing activity identified within the considered minute. The effective time span in seconds is displayed in the fourth column.

The non-biophonic contributions are split into road traffic, trains, airplanes and other technophonic noise sources. The former is subdivided into two categories, level and type of traffic, while the last three are associated with the fifth parameter if they are present and with the third one if absent (see below). Finally, geophonies such as rain and wind were not considered due to their negligible contribution to the soundscape during the measurement campaign.

2.5. The Soundscape Ranking Index

We are interested in quantifying the quality of the local environment sound by means of a soundscape ranking index, SRI, as proposed recently [38]. In the following, we briefly discuss how to evaluate the SRI, aimed at describing the local soundscape, at a given site j in the network, in an average sense. The value of SRI(j, t) depends on the time interval t, in our case ($t = 1,\ 70$), listened during the aural survey at site j. For each time interval t and site j, we determine the event function, $N(i, j, t)$ for the ith sound category ($i = 1,\ 19$), described in Table 1. The event function can be either $N = 1$ or $N = 0$, depending on whether the event i is present or not. The SRI can then be obtained as the sum

$$\mathrm{SRI}(j,t) = \sum_{i=1}^{19} N(i,j,t)\, w(i), \tag{1}$$

where the weights, $w(i) = P(n_i)\, c(n_i)$, with $n_i = (1, 2, 3, 1, 2, 3, 1, 1, 1, 1, 1, 2, 4, 5, 2, 4, 5, 5, 5)$ (cf. Table 1), while the additional factor, $c(n_i) = (1.0, 0.75, 0.50, 0.25, 0.0)$ for $i = (7, 8, 9, 10, 11)$, respectively, and $c(n_i) = 1$, otherwise. For instance, $w(6) = P(3)\, c(6) = P(3)$, while $w(7) = P(1)\, c(7) = 0$, and $w(8) = P(1)\, 0.25$, etc.

The choice of the values for the parameter $P(n)$ is rather arbitrary of course, but we can stick to the following simple considerations in order to make up a representative and useful picture of the soundscape. We follow our previous work [38] and assume that $P(n) > 0$ if the sound is associated with a natural source, while we take $P(n) < 0$ if it is of anthropogenic origin. Given the fact that we consider five different situations, we chose the values reported in Table 2 to start with.

Now, we have all the ingredients to calculate the $\mathrm{SRI}(j,t)$, using Equation (1), once the event function, $N(i,j,t)$, is known for all the subcategories i, at site j and time interval t. SRI is expected to fluctuate as a function of t, reflecting the ever changing environmental conditions of the varying soundscape. However, we may average the index over time in order to get a mean value, $\langle \mathrm{SRI}(j) \rangle$, at site j, which should provide us with a quantitative element to estimate the *quality* of the local environmental sound. By quality, we mean that $\langle \mathrm{SRI}(j) \rangle$ should be large for a natural soundscape and small for a poor one affected by strong anthropogenic perturbations. At this point, and in order to fix the ideas, we suggest a simple classification of the mean quality index, $\langle \mathrm{SRI} \rangle$, as displayed in Table 3.

Table 2. The starting parameters $P(n)$ $(n = 1, 5)$ associated with each sound category to be used in Equation (1) (see also [38]).

n	P(n)	Identification
1	2	Many birds/species
2	1	Few birds/no traffic
3	0	No birds/no other sources
4	−1	Continuous/low traffic
5	−2	Intermittent/high traffic/other sources

Table 3. Quality intervals for the mean soundscape ranking index, $\langle \mathrm{SRI} \rangle$.

Poor Quality	Medium Quality	Good Quality
SRI < 0	$0 \le \mathrm{SRI} \le 2$	SRI > 2

2.6. Optimization Procedure for $P(n)$

The values of $P(n)$ reported in Table 2 can be seen as our starting guess of the unknown parameters and are therefore arbitrary. As we discuss in Section 3, however, they appear to build a quite robust starting set from which one can search for new "optimized" values. In addition, we find that the optimized set is closer to the prescription in Table 3 than the original set. We show that this different behavior is a result of the self-consistent optimization method we have developed for obtaining the parameters $P(n)$.

The basic quantity in our approach is the probability distribution function, $H_j(\mathrm{SRI})$, of the set of SRI values obtained at a given site j, at different time intervals t, using Equation (1). Based on the distribution functions obtained for all active sites, we can define a quantity representing how "dissimilar" two distributions are. The dissimilarity between say, sites i and j, denoted as $D_{i,j}$, is here defined by the following relation:

$$D_{i,j} = 1 - \frac{1}{A_i\, A_j} \int_{-\infty}^{\infty} dx\; H_i(x)\, H_j(x), \quad \text{with} \quad A_i^2 = \int_{-\infty}^{\infty} dx\; H_i^2(x), \tag{2}$$

where the normalizing factor A_i ensures that $D_{i,i} = 0$. The integral form used in Equation (2) can be formally seen as the internal product of two vectors, $H_i(m)$ and $H_j(m)$, where $m \in Z$ are the coordinates in a high-dimensional space. Specifically, if we define the histograms $H_i(x)$ on a set of discrete coordinates, $x = m\,b$, where b is the bin size used to build the histograms, we may regard $H_i(m)$ as the mth component of the vector $\bar{H}_i$. This internal product is reminiscent of a similar form, typically used in financial studies, to define a distance between time series in terms of their cross-correlations [39]. Intuitively, the larger $D_{i,j}$ is, the larger the soundscape dissimilarity between sites i and j. In other words, Equation (2) represents a measure that can be used to estimate a soundscape distance between sites. We can loosely say that two sites displaying very different SRI distribution functions are very distant in soundscape space.

Actually, sites i and j are located at well defined positions in space; in particular, they are at a fixed spatial distance, $R_{i,j}$, within the network (cf. Figure 1). Therefore, the question arises of whether both distances, $D_{i,j}$ and $R_{i,j}$, are somehow related to each other. Intuitively, one would expect that dissimilarity increases with distance. In other words, we are interested in finding how the local soundscape changes as a function of spatial distance R within the network. To do this, we assume, as our working hypothesis, a simple relation of the form

$$D_{i,j} \simeq a\; (R_{i,j}[\mathrm{m}])^{\alpha}, \tag{3}$$

where the distances $R_{i,j}$ are expressed in meters, and the exponent $\alpha > 0$ needs to be determined empirically. We expect that Equation (3) should be valid at least in an average sense. Note that it can be written as $D \sim (R/R_0)^{\alpha}$, where R_0 is an effective length scale associated with the network soundscape. This in keeping with the fact that D is actually dimensionless. Numerically, the constants a and R_0 are related to each other by, $R_0 = a^{-1/\alpha}$. Note that R_0 is given in meters (see also Equation (4)). To simplify the notation, in what follows, we simply write $R_{i,j}^{\alpha}$, meaning $(R_{i,j}[\mathrm{m}])^{\alpha}$.

As a matter of fact, α becomes the sixth parameter in our approach. Notice, however, that $P(3) = 0$ in all cases, so we are effectively dealing with only five parameters: $P(1)$, $P(2)$, $P(4)$, $P(5)$ and α. The five parameters are obtained by a least-square fit that minimizes the total deviation of the data from the fit, i.e.,

$$\Sigma^2 = \frac{1}{120} \sum_{i=1,j>i}^{16} \left(D_{i,j} - a\; R_{i,j}^{\alpha} \right)^2, \quad \text{with} \quad a = \langle D_{i,j} R_{i,j}^{\alpha} \rangle / \langle R_{i,j}^{2\alpha} \rangle \equiv R_0^{-\alpha}. \tag{4}$$

The constant a is obtained by requiring $\partial\Sigma/\partial a = 0$, and as a result, it is a function of all the five unknown parameters. We note that in the expression for a, the symbols $\langle\rangle$ represent averages over the 120 distinct site pairs (i,j) in the network. The effective soundscape length R_0 now becomes

$$R_0 = [\langle R_{i,j}^{2\alpha} \rangle / \langle D_{i,j} R_{i,j}^{\alpha} \rangle]^{1/\alpha}. \tag{5}$$

3. Results

The results presented in this section refer to the audio files recorded on 25 May 2015, from 06:30 a.m. to 10:00 a.m., and based on the aural survey discussed in Section 2.4.

3.1. Initial $P(N)$ Values

We start the analysis of the soundscape using the initial values of the parameters reported in Table 2. As discussed in Section 2.5, we evaluate the SRI values using Equation (1) at the 70 time intervals for each of the 16 active sites. The corresponding histograms vs. SRI are reported in Figure 2.

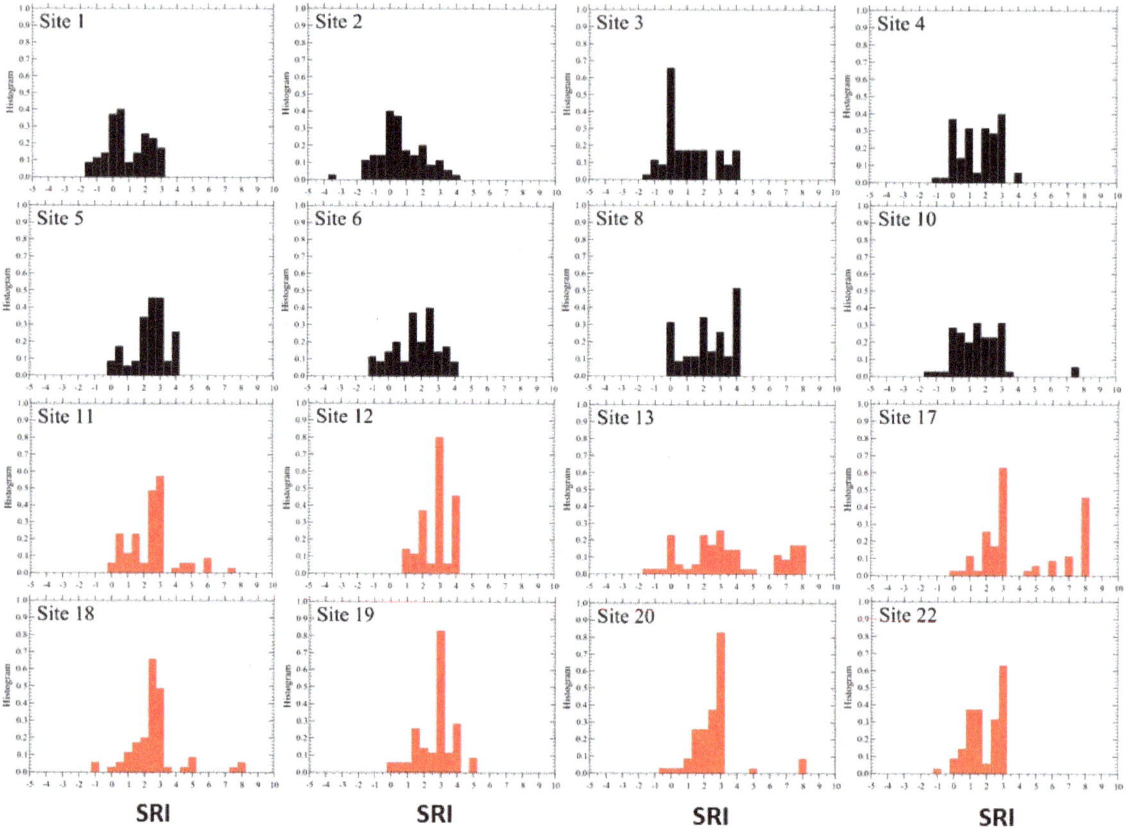

Figure 2. Variability of SRI at each active site. The distribution functions H(SRI) were obtained using Equation (1), from the initial values of parameters $P(n) = (2, 1, 0, -1, -2)$. The scale for SRI is fixed for all sites in the range $(-5, 10)$. The two sets of colors (black on top and red on bottom panel) are for convenience but indicate that the former are centered to lower values of SRI while the latter are centered to more positive ones, suggesting an underlying organization of the sites into two main groups.

Once the distribution functions, H(SRI), for each active site have been obtained (Figure 2), we can evaluate, using Equation (2), the dissimilarity distances between pairs of sites. We note that, so far, the exponent α has not been required. Indeed, it can be read off from the plot of $D_{i,j}$ values versus the corresponding inter-site distances $R_{i,j}$, as shown in Figure 3. We notice in the figure a pronounced scattering of the empirical data from the behavior expected from Equation (3). Despite this fact, we can still recognize an increasing trend of D vs. R, as visualized by the obtained least-square fit representing the expected power-law featured in Equation (3). From the latter, we find $\alpha \simeq 1/3$, which is rather small, but significant enough to conclude that the assumed power-law (Equation (3)) may be acceptable.

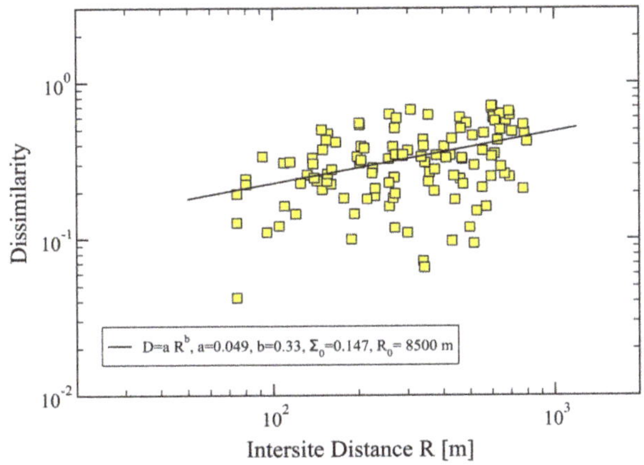

Figure 3. Dissimilarity, $D_{i,j}$, between two sites, (i, j) vs. inter-site distance $R_{i,j}$, obtained from Equation (2) using the initial set of parameters $P(n) = (2, 1, 0, -1, -2)$. The full squares are the empirical data, and the straight line is a least-square fit with the power law $D = aR^{\alpha}$, yielding $a = 0.049$, corresponding to $R_0 = 8500$ m, and $\alpha \simeq 1/3$. Specifically, we find $\langle DR^{\alpha} \rangle \simeq 2.354$ and $\langle R^{2\alpha} \rangle \simeq 48.03$. The total error of the fit is $\Sigma_0 = 0.147$. The mean inter-site distance in the network is $\langle R_{i,j} \rangle = 353.64$ m, the minimum and maximum ones: $R_{\min} = 75$ m and $R_{\max} = 800$ m, respectively.

The associated distribution function of dissimilarity is shown in Figure 4, together with a fitting function of a simple analytical form, which should be useful for making a quantitative comparison with the case of optimized parameters discussed below.

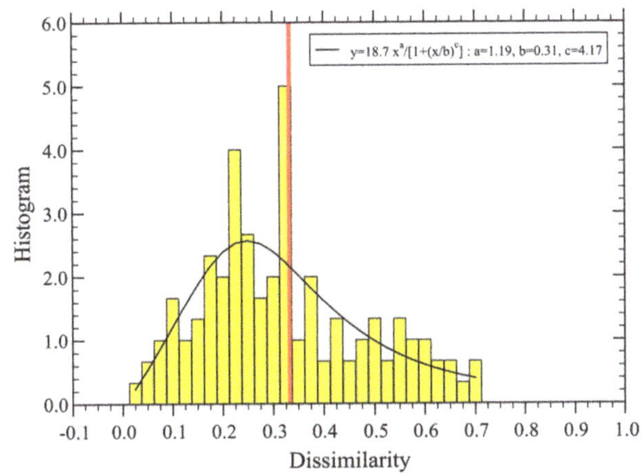

Figure 4. Probability density distribution of dissimilarity, corresponding to the values used in Figure 3. We find $\langle D \rangle \simeq 0.333$ (vertical red line). The continuous line is a fit with the form $y = Ax^a/[1 + (x/b)^c]$, with the normalization constant $A = 19$, and the fit parameters $a = 1.19$, $b = 0.31$ and $c = 4.17$.

Finally, we plot in Figure 5 the mean SRI values at each site, obtained from Equation (1) and corresponding to the parameter set: $P(n) = (2, 1, 0, -1, -2)$ and $\alpha = 0.33$. We note that all mean values turn out to be positive, $\langle \text{SRI} \rangle > 0$, and they are not quite consistent with the scenario expected from Table 3. This observation actually motivated us to search for a way to improve on this quality scenario results. The idea is then to search for a "better" set of parameters that can bring us beyond our initial guess, thus eliminating the arbitrariness endowed in our parameter values.

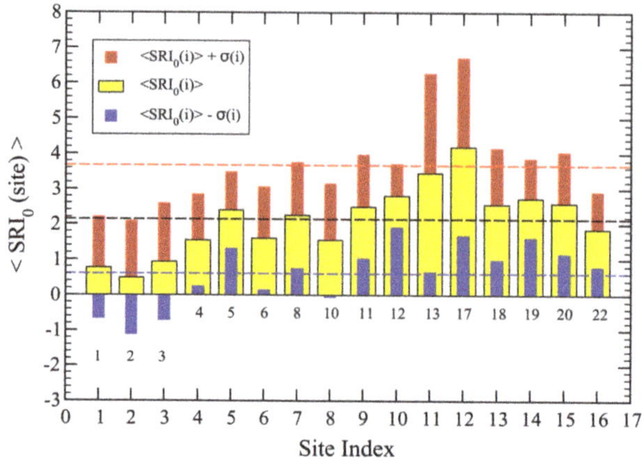

Figure 5. Mean $\langle \mathrm{SRI}_0 \rangle$ obtained by averaging the index over the 70 temporal values for each active site (yellow bars). The red and blue bars display the one $\sigma(i)$ variations of the data at each site i, above and below the local mean values, respectively. The dashed lines represent averages over all sites of the three set of data. We find $<\mathrm{SRI}_0> = 2.14$, $<\mathrm{SRI}_0>+\sigma = 3.68$, $<\mathrm{SRI}_0>-\sigma = 0.60$, with $\sigma = 1.54$. The actual site identification numbers are reported just below the zero line: (1–6, 8, 10–13, 17–20, 22).

3.2. Optimized $P(N)$ Values

In order to improve on the previous results, we let all the five parameters, $P(1)$, $P(2)$, $P(4)$, $P(5)$ and α, vary independently and search for the "best" set that minimizes the function Σ in Equation (4). For each set of new parameters, we need to evaluate the SRI given by Equation (1), followed by the calculation of the dissimilarity distances, Equation (2), once the corresponding histograms of the SRI have been obtained. To speed up the search, we start varying each parameter value within a given range, subdivided into few equidistant smaller intervals. Once a minimum has been found, we choose new ranges centered around the putative minimum parameter values. In this way, we can refine the search more efficiently and converge fast towards a solution. Indeed, few such iterations are needed to get a stable result.

It is convenient to check the convergence of the method by first keeping α fixed and letting the $P(n)$ values adjust themselves so as to minimize Σ. We tried both, $\alpha = 0.33$ and $\alpha = 1$, and from each of the found sets, we minimized Σ with respect to α. We obtained similar values for α in both cases, yielding $\alpha \approx 0.5 \pm 0.2$. We then performed a full search with all five parameters, finding indeed $\alpha \simeq 0.5 \pm 0.05$, suggesting that we may take $\alpha = 1/2$ as our best value for this unknown exponent. Finally, we set $\alpha = 1/2$ and optimized the search for our final set of parameters, yielding

$$P(1) = 2.290, \quad P(2) = 0.766, \quad P(4) = -1.528, \quad P(5) = -2.262, \tag{6}$$

with $P(3) = 0$, and a final error, $\Sigma = 0.139$, for dissimilarity versus inter-site distance, as shown in Figure 6.

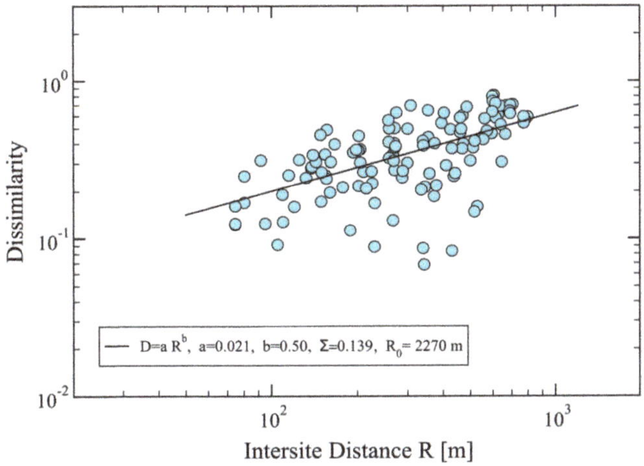

Figure 6. Dissimilarity, $D_{i,j}$, between two sites, (i, j) vs. inter-site distance $R_{i,j}$, obtained from the optimized parameters $P(n) = (2.29, 0.766, 0, -1.528, -2.262)$, corresponding to the best value $\alpha = 1/2$. The full circles are the empirical data, and the straight line is a least-square fit with the power law $D = aR^\alpha$, yielding $a = 0.021$, corresponding to $R_0 = 2270$ m, with $\langle DR^\alpha \rangle \simeq 7.39$ and $\langle R^{2\alpha} \rangle \simeq 353.6$. The total error of the fit is $\Sigma = 0.139$, which is smaller than the one obtained for the initial parameter values, $\Sigma_0 = 0.147$ (cf. Figure 3).

The distribution function of the newly obtained dissimilarity distances, $D_{i,j}$, is displayed in Figure 7. Now, the mean dissimilarity, $\langle D \rangle \simeq 0.375$, becomes a bit larger than for the non-optimized parameters (cf. Figure 4), while the new distribution function becomes flatter and extends to larger values of D.

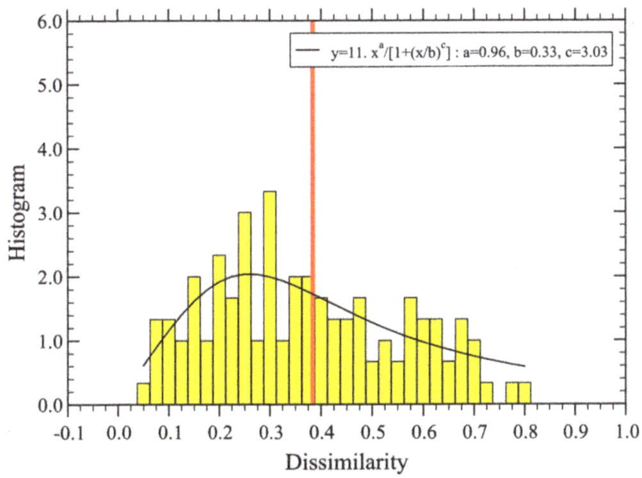

Figure 7. Probability density distribution of dissimilarity, corresponding to the values used in Figure 6. We find $\langle D \rangle \simeq 0.375$ (vertical red line). The continuous line is a fit with the form $y = Ax^a/[1 + (x/b)^c]$, with the normalization constant $A = 11$ and the fit parameters $a = 0.96$, $b = 0.33$ and $c = 3.03$.

The distribution functions of the SRI for each site are displayed in Figure 8, and their mean values are shown in Figure 9.

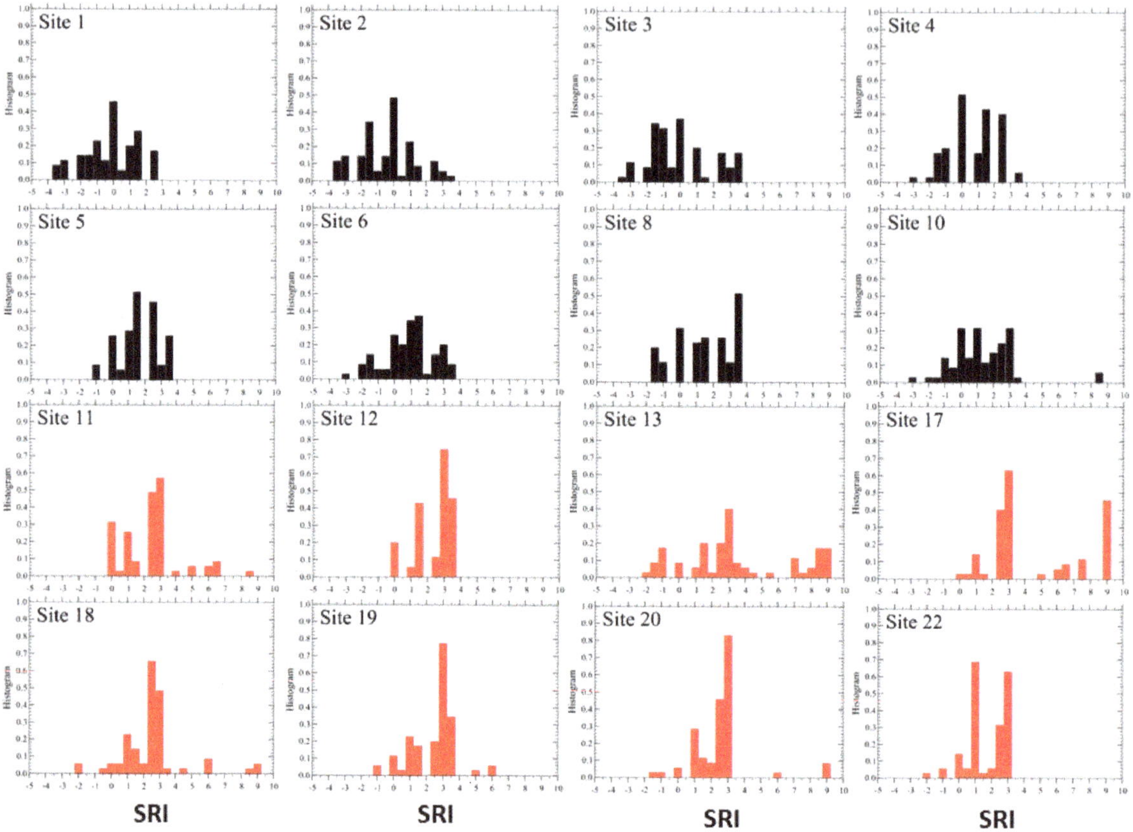

Figure 8. The distribution functions H(SRI) for the optimized parameters displayed in Equation (6). The scale for SRI is the same used in Figure 2. Here again, we have used two sets of colors for the histograms.

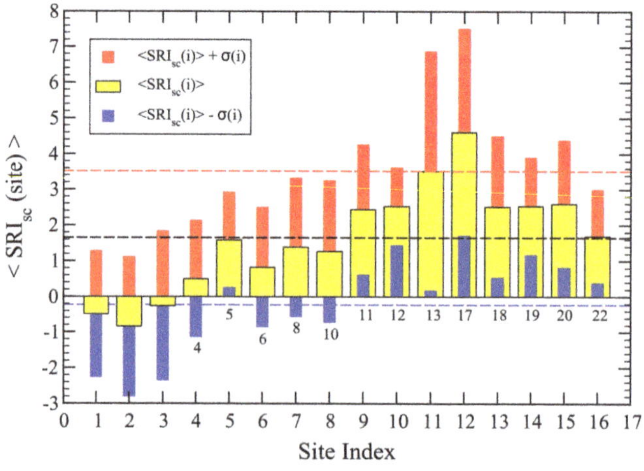

Figure 9. Same as in Figure 5 for the self-consistent set of parameters (Equation (6)). Here, we find <SRI> = 1.65 (2.14), <SRI>+σ = 3.53 (3.68), <SRI>-σ = −0.23 (0.60), with σ = 1.88 (1.54). The corresponding values from Figure 5 are reported in parenthesis.

4. Discussion and Conclusions

It is convenient to summarize the main results of our work as shown in Table 4. Let us start with the exponent α, expressing the way that dissimilarity decays with inter-site distance (Equation (3)). Quantitatively, the values indicate that for the initial set of parameters, dissimilarity decays more slowly (1/3) than for the optimized one (1/2), suggesting a more persistent behavior of the soundscape. The optimized values for the parameters $P(n)$ display the same trend as the initial ones, differing by at most a 50 % as for $P(4)$. Despite this, crucial differences develop when we analyze the remaining quantities.

Table 4. Summary of the main results for the exponent α in Equation (3), the parameters $P(n)$ $(n = 1,5)$, the mean value of the SRI in the network, the dispersion of the SRI values σ (cf. Figures 5 and 9), the mean value of dissimilarity $\langle D \rangle$ among all sites, the effective scale distance R_0 (cf. Equation (3)) and the total error of the fit Σ (Equation (4)).

	Initial Parameters	Optimized Parameters
α	1/3	1/2
$P(1)$	2.0	2.290
$P(2)$	1.0	0.766
$P(3)$	0.0	0.000
$P(4)$	−1.0	−1.528
$P(5)$	−2.0	−2.262
<SRI>	2.14	1.65
σ	1.54	1.88
$\langle D \rangle$	0.333	0.375
R_0	8500 m	2270 m
Σ	0.147	0.139

The SRI behaves quite differently. Indeed, the optimized set yields a much smaller mean SRI value for the whole network, about 23 %, than for the original non-optimized set of parameters. This again is consistent with the behavior of dissimilarity, indicating a lower soundscape quality than the one predicted by the original set. The dispersion of the SRI at each site, $\sigma(i)$, increases for the optimized parameter set, consistent with a more extended variability of dissimilarity (cf. Figures 4 and 7), also reflected in their mean values, $\langle D \rangle$. The effective length scale R_0 also manifests this difference as it is larger for the original set, suggesting that dissimilarity is smaller than for the case of optimized parameters. Finally, as expected, the optimized set yields a smaller dispersion of the data from the assumed power-law decay of dissimilarity as given by its smaller value of Σ.

We may conclude that in our case, the initial guess for the parameter values over-estimate the sound quality in the urban habitat we studied. This conclusion is based on the smaller dispersion of the data obtained from the assumed empirical relation between dissimilarity and intersite distance, as shown in Equation (3). We may expect that in other habitats, the self-consistent parameters would be different from the initially chosen values in either way, i.e., the soundscape quality may be either smaller or larger than initially predicted. In both cases, the present method suggests which of the scenarios is more likely to be correct, as it provides us with a recipe to check for the self-consistency of the empirical data. From a more fundamental perspective, the problem of deriving the basic relation in Equation (3) should be considered for future studies.

In view of these results, we may suggest that sound dissimilarity between sites decays with their inter-site distance in space with an exponent $\alpha = 1/2$, yielding

$$D_{i,j} \simeq [R_{i,j}/R_0]^{1/2}, \tag{7}$$

where the effective length scale in the network is given by,

$$R_0 = \langle R_{i,j} \rangle^2 / \langle D_{i,j} \sqrt{R_{i,j}} \rangle^2. \quad (8)$$

These relations can be tested in other cases, and if confirmed, they may represent a useful technique to estimate the internal correlations of the soundscape in the area of interest. Indeed, the set of parameters $P(n)$, obtained in a self-consistent fashion with the assumed decay of dissimilarity, allows us to classify the sites according to our scheme discussed in Table 3.

According to the results in Figure 9, we find it quite remarkable that the local SRI mean values are consistent with our simple quality rules displayed in Table 3. We have therefore classified them according to these rules and plotted the corresponding set of sites with different colors, as shown in Figure 10. The good quality sites are in green color, the medium quality ones in yellow and the poor quality sites in red. There is a "borderline" site, number 4, which has a mean SRI close to 0. We have therefore indicated it as a full yellow circle with a red contour.

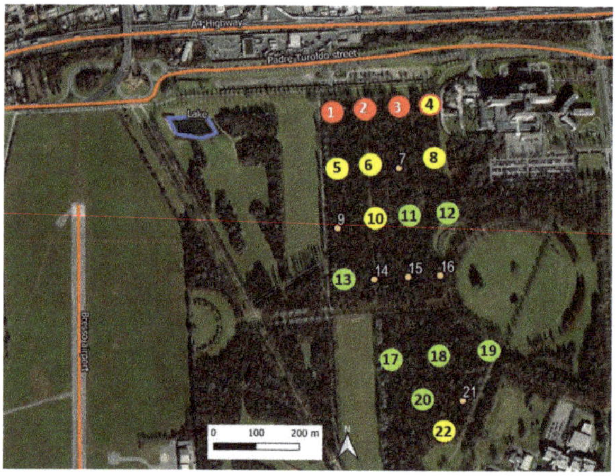

Figure 10. Final classification of the active sensors: (red circles) $\langle \text{SRI} \rangle < 0$, (yellow circles) $0 < \langle \text{SRI} \rangle < 2$ (sensor 4 has been indicated with a red border line since its SRI$\simeq 0$), and (green circles) $\langle \text{SRI} \rangle > 2$. The latter represents the best quality locations in the network. The value of the optimized parameters are displayed in Table 4. The inactive sensors (7, 9, 14, 15, 16, 21) are depicted by the small circles.

Environmental sound represents an important phenomenon that is fully integrated in the ecological system as a whole [40]. This feature is strongly interconnected with the ecological processes and patterns driven by biotic and abiotic relationships [40,41]. For this reason, environmental sounds may represent measurable indicators of ecological relationships and environmental degradation [42,43]. In addition, a simple but effective classification scheme to visualize the internal "structure" of the soundscape inside a habitat should be useful for many applications, including helping the monitoring of a given area for implementing better environmental policies.

One may also attempt to apply pattern recognition algorithms to analyze larger audio recorders in a systematic way, in order to extract the sound subcategories (and possibly other ones) considered in Table 1. This could be of great help, and one can rely on human intervention only to validate the fully automatized algorithm. Our approach represents a methodological effort to describe the environmental sound quality on the basis of a receiver perception. For this reason, although we used short recording periods (3.5 h), obtained from an area of approximately 22 hectares, and some of the geophonic categories were not represented in the audio recordings (wind, rain), we expect this study to provide a useful measure of a habitat soundscape quality in terms of a simple SRI index.

Finally, we expect these results to be useful also in connection with more elaborated approaches based on ecoacoustic indices and associated time series, analyzed using artificial intelligence (AI) methods, as they may serve as simpler qualitative indicators of the soundscape quality to complement the AI results. Work in this direction is in progress.

Author Contributions: Conceptualization, H.E.R., R.B. and G.Z.; methodology, H.E.R. and R.B.; formal analysis, H.E.R.; investigation, R.B., H.E.R. and A.A.; resources, G.Z.; data curation, A.P.; writing—original draft preparation, H.E.R.; writing—review and editing, H.E.R. and R.B.; visualization, R.B., H.E.R. and A.P.; funding acquisition, G.Z. All authors have read and agreed to the published version of the manuscript.

Funding: This research received no external funding.

Institutional Review Board Statement: Not applicable.

Informed Consent Statement: Not applicable.

Data Availability Statement: Data are available upon request.

Conflicts of Interest: The authors declare no conflict of interest.

References

1. Schafer, R.M. *The Soundscape: Our Sonic Environment and the Tuning of the World*; Simon and Schuster: New York, NY, USA, 1993.
2. Dumyahn, S.L.; Pijanowski, B.C. Soundscape conservation. *Landsc. Ecol.* **2011**, *26*, 1327–1344. [CrossRef]
3. Barber, J.R.; Crooks, K.R.; Fristrup, K.M. The costs of chronic noise exposure for terrestrial organisms. *Trends Ecol. Evol.* **2010**, *25*, 180–189. [CrossRef]
4. Lawson, G.M. Networks cities and ecological habitats. In *Networks Cities*; Qun, F., Brearley, J., Eds.; China Architecture and Building Press: Beijing, China, 2011; pp. 250–253. Available online: https://eprints.qut.edu.au/40229/(accessed on 15 March 2023).
5. Francis, C.D.; Newman, P.; Taff, B.D.; White, C.; Monz, C.A.; Levenhagen, M.; Petrelli, A.R.; Abbott, L.C.; Newton, J.; Burson, S.; et al. Acoustic environments matter: Synergistic benefits to humans and ecological communities. *J. Environ. Manag.* **2017**, *203*, 245–254. [CrossRef]
6. Doser, J.W.; Hannam, K.M.; Finley, A.O. Characterizing functional relationships between technophony and biophony: A western New York soundscape case study. *Landsc. Ecol.* **2020**, *35*, 689–707. [CrossRef]
7. Sayuri Moreira, S.L.; Freire, S.T.S.; Wagner, R.J., Jr.; Diego, L. Terrestrial Passive Acoustic Monitoring: Review and Perspectives. *BioScience* **2019**, *69*, 15–25. [CrossRef]
8. Mellinger, D.K.; Stafford, K.M.; Moore, S.E.; Dziak, R.P.; Matsumoto, H. An overview of fixed passive acoustic observation methods for cetaceans. *Oceanography* **2007**, *20*, 36–45. Available online: https://www.jstor.org/stable/24860138 (accessed on 15 March 2023). [CrossRef]
9. Ribeiro, J.W.; Sugai, L.S.M.; Campos-Cerqueira, M. Passive acoustic monitoring as a complementary strategy to assess biodiversity in the Brazilian Amazonia. *Biodivers. Conserv.* **2017**, *26*, 2999–3002. [CrossRef]
10. Browning, E.; Gibb, R.; Glover-Kapfer, P.; Jones, K.E. *Passive Acoustic Monitoring in Ecology and Conservation*; WWF: Woking, UK, 2017; p. 76. Available online: https://repository.oceanbestpractices.org/handle/11329/1370 (accessed on 15 March 2023).
11. Sueur, J.; Farina, A.; Gasc, A.; Pieretti, N.; Pavoine, S. Acoustic indices for biodiversity assessment and landscape investigation. *Acta Acust. United Acust.* **2014**, *100*, 772–781. [CrossRef]
12. Krause, B. The Loss of Natural Soundscapes. *Earth Isl. J.* **2002**, *17*, 27–29. Available online: www.earthisland.org/journal/index.php/magazine/archive (accessed on 15 March 2023).
13. Pijanowski, B.C.; Farina, A.; Gage, S.H.; Dumyahn, S.L.; Krause, B.L. What is soundscape ecology? An introduction and overview of an emerging new science. *Landsc. Ecol.* **2011**, *26*, 1213–1232. [CrossRef]
14. Yang, W.; Kang, J. Soundscape and sound preferences in urban squares: A case study in Sheffield. *J. Urban Des.* **2005**, *10*, 61–80. [CrossRef]
15. Pavan, G. Fundamentals of Soundscape Conservation. In *Ecoacoustics: The Ecological Role of Sounds*; Farina, A., Gage, S.H., Eds.; Wiley & Sons: Hoboken, NJ, USA, 2017; pp. 235–258. [CrossRef]
16. Sethi, S.S.; Jones, N.S.; Fulcher, B.D.; Picinali, L.; Clink, D.J.; Klinck, H.; Orme, C.D.L.; Wrege, P.H.; Ewers, R.M. Characterizing soundscapes across diverse ecosystems using a universal acoustic feature set. *Proc. Natl. Acad. Sci. USA* **2020**, *117*, 17049–17055. [CrossRef] [PubMed]
17. Liu, J.; Kang, J.; Luo, T.; Behm, H. Landscape effects on soundscape experience in city parks. *Sci. Total. Environ.* **2013**, *454–455*, 474–481. [CrossRef]

18. Raimbault, M.; Lavier, C.; Bérengier, M. Ambient sound assessment of urban environments: Field studies in two French cities *Appl. Acoust.* **2003**, *64*, 1241–1256. [CrossRef]
19. Uebel, K.; Rhodes, J.R.; Wilson, K.; Dean, A.J. Urban park soundscapes: Spatial and social factors influencing bird and traffic sound experiences. *People Nat.* **2022**, *4*, 1616–1628. [CrossRef]
20. Brambilla, G.; Gallo, V.; Asdrubali, F.; D'Alessro, F. The perceived quality of soundscape in three urban parks in Rome. *J. Acoust. Soc. Am.* **2013**, *134*, 832–839. [CrossRef]
21. Szeremeta, B.; Zannin, P.H.T. Analysis and evaluation of soundscapes in public parks through interviews and measurement of noise. *Sci. Total. Environ.* **2009**, *407*, 6143–6149. [CrossRef]
22. Uebel, K.; Marselle, M.; Dean, A.J.; Rhodes, J.R.; Bonn, A. Urban green space soundscapes and their perceived restorativeness. *People Nat.* **2021**, *3*, 756–769. [CrossRef]
23. Lellouch, L.; Pavoine, S.; Jiguet, F.; Glotin, H.; Sueur, J. Monitoring temporal change of bird communities with dissimilarity acoustic indices. *Methods Ecol. Evol.* **2014**, *5*, 495–505. [CrossRef]
24. Kasten, E.P.; Gage, S.H.; Fox, J.; Joo, W. The remote environmental assessment laboratory's acoustic library: An archive for studying soundscape ecology. *Ecol. Inform.* **2012**, *12*, 50–67. [CrossRef]
25. Boelman, N.T.; Asner, G.P.; Hart, P.J.; Martin, R.E. Multitrophic invasion resistance in hawaii: Bioacoustics, field surveys, and airborne remote sensing. *Ecol. Appl.* **2007**, *17*, 2137–2144. [CrossRef] [PubMed]
26. Sueur, J.; Pavoine, S.; Hamerlynck, O.; Duvail, S. Rapid acoustic survey for biodiversity appraisal. *PLoS ONE* **2008**, *3*, e4065. [CrossRef]
27. Pieretti, N.; Farina, A.; Morri, D. A new methodology to infer the singing activity of an avian community: The Acoustic Complexity Index (ACI). *Ecol. Indic.* **2011**, *11*, 868–873. [CrossRef]
28. Eldridge, A.; Guyot, P.; Moscoso, P.; Johnston, A.; Eyre-Walker, Y.; Peck, M. Sounding out ecoacoustic metrics: Avian species richness is predicted by acoustic indices in temperate but not tropical habitats. *Ecol. Indic.* **2018**, *95*, 939–952. [CrossRef]
29. Benocci, R.; Brambilla, G.; Bisceglie, A.; Zambon, G. Eco-Acoustic Indices to Evaluate Soundscape Degradation Due to Human Intrusion. *Sustainability* **2020**, *12*, 10455. [CrossRef]
30. Bertucci, F.; Parmentier, E.; Berten, L.; Brooker, R.M.; Lecchini, D. Temporal and spatial comparisons of underwater sound signatures of different reef habitats in Moorea Island, French Polynesia. *PLoS ONE* **2015**, *10*, e0135733. [CrossRef]
31. Harris, S.A.; Shears, N.T.; Radford, C.A. Ecoacoustic indices as proxies for biodiversity on temperate reefs. *Methods Ecol. Evol.* **2016**, *7*, 713–724. [CrossRef]
32. Shonfield, J.; Bayne, E.M. Autonomous recording units in avian ecological research: Current use and future applications. *Avian Conserv. Ecol.* **2017**, *12*, 14. [CrossRef]
33. Pérez-Granados, C.; Traba, J. Estimating bird density using passive acoustic monitoring: A review of methods and suggestions for further research. *Ibis* **2021**, *163*, 765–783. [CrossRef]
34. Benocci, R.; Roman, H.E.; Bisceglie, A.; Angelini, F.; Brambilla, G.; Zambon, G. Eco-acoustic assessment of an urban park by statistical analysis. *Sustainability* **2021**, *13*, 7857. [CrossRef]
35. Benocci, R.; Potenza, A.; Bisceglie, A.; Roman, H.E.; Zambon, G. Mapping of the Acoustic Environment at an Urban Park in the City Area of Milan, Italy, Using Very Low-Cost Sensors. *Sensors* **2022**, *22*, 3528. [CrossRef]
36. Truax, B. *Handbook for Acoustic Ecology*; A.R.C. Publications: Vancouver, BC, Canada, 1978.
37. Pijanowski, B.C.; Villanueva-Rivera, L.J.; Dumyahn, S.L.; Farina, A.; Krause, B.L.; Napoletano, B.M.; Gage, S.H.; Pieretti, N. Soundscape ecology: The science of sound in the landscape. *Bioscience* **2011**, *61*, 203–216. [CrossRef]
38. Benocci, R.; Roman, H.E.; Bisceglie, A.; Angelini, F.; Brambilla, G.; Zambon, G. Auto-correlations and long time memory of environment sound: The case of an Urban Park in the city of Milan (Italy). *Ecol. Indic.* **2022**, *134*, 108492. [CrossRef]
39. Roman, H.E.; Albergante, M.; Colombo, M.; Croccolo, F.; Marini, F.; Riccardi, C. Modeling cross correlations within a many-assets market. *Phys. Rev. E* **2006**, *73*, 036129. [CrossRef]
40. Gage, S.H.; Farina, A. *Ecoacoustics: The Ecological Role of Sounds*; John Wiley & Sons: Hoboken, NJ, USA, 2017.
41. Sueur, J.; Farina, A. Ecoacoustics: The ecological investigation and interpretation of environment sound. *Biosemiotics* **2015**, *8*, 493–502. [CrossRef]
42. Farina, A. *Soundscape Ecology*; Springer: Dordrecht, The Netherlands, 2014.
43. Krause, B.; Farina, A.Using ecoacoustic methods to survey the impacts of climate change on biodiversity. *Biol. Conserv.* **2016**, *195*, 245–254. [CrossRef]

sensors

MDPI

Article

Advanced Noise Indicator Mapping Relying on a City Microphone Network

Timothy Van Renterghem [1,*], Valentin Le Bescond [2], Luc Dekoninck [1] and Dick Botteldooren [1]

1 WAVES Research Group, Department of Information Technology, Ghent University, Technologiepark 126, B 9052 Gent-Zwijnaarde, Belgium
2 Joint Research Unit in Environmental Acoustics (UMRAE), Centre for Studies on Risks, Mobility, Land Planning and the Environment (CEREMA) and University Gustave Eiffel, F-44344 Bouguenais, France
* Correspondence: timothy.vanrenterghem@ugent.be

Abstract: In this work, a methodology is presented for city-wide road traffic noise indicator mapping. The need for direct access to traffic data is bypassed by relying on street categorization and a city microphone network. The starting point for the deterministic modeling is a previously developed but simplified dynamic traffic model, the latter necessary to predict statistical and dynamic noise indicators and to estimate the number of noise events. The sound propagation module combines aspects of the CNOSSOS and QSIDE models. In the next step, a machine learning technique—an artificial neural network in this work—is used to weigh the outcomes of the deterministic predictions of various traffic parameter scenarios (linked to street categories) to approach the measured indicators from the microphone network. Application to the city of Barcelona showed that the differences between predictions and measurements typically lie within 2–3 dB, which should be positioned relative to the 3 dB variation in street-side measurements when microphone positioning relative to the façade is not fixed. The number of events is predicted with 30% accuracy. Indicators can be predicted as averages over day, evening and night periods, but also at an hourly scale; shorter time periods do not seem to negatively affect modeling accuracy. The current methodology opens the way to include a broad set of noise indicators in city-wide environmental noise impact assessment.

Keywords: noise monitoring networks; microphones; road traffic noise; environmental noise mapping; noise indicators

Citation: Van Renterghem, T.; Le Bescond, V.; Dekoninck, L.; Botteldooren, D. Advanced Noise Indicator Mapping Relying on a City Microphone Network. *Sensors* **2023**, 23, 5865. https://doi.org/10.3390/s23135865

Academic Editor: Hector Eduardo Roman

Received: 25 May 2023
Revised: 16 June 2023
Accepted: 21 June 2023
Published: 24 June 2023

1. Introduction

Road traffic is commonly the main source of exposure to environmental noise in European cities [1]. A basic step in road traffic noise mapping is gaining access to traffic parameters such as traffic intensity, vehicle speed, acceleration and traffic composition on each road segment in the network [2]. However, most traffic models focus on major roads only to perform congestion analysis during rush hours [3]. On smaller roads, in contrast, traffic data are most often lacking. Although this might be in line with the Environmental Noise Directive [4] in Europe, stipulating that noise maps should only be produced from 55 dB(A) L_{den} on, this is nevertheless problematic in view of city-wide noise mapping. When assessing human sleep disturbance due to noise, exposure mapping becomes even more challenging and should go down to levels as low as 40 dB(A) L_{night}.

Knowledge of less exposed zones is relevant as well since these zones should be of primary interest for future residential developments and are potentially restorative places in a city. Furthermore, only mapping exposure in part of a city could introduce bias in environmental justice studies, an important concern nowadays when making sustainable cities (see, e.g., [5]).

An interesting line of research showed that street categorization in a city is able to estimate street-side exposure levels reasonably well [6–11], possibly accompanied with limited sets of snapshot measurements, where efforts can be minimized by suited sampling strategies [6,12]. Similarly, roadside noise measurements were shown to be able to adequately predict the underlying road traffic parameters such as vehicle speed, traffic intensity and the share of heavy vehicles [13]. In the open GIS initiative Open Street Map (OSM), every road in a city is present and assigned a specific category. This opens possibilities for full city noise mapping, including low(er) exposure zones.

Linking noise exposure maps to human health effects is currently not very successful. For an important noise policy indicator such as self-reported noise annoyance, less than 30% of the observed variance found in a surveyed population is actually captured [14]. A possible reason for this low predictive power is that current noise mapping initiatives focus on long-term equivalent sound pressure levels only. This undermines a noise map as an efficient urban sound planning instrument. Only recently, the shortcomings of the commonly used energetically equivalent levels have been officially acknowledged [15].

The way people perceive environmental noise is much more complex than what can be quantified with these standard energetically averaged sound pressure levels. A wider set of noise indicators and psycho-acoustical indicators have long been used in other contexts, e.g., in product design [16] and soundscape studies [17,18]. Essentially, people are very good listeners, and even subtle changes in the spectro-temporal content of a sound might impact the perception and reaction to it. Noise indicators of potential interest are statistical sound pressure levels, the number of events and indicators describing the dynamic nature of urban sound. Currently, city-wide mapping of such noise indicators is very scarce. A few measurement-based initiatives can be found, where walkers equipped with microphones scan a particular city quarter [19–21]. The use of these more advanced indicators to better predict the impacts of environmental noise is currently underexplored.

In this work, a methodology is described to predict both equivalent sound pressure levels and a wide range of other noise indicators, by means of deterministic noise modeling, where the accuracy of the predictions is improved by fitting to long-term measurements of a city noise monitoring network in a final step. The state-of-the-art deterministic noise modeling procedure, facilitating the calculation of dynamic noise indicators, is described in brief in Section 2. The proposed methodology is illustrated for the city of Barcelona (Spain) in Section 3, where a city-wide microphone network has been operational for more than a decade.

2. Deterministic Noise Mapping Procedure

2.1. Linking Traffic Data and Open Street Map Road Categorization

Street categorization data were directly used from Open Street Map (OSM). Each street category was assigned a set of plausible traffic parameters (more precisely traffic intensity, vehicle speed and share of heavy vehicles). This assignment starts from existing (highway) traffic count databases, and it is ensured that the expected logics such as a lower traffic intensity, lower vehicle speed and a lower share of heavy vehicles on minor streets compared to major streets are present. Depending on the deterministically predicted noise indicators corresponding to a given scenario, additional scenarios were manually added. In total, 15 scenarios were used (see Appendix A for an overview of the parameter settings). In a final step, the calculated outcomes for a wide set of noise indicators are weighted to minimize the difference with measurement from the microphone network, as will be discussed in Section 3.2.

2.2. Dynamic Traffic Model

Simplified vehicle movements are modeled based on the hourly averaged number of vehicles and their speeds. Vehicles are launched on a road segment at a fixed speed. When reaching the end of that segment, the vehicle is removed from the simulation, meaning there is no vehicle transfer from one segment to another. The inter-vehicle times respect a Poisson distribution, and vehicle speeds of the different cars follow a normal distribution. At the end of the simulated hour, at each road segment, the (static) vehicle counts and average speeds are respected. More information on this simplified micro-simulation traffic procedure can be found in [22]. A time step of 1 s was considered in this work.

2.3. Traffic Noise Emission Model

Vehicle category, number of vehicles per hour, and average speed are used as input for the CNOSSOS [23] road traffic acoustic emission model. These traffic-related inputs come from the dynamic traffic modeling procedure described in Section 2.2.

2.4. Sound Propagation Model

The sound propagation modeling procedure combines the CNOSSOS sound propagation model [23] with aspects from the QSIDE urban sound propagation model [24]. Vehicles close to a receiver (within a radius of 500 m) are treated differently from those further away (between 500 m and at maximum 2000 m).

At close distance, and when a direct line-of-sight propagation path is possible between a source and a receiver, geometrical divergence, ground effect and atmospheric absorption are included following the CNOSSOS sound propagation model, implemented in the open access NoiseModelling framework [25,26]. Only in the absence of a line-of-sight propagation path, diffractions around horizontal edges and reflections on vertical objects are accounted for. The maximum reflection order is 2, and the maximum source-reflection distance is 50 m (which are standard settings; see, e.g., [27]). A standard noise mapping receiver height of 4 m is used.

Scattering on atmospheric turbulence is added to the attenuation factors to avoid levels becoming unrealistically low, especially behind objects. The QSIDE engineering scattering approach [28] was used, providing an easy-to-evaluate expression adapted to the urban environment, accounting for sound frequency, propagation distance, street canyon geometry and turbulence strength. Although the model could include local street canyon geometry in detail, standard building heights and widths were used to avoid time-demanding retrieval from geographical input data. Turbulence structure parameters C_v^2 and C_T^2 (see Table 1) depend on whether the propagation occurs in rural/suburban settings or in the dense urban fabric, and during the daytime or at night. These turbulence parameters are based on long-term observations over flat rural zones with dispersed smaller cities [29]; in the dense urban environment, turbulence strength is doubled in a simplified approach.

Table 1. Overview of the sound propagation models and parameter settings.

	Close-by Traffic		Far Traffic	
Radius around receiver (in m)	<500		≥500 and <2000	
Traffic (noise emission) modeling	Simplified dynamic traffic modeling following [22], at a 1 s time interval.		Aggregated traffic at discrete emission points. Number of emission points minimized by NoiseModelling [26]	
If a direct line-of-sight path is possible	CNOSSOS sound propagation model [23] without reflections on vertical objects, without diffractions, and in a non-refracting atmosphere.			
Only obstructed sound paths are present	CNOSSOS sound propagation model [23] including reflections on vertical objects (reflection order 2, maximum source-reflection distance 50 m) and including diffractions on horizontal edges. Downward refraction ("favorable conditions") is assumed with 50% occurrence in any direction.			
	Turbulent scattering model [28]			
	Rural/suburban		Dense urban fabric	
Distance to façade (m)	Not applicable		5	
City canyon width (m)	Not applicable		15	
Building height (m)	8		20	
	Day	Night	Day	Night
C_v^2 ($m^{4/3}/s^2$)	0.4	0.2	0.8	0.4
C_T^2 ($K^2/m^{2/3}$)	0.7	0.04	1.4	0.08

In order to capture dynamic noise indicators and noise events, considering individual nearby vehicles is essential. This is not the case anymore for road traffic further away contributing mainly to the background noise at a receiver. This allows bundling the acoustic energy of cars in a limited number of emission points as optimized by the NoiseModelling framework. For the propagation simulations, an approach similar to that for nearby traffic is followed, so depending on whether a line-of-sight path is possible or not.

The CNOSSOS favorable sound propagation approach (i.e., downward refraction) is only considered in the absence of line-of-sight paths. In the CNOSSOS simplified curved ray approach, the difference between refraction/no refraction is mainly relevant in the case of propagation over objects. The probability of favorable sound propagation is then set to 50% in any propagation direction.

3. The Barcelona Microphone Sensor Network

3.1. Measurements and Data Handling

The Barcelona microphone measurement network is unique in its kind due to its size (roughly 250 monitoring points spread over the city) and since it has been operational for more than a decade. The network contains both fixed sensors and sensors that are repositioned period-wise to maximize the zone monitored. Together with the fact that individual sensors are prone to accidental failure, the dataset is rather discontinuous in nature. Nevertheless, at some fixed sensors, continuous sound pressure level measurements over several years are present.

The sensors are opportunistically positioned, e.g., directly attached to window sills or near balconies. This means that the extent to which façade reflections impact the sound pressure level measurements is not fixed (see Section 4). Microphones are always facing the streets and are representative of the most exposed building side.

To limit the impact of changes in the traffic network infrastructure and its management (such as limiting traffic in specific streets, changing the direction of circulation, ban of heavy traffic), a 3-year period was selected, which was a compromise between keeping this period as short as possible and having a sufficient amount of data while keeping as many sensor locations as possible for processing. Nevertheless, changes in the traffic network cannot be fully avoided within this time frame, and if this was the case, the measurements were then the average between the two different traffic situations. Note that convergence must still be reached at such locations (see next paragraph) for a microphone position to be used.

The processing of the measurement sensors was performed as follows. A basic time period of 15 min was chosen for all indicators. Previous research [12] showed that this is a suitable time frame in road traffic noise-dominated urban environments. Shorter periods could lead to difficulties in stabilizing the noise indicators, giving too much emphasis on momentary variations. When extending to longer periods, the temporal variations in the sonic environment might not be captured sufficiently.

For a sensor location to be considered in further analysis, at least 3 weeks of data (not necessarily continuous) should be available. Weekends were excluded to avoid uncommon traffic situations. As a simplified convergence criterion, the difference between taking 80% of the data and all available data (in a chronological way) should be less than 1 dB when (linearly) averaging a noise indicator that uses a decibel scale. For event-based indicators, this criterion is set to five events, and for the intermittency ratio set to 5% (see further). If this condition is not met, this sensor location is disregarded at least for a specific time period. Removing sensor data during the day period, e.g., does not necessarily mean that the sensor location is also disregarded during the evening and night periods.

The measurement network contains sensors with two levels of detail. Most microphone stations report total A-weighted sound pressure levels with a basic integration period of 1 min. These data were available at 93 sensors in the current study (see Section 3.3), during the period 2020–2021–2022. Secondly, measurement stations logging 1/3-octave bands with a basic integration period of 1 s were used during the period 2016–2017–2018. These more detailed data were available at 23 stations and allowed calculating more advanced noise indicators as discussed in Section 3.4.

3.2. Machine Learning Fitting Procedure

As an example supervised machine learning fitting algorithm, an artificial neural network was used, as implemented in Matlab [30]. A standard split into training, validation and test sets using 70%, 15% and 15% of the data, respectively, was chosen. The training algorithm "Levenberg–Marquardt backpropagation" was used, which is recommended as a fast, first-choice procedure [30]. To prevent overfitting, only five neurons were used, with a single hidden layer [31]. Given the random split into training, validation and test datasets, models were repeatedly constructed, allowing the use of averaged model predictions and giving an indication of confidence intervals on repeated predictions.

In the case of predicting A-weighted equivalent sound pressure levels (see Section 3.3), the input consists of 93 locations × 15 traffic scenarios; there are 93 (locations) × 3 (day, evening and nightly averaged) or 93 (locations) × 24 (hourly averaged) outputs. In case more advanced noise indicators were included (see Section 3.4), 23 (locations) × 15 (traffic scenario) × 29 (indicators) inputs were used to predict 23 (locations) × 29 (indicators) × 3 (day, evening and nightly averaged) outputs.

This work does not aim at finding the most accurate or fastest machine learning approach for this specific application, but rather showcases what can be achieved with a standard and well-established supervised machine learning fitting approach. Similarly, further optimization of the neural network settings is also beyond the scope of the current work.

3.3. Predicting A-Weighted Equivalent Sound Pressure Levels

Using the basic $L_{Aeq,1min}$ values, an integration is performed to 15 min. In the next step, $L_{Aeq,15min}$ data are linearly averaged over day (7:00–19:00), evening (19:00–23:00) and night (23:00–7:00) periods, thus providing a typical value in each period, and form the basis for the artificial neural network predictions. In a second set of predictions, hourly averaged $L_{Aeq,15min}$ data are used as well.

Figures 1–3 depict the 15 deterministic predictions at each sensor location that formed the basis for the weighting by the artificial neural network, together with the measured values (i.e., the ground truth), the mean predicted values and the 90th and 10th percentiles based on repeated model constructions. On the horizontal axis, the location ID number is used, which is an arbitrary number but easily allows assessing changes from location to location, both in measurements and predictions. Figures are shown for the daily, evening and nightly averaged $L_{Aeq,15min}$. Note that sensors with an insufficient number of data points or sensors not leading to converged indicators (see Section 3.1) were obviously not used during the construction of the machine learning model. Once the model was constructed, predictions with the model were performed at all 93 sensor locations. Clearly, only locations with both measurements and predictions were considered in the subsequent accuracy analysis.

In Figures 4–6, the measured data are plotted on the street map of Barcelona, complying with the selection criteria (see Section 3.1), the (mean) predictions at (all) sensor locations, and the difference between measurements and predictions where possible (as root-mean-square error, RMSE). As an example, daytime data only are shown. At most locations, prediction errors are limited, although a few points give rise to larger errors.

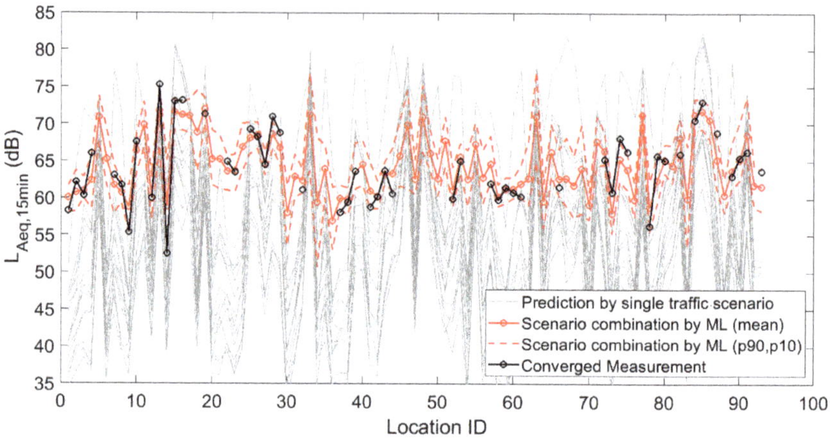

Figure 1. Deterministic predictions of $L_{Aeq,15min}$ for each single traffic scenario, by the artificial neural network (showing the mean prediction and the 90th and 10th percentiles, based on repeated model constructions), together with the converged measurements, at each of the 93 locations where a sensor node is/was operational. Data shown here are the linearly averaged $L_{Aeq,15min}$ during the daytime.

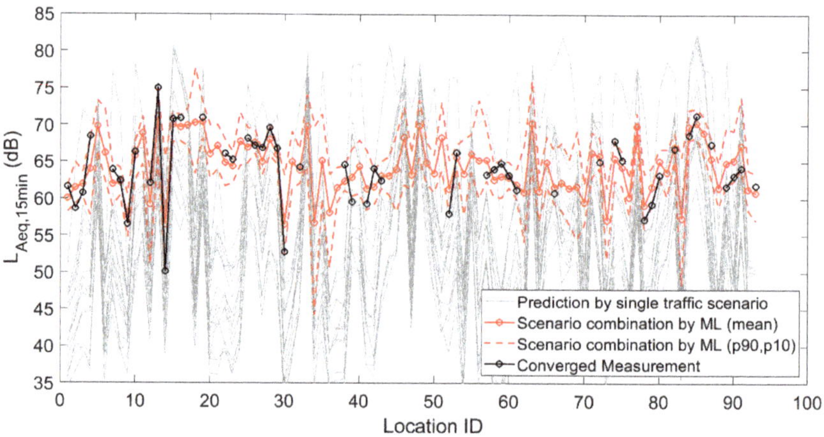

Figure 2. Deterministic predictions of $L_{Aeq,15min}$ for each single traffic scenario, by the artificial neural network (showing the mean prediction and the 90th and 10th percentiles, based on repeated model constructions), together with the converged measurements, at each of the 93 locations where a sensor node is/was operational. Data shown here are the linearly averaged $L_{Aeq,15min}$ during the evening.

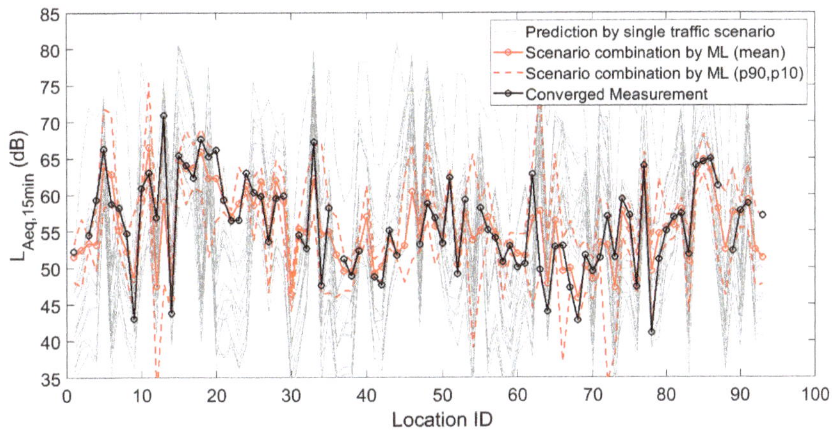

Figure 3. Deterministic predictions of $L_{Aeq,15min}$ for each single traffic scenario, by the artificial neural network (showing the mean prediction and the 90th and 10th percentiles, based on repeated model constructions), together with the converged measurements, at each of the 93 locations where a sensor node is/was operational. Data shown here are the linearly averaged $L_{Aeq,15min}$ during the night.

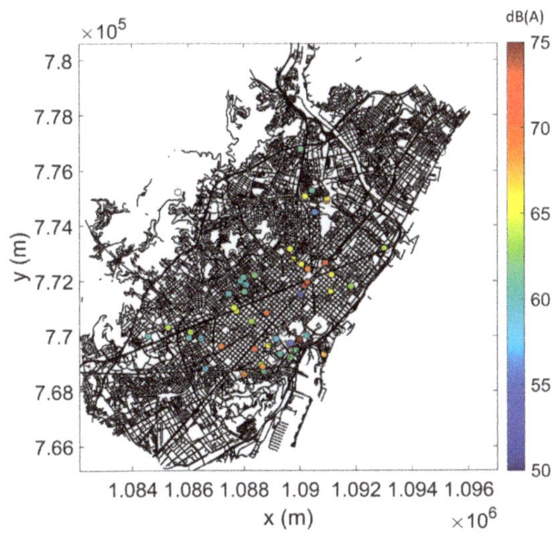

Figure 4. Linearly averaged $L_{Aeq,15min}$ from measurements during daytime. Only sensor locations with converged measurements and data falling within the pre-selected timeframe are shown.

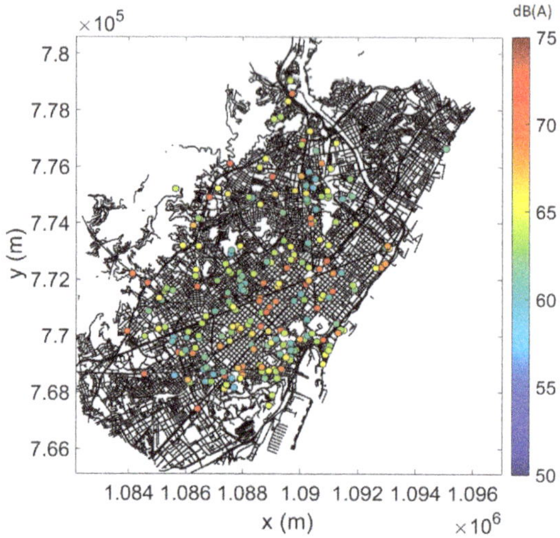

Figure 5. Mean $L_{Aeq,15min}$ predictions during daytime, at 93 spots where a sensor node is/was operational.

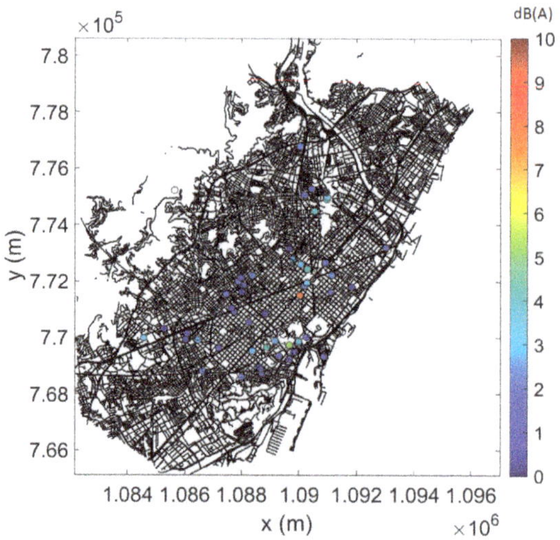

Figure 6. Root-mean-square error (RMSE) between measured and (mean) predicted $L_{Aeq,15min}$ during daytime. Only sensor locations with converged measurements and data falling within the pre-selected timeframe were used for this analysis.

The histograms in Figure 7 depict the actual differences between measurements and predictions, showing that the zero error class is most populated in all time periods considered. During the night, the distribution is still symmetrical, but the spread seems somewhat larger. The RMSEs are 2.0 dB(A) during the daytime, 2.1 dB(A) during the evening and 3.3 dB(A) during the night.

Results for hourly predictions are depicted in Figure 8, shown as temporal patterns over 24 h periods at each sensor location. The measured temporal patterns are shown as well. This figure does not allow the comparison of measurements and predictions at any individual sensor location, but it nicely shows that the bulk of the temporal patterns are well predicted. Both locations with a rather flat pattern and those with stronger level drops during the night hours can be distinguished, both in the measurements and predictions. Directly related to Figure 8, Figure 9 shows the hourly RMSEs. Minimum values are found around noon, near 2 dB(A), and increase slightly above 3 dB(A) between 3 and 4 o'clock at night.

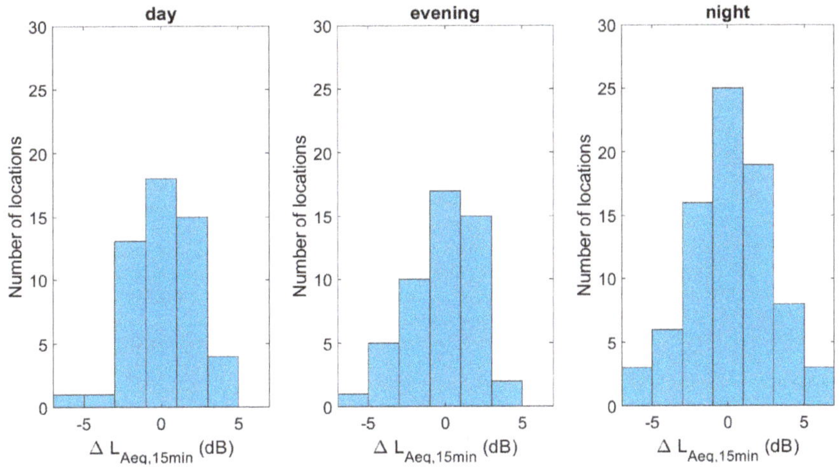

Figure 7. Histograms showing the difference between the (mean) predicted and measured $L_{Aeq,15min}$, linearly averaged over daytime, evening and night hours.

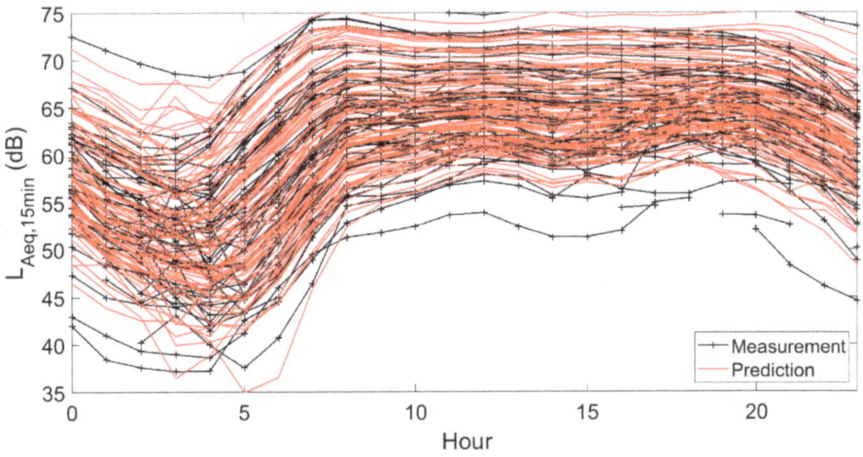

Figure 8. Hourly temporal patterns of $L_{Aeq,15min}$ at all 93 measurement locations. The measurements are shown together with the mean predictions based on repeated model construction. Data shown here are the linearly averaged $L_{Aeq,15min}$ during a specific hour. Interrupted lines indicate hours where measurements are not converged due to an insufficient amount of data.

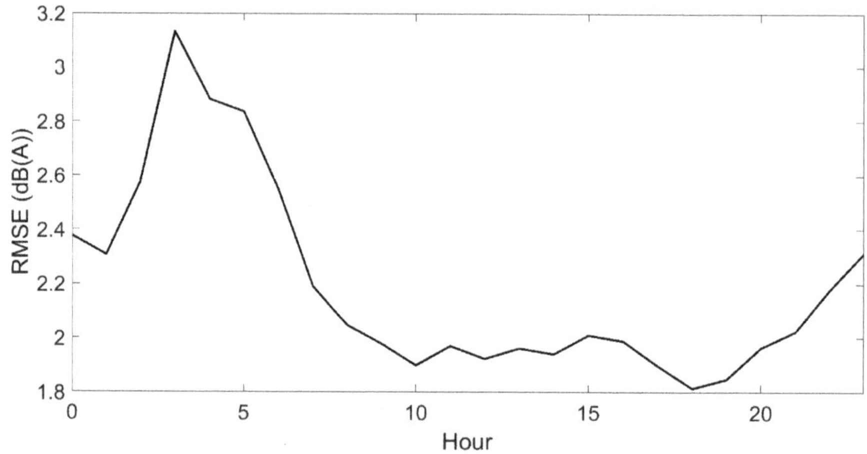

Figure 9. Root-mean-square error (RMSE) between measured and (mean) predicted $L_{Aeq,15min}$ over all locations on an hourly basis. Only sensor locations with converged measurements and data falling within the pre-selected timeframe are considered.

3.4. Predicting Advanced Noise Indicators

The spectro-temporal detail of the more detailed measurement stations is 1 s and in 1/3-octave bands. All indicators are first evaluated over a 15 min period. Calculating over longer time frames (hourly, or day/evening/night period) is performed by linearly averaging the 15 min indicators.

In Table 2, an overview is given of the noise indicators that were considered in this work. It concerns the basic equivalent sound pressure levels, either non-weighted (L_{eq}), A-weighted (L_{Aeq}) or C-weighted (L_{Ceq}). For the calculation, the basic 1 s 1/3-octave bands are first frequency-weighted and integrated to a total sound pressure level, and in the next step, they are integrated to 15 min.

Table 2. Overview of the 29 noise indicators considered in this work.

Equivalent Sound Pressure Levels	Statistical Sound Pressure Levels	Number of Events above x dB(A)	Number of Events above a Specific Indicator	Sound Dynamics Indicators	Intermittency Ratio
L_{eq}	L_{A01}	EN55	ENL_{A10}	σ_{AS}	IntRatio
L_{Aeq}	L_{A05}	EN60	ENL_{A50}	σ_{CS}	
L_{Ceq}	L_{A10}	EN65	ENL_{A50} + 3	L_{A10}–L_{A90}	
	L_{A50}	EN70	ENL_{A50} + 10	L_{C10}–L_{C90}	
	L_{A90}	EN75	ENL_{A50} + 15		
	L_{A95}	EN80	ENL_{A50} + 20		
	L_{A99}		ENL_{Aeq} + 10		
			ENL_{Aeq} + 15		

Statistical level L_{An} denotes the A-weighted sound pressure level exceeded n% of the time, where L_{A10}, e.g., is representative of peak levels, while L_{A90}, e.g., is representative of background noise levels. The percentiles are calculated over a period of 15 min based on the 1 s A-weighted total levels.

ENx is the number of events above a fixed L_{Aeq} of x dB. An event is defined by a peak in the time history of total A-weighted levels, lasting for at least 1 s. The number of occurrences is counted over a 15 min period. Similarly, the number of events above a statistical noise level, or a statistical noise level plus x dB, is considered as well.

Next, a number of sound dynamics indicators are considered. The difference between L_{A10} and L_{A90} is a commonly used metric in this respect; a large value means a strong variation in exposure level. $L_{C10}-L_{C90}$ is a similar indicator; in theory, it is applicable to higher absolute exposure levels, but it is often used to study dynamics with a focus on the low sound frequency range. These indicators are calculated as the differences between statistical levels assessed over the 15 min period.

Indicators σ_{AS} and σ_{CS} are the standard deviations on either the A-weighted or C-weighted 1 s total sound pressure levels, calculated over a period of 15 min.

The intermittency ratio IntRatio [32] is defined as the acoustic energy present in peaks relative to the total amount of energy in a particular time period, expressed as a percentage. The threshold to detect peaks is set to 3 dB (above the $L_{eq,15min}$) as proposed in [32].

When having access to detailed spectro-temporal data, the list of noise indicators that could be calculated is clearly not limited to the current selection. The current selection could especially be relevant to acoustically characterize urban traffic noise.

In Figures 10–12, fifteen deterministic predictions for (a selection of) statistical sound pressure levels at each sensor location are depicted, forming the basis for the weighting by the artificial neural network. These graphs further show the measured values, the mean predicted values and the 90th and 10th percentiles based on repeated model constructions. Figures are shown for the daily averaged indicators only for brevity.

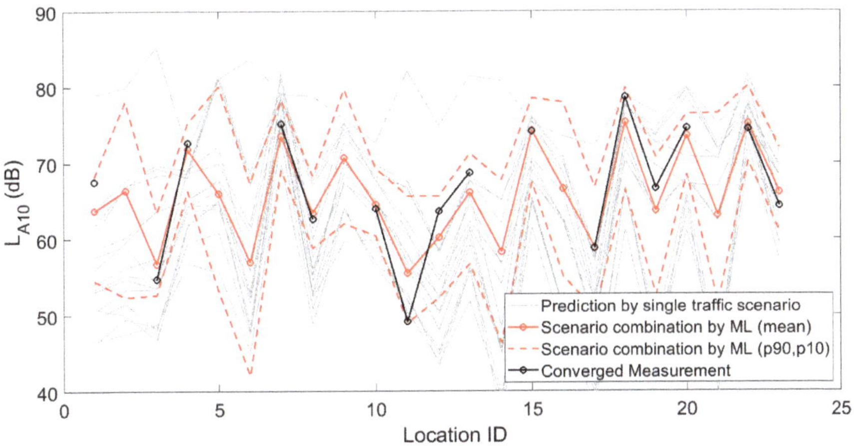

Figure 10. Deterministic predictions of the statistical sound pressure level L_{A10} for each single traffic scenario, by the artificial neural network (showing the mean prediction and the 90th and 10th percentiles, based on repeated model constructions), together with the converged measurements, at each of the 23 locations where a sensor node is present with detailed logging capabilities. Data shown here are linearly averaged over 15 min periods during daytime.

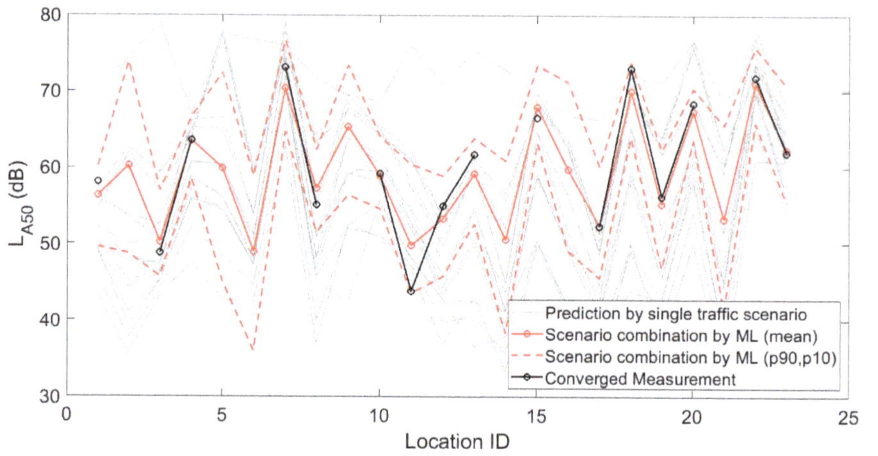

Figure 11. Deterministic predictions of the statistical sound pressure level L_{A50} for each single traffic scenario, by the artificial neural network (showing the mean prediction and the 90th and 10th percentiles, based on repeated model constructions), together with the converged measurements, at each of the 23 locations where a sensor node is present with detailed logging capabilities. Data shown here are linearly averaged over 15 min periods during daytime.

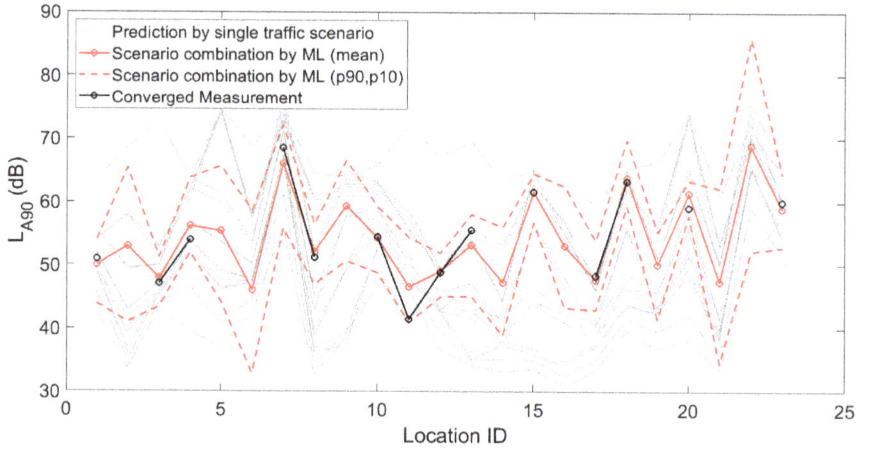

Figure 12. Deterministic predictions of the statistical sound pressure level L_{A90} for each single traffic scenario, by the artificial neural network (showing the mean prediction and the 90th and 10th percentiles, based on repeated model constructions), together with the converged measurements, at each of the 23 locations where a sensor node is present with detailed logging capabilities. Data shown here are linearly averaged over 15 min periods during daytime.

As another example, intermittency ratio predictions can be evaluated based on Figure 13 during the daytime. An overview of the RMSEs of all 29 indicators, averaged over day, evening and night periods, is given in Table 3.

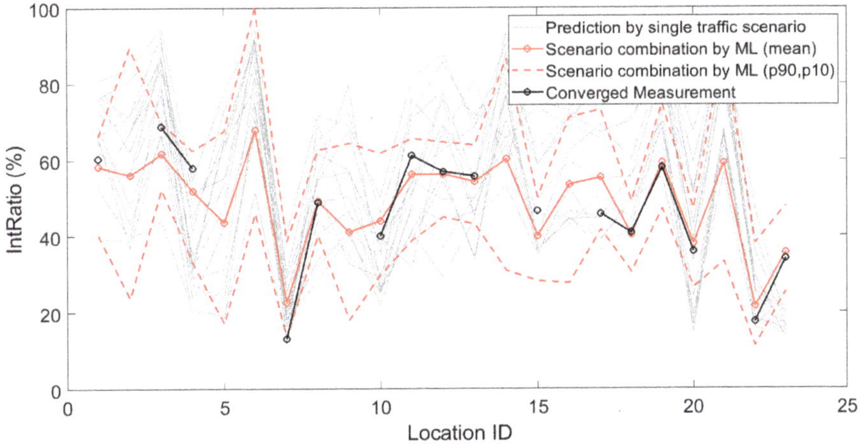

Figure 13. Deterministic predictions of the intermittency ratio for each single traffic scenario, by the artificial neural network (showing the mean prediction and the 90th and 10th percentiles, based on repeated model constructions), together with the converged measurements, at each of the 23 locations where a sensor node is/was operational with detailed logging. Data shown here are linearly averaged over 15 min periods during daytime.

Table 3. Root-mean-square errors (RMSEs) for each noise indicator considered over all sensor locations, split into day, evening and night periods.

Indicator	Day	Evening	Night
L_{eq} (dB)	1.9	2.2	1.8
L_{Aeq} (dB)	2.1	2.1	1.9
L_{Ceq} (dB)	2.1	2.3	2.0
L_{A01} (dB)	2.2	2.2	2.3
L_{A05} (dB)	2.4	2.4	2.1
L_{A10} (dB)	2.6	2.5	1.9
L_{A50} (dB)	2.2	2.3	2.3
L_{A90} (dB)	1.9	2.5	2.6
L_{A95} (dB)	2.1	2.6	2.0
L_{A99} (dB)	2.1	2.7	2.0
σ_{AS} (dB)	0.9	1.0	1.1
σ_{CS} (dB)	0.8	0.8	0.9
L_{A10}–L_{A90} (dB)	2.3	2.4	2.5
L_{C10}–L_{C90} (dB)	1.6	1.9	1.9
EN55 (n.o.e.) [1]	4.3	6.5	5.3
EN60 (n.o.e.)	7.4	7.0	8.4
EN65 (n.o.e.)	10.6	7.8	9.4
EN70 (n.o.e.)	12.3	10.9	2.7
EN75 (n.o.e.)	3.4	3.9	4.2
EN80 (n.o.e.)	3.2	3.1	1.6
ENL_{A10} (n.o.e.)	4.9	4.8	4.8
ENL_{A50} (n.o.e.)	8.3	9.8	8.6
ENL_{A50} + 3 (n.o.e.)	7.8	8.8	6.2
ENL_{A50} + 10 (n.o.e.)	3.9	4.2	3.9
ENL_{A50} + 15 (n.o.e.)	1.8	2.1	2.5
ENL_{A50} + 20 (n.o.e.)	0.9	0.8	1.7
ENL_{Aeq} + 10 (n.o.e.)	1.0	1.0	1.3
ENL_{A10} (n.o.e.)	0.5	0.5	0.6
IntRatio (%)	4.9	6.1	5.5

[1] n.o.e. stands for "number of events".

The A-weighted statistical levels, averaged over the three periods considered, only score slightly worse (2.3 dB) than the equivalent sound pressure levels (2.0 dB). The σ_S indicators, either A-weighted or C-weighted, have an RMSE lower than 1 dB, while L_{10}–L_{90} is predicted within 2 dB.

The errors in the number of noise events shown in Table 3 should be seen relative to the total number of events at a particular location. The larger errors are typically for locations and indicators with more events. When expressing the RMSEs relative to the (measured) total number of events, values equal to 26%, 28% and 30% are found for day, evening and night periods, respectively. The IntRatio is predicted with an RMSE of about 5–6%.

4. Discussion

The current work shows that a basic set of deterministic noise mapping simulations relying on street categorizations, forms a good basis for predicting sound pressure levels and related noise indicators. RMSEs with relation to 15 min equivalent sound pressure levels, averaged over day, evening and night time, were 2.0 dB(A), 2.1 dB(A) and 3.3 dB(A) respectively, for the case study of Barcelona, relying on a subset of 93 sensor locations with data over a period of 3 years. The histograms of differences between (fitted) simulations and measurements show that most data fall in the 0 dB(A) error class. Using this same approach, linearly averaged hourly predictions are possible, showing an RMSE below 2 dB(A) around noon, increasing to 3 dB(A) in the middle of the night. However, at a few sensor locations, larger errors are observed. Averaged daily temporal patterns of $L_{Aeq,15min}$ can be predicted too, showing logical results, and clearly distinguish between sensor locations with a rather flat pattern and those with a stronger level drop during the night hours.

This same procedure also works well for the more advanced noise indicators considered here. Some care is needed since this analysis is based on a more limited number of sensors Nevertheless, similar RMSEs were obtained, generally between 2 and 3 dB for indicators expressed in decibel units. Predicting a set of statistical sound pressure levels leads to errors similar to those for equivalent sound pressure levels. Capturing sound dynamics by means of the standard deviation, calculated based on 1 s L_{Aeq} or L_{Ceq}, leads to prediction errors near 1 dB. The intermittency ratio is predicted with an RMSE of about 5–6%.

The number of events, an indicator that might be especially relevant for assessing sleep disturbance by environmental noise, can be reasonably well predicted too. The larger RMSEs are typically for locations and indicators with more events. Note that at a location with a large number of events, missing a few events will not change the perception of the sonic environment. In contrast, if there are only a few events, each individual event has a bigger importance. Median relative errors are near 30%.

The artificial neural network fitting procedure uses a random split into training, validation and test subsets. Various model constructions allow deriving prediction ranges per location as a measure for the modeling uncertainty, while averaging model responses stabilizes results. For the $L_{Aeq,15min}$ predictions, using 93 sensor locations, the 10th and 90th percentiles on the predictions cover rather small ranges. For the more advanced noise indicator predictions (23 sensors), these uncertainty ranges are extensive and could indicate a lack of a sufficient amount of data. Notwithstanding these concerns, the linearly averaged fitted predictions make sense and do not seem to deviate more from the measurements than the artificial network trained on 93 sensor locations. Although the fitting procedure does not (explicitly) impose boundaries during the training, the average predictions nicely fit within these ranges. This indicates that the translation of street categories to traffic parameters, performed iteratively based on expert judgment, is performed adequately, at least for the specific sensor locations considered.

These observed errors should further be seen in view of façade reflections that play a significant role in inner-city street-side measurements. In the current dataset, the actual distance of the microphones relative to the façades is not fixed. Since exterior building surfaces are predominantly rigid, this will lead to strong reflections; consequently, standing waves will appear in front of a façade, characterized by frequency ranges with constructive interferences. At other frequencies, pronounced destructive interference dips appear. Measurements and numerical simulations, focusing on road traffic noise sources, showed that the increase in sound pressure level, relative to the free field, is typically between 3 and 6 dB [33–36]. Interference effects are not accounted for in the noise mapping propagation module. This difference in 3 dB could therefore be considered as a random factor during the fitting procedure. In addition, the distance relative to the driving lane is also relevant in relation to the magnitude of the façade reflections [37,38].

The deterministic modeling process is based on the CNOSSOS [23] and QSIDE models [24]. CNOSSOS is currently the recommended model for noise mapping in the European Union and basically stems from the ISO9613-2 model [39]. The model captures the basic physics of sound propagation in the outdoor environment. The focus is on rapid evaluation rather than achieving high accuracy, making the model suited for strategic noise mapping in large zones. The QSIDE model was designed specifically to model exposure at non-directly exposed building sides in urban environments. Only the QSIDE turbulent scattering formula [28] is used in the current context. The parameter settings and model choices aim at balancing the expected physical accuracy and computational cost.

In this respect, simplifications and inaccuracies arising from the deterministic modeling process might be—at least partly—corrected for by the fitting procedure, aiming primarily at weighting the level predictions of the different traffic parameter scenarios. In the current deterministic simulations, the reflection order on vertical objects is limited to 2. In a street canyon, however, a much larger number of reflections is needed; convergence in sound pressure level might need an order of 20 or more [40]. Despite this strong limitation during the deterministic modeling, street-side levels are still adequately predicted. A possible explanation is that a constant factor ("street amplification") is implicitly added during the fitting. Research found that the many sound reflections in a street build up a reverberant field that can be captured by a "building correction" [37] or "reflection ratio" [41], mainly depending on street width [41]. Logically, street width depends on street category. As another example, the local vehicle fleet might not fully align with the standard sound emission model and might depend on the average age and maintenance degree of the cars and the popularity of specific engine types [42].

In addition, the diffraction formula used in CNOSSOS is a strong simplification of a complex physical process (see, e.g., [43]). Especially in the case of realistic urban environments, characterized by consecutive diffractions over roof edges, accuracy will further degrade. The minimum level set by accounting for turbulent scattering could at least avoid levels becoming unrealistically low, which is a common issue with simplified diffraction modeling over buildings [24]. Note, however, that concerns on modeling sound propagation towards shielded building sides might not be a main problem in the current work since most sensors are positioned at the street side. Application of the current methodology to shielded building façades, of high relevance, e.g., with respect to the promotion of quiet sides [44], needs further study. Indeed, sound fields in non-directly exposed zones in the urban environment might be strongly different, e.g., in relation to their dynamics [45].

The methodology presented in this work needs an extensive microphone measurement network for weighting the traffic scenarios and therefore is not readily applicable to any city. Although such (permanent) networks are still scarce, they are increasingly being deployed in bigger cities all over the world such as Barcelona [46], New York [47], and Paris [48], just to name a few. Although high-end network-based microphone systems are possible, more affordable options using consumer electronics microphones exist. Such sensors have become very cheap due to mass production. Although such sensors are not primarily intended as measurement devices, they can measure sound pressure levels reasonably well. It was shown before that cheap microphones that highly correlate to reference type-1 microphones even in harsh outdoor conditions can be identified; when the deviations are expressed in total A-weighted (road traffic) noise levels, values of less than 1 dB are obtained, in excess of the deviation amongst reference microphones themselves [49]. More recent developments and experience with MEMS microphones [50–52] could further boost the deployment of city-wide microphone networks. The implicit traffic data retrieval in the current work could further benefit from including computer vision technologies [53]. In [54], e.g., camera images were directly used for noise mapping using machine learning.

A relevant question is whether the trained fitting network could be transferable to other cities. It is expected that this is unlikely, since the link between street categories and traffic parameters could be strongly locally dependent. In addition, as discussed before, the network might not only weight the traffic scenarios, but also traffic and propagation-related aspects are corrected for. Examples are the typical street widths and number of lanes corresponding to a specific street category in a city, traffic management policy, and preferred road surfaces and their maintenance.

5. Conclusions

The proposed advanced noise indicator mapping procedure, using a set of deterministic predictions combined with data from a city microphone measurement network, has been shown to be an approach with high potential. Both equivalent sound pressure levels and more advanced noise indicators expressed in decibel units lead to RMSEs between 2 and 3 dB. These deviations should be positioned relative to the 3 dB variation in street-side urban road traffic noise exposure measurements when the microphone positioning relative to the façade is not fixed. The current work further shows that city-wide noise mapping without access to direct traffic data is feasible on the condition that a microphone network is available, and at the same time, systematic inaccuracies occurring at any stage during the deterministic modeling process might be implicitly corrected for, at least to some extent. Continued research and more case studies are needed to see whether the current concept can grow to a mature urban traffic noise mapping methodology.

Author Contributions: Conceptualization, D.B. and T.V.R.; methodology, D.B., T.V.R., L.D. and V.L.B.; software, V.L.B., T.V.R., D.B. and L.D.; validation, T.V.R.; formal analysis, T.V.R.; investigation, V.L.B., D.B., T.V.R. and L.D.; resources, D.B. and T.V.R.; data curation, T.V.R.; writing—original draft preparation, T.V.R.; writing—review and editing, T.V.R., V.L.B., D.B. and L.D.; visualization, T.V.R.; supervision, T.V.R. and D.B.; project administration, D.B.; funding acquisition, D.B. All authors have read and agreed to the published version of the manuscript.

Funding: We acknowledge the funding from the European Union's Horizon 2020 Research and Innovation Programme for the project "Equal life", as part of the European Human Exposome Network, under grant agreement No. 87474.

Institutional Review Board Statement: Not applicable.

Informed Consent Statement: Not applicable.

Data Availability Statement: This study made use of third-party data and software that are partly open. The new data created in this work is available upon request.

Acknowledgments: We are grateful to Javier Casado Novas and Julia Camps Farres for enabling access to the historical data from Barcelona's microphone network.

Conflicts of Interest: The authors declare no conflict of interest.

Appendix A

In Table A1, the link between the Open Street Map categories and the traffic intensity, vehicle speed and share of heavy vehicles is shown, for the 15 scenarios that were used in this work.

Table A1. Traffic parameters assigned to the Open Street Map street categories that were explicitly calculated with the deterministic noise mapping methodology. Vehicle intensities (VIs) are expressed in cars per hour, the share of heavy vehicles (SHV) in %.

Period	Road Type	Vehicle Speed (km/h)	VI (SHV) Scenario 1	VI (SHV) Scenario 2	VI (SHV) Scenario 3	VI (SHV) Scenario 4	VI (SHV) Scenario 5	VI (SHV) Scenario 6
Day	motorway	130	20,400 (15%)	10,200 (15%)	5100 (15%)	20,400 (15%)	20,400 (20%)	20,400 (15%)
	trunk	110	8400 (15%)	4200 (15%)	2100 (15%)	8400 (15%)	33,600 (20%)	16,800 (15%)
	primary	80	4800 (0%)	2400 (0%)	1200 (0%)	4800 (0%)	19,200 (5%)	9600 (0%)
	secondary	80	3300 (0%)	3300 (0%)	3300 (0%)	1750 (0%)	26,400 (5%)	6600 (0%)
	tertiary	50	350 (0%)	350 (0%)	350 (0%)	175 (0%)	8400 (0%)	2100 (0%)
	residential	30	175 (0%)	175 (0%)	175 (0%)	85 (0%)	350 (0%)	1400 (0%)
	service	30	80 (0%)	80 (0%)	80 (0%)	42 (0%)	175 (0%)	175 (0%)
Evening	motorway	130	20,400 (11%)	10,200 (11%)	5100 (11%)	20,400 (11%)	20,400 (16%)	20,400 (11%)
	trunk	110	1600 (11%)	800 (11%)	400 (11%)	1600 (11%)	12,800 (16%)	3200 (11%)
	primary	80	1000 (0%)	500 (0%)	250 (0%)	1000 (0%)	8000 (5%)	2000 (0%)
	secondary	80	600 (0%)	600 (0%)	600 (0%)	300 (0%)	9600 (5%)	1200 (0%)
	tertiary	50	100 (0%)	100 (0%)	100 (0%)	50 (0%)	2400 (0%)	600 (0%)
	residential	30	50 (0%)	50 (0%)	50 (0%)	25 (0%)	100 (0%)	400 (0%)
	service	30	25 (0%)	25 (0%)	25 (0%)	12 (0%)	50 (0%)	50 (0%)
Night	motorway	130	20,400 (32%)	10,200 (32%)	5100 (32%)	20,400 (32%)	20,400 (37%)	20,400 (32%)
	trunk	110	800 (32%)	400 (32%)	200 (32%)	800 (32%)	6400 (37%)	1600 (32%)
	primary	80	640 (0%)	320 (0%)	160 (0%)	640 (0%)	5120 (5%)	1280 (0%)
	secondary	80	360 (0%)	180 (0%)	180 (0%)	160 (0%)	5760 (0.5%)	720 (0%)
	tertiary	50	50 (0%)	50 (0%)	50 (0%)	25 (0%)	1200 (0%)	300 (0%)
	residential	30	25 (0%)	25 (0%)	25 (0%)	12 (0%)	50 (0%)	50 (0%)
	service	30	12 (0%)	12 (0%)	12 (0%)	6 (0%)	25 (0%)	100 (0%)

Period	Road Type	VI (SHV) Scenario 7	VI (SHV) Scenario 8	VI (SHV) Scenario 9	VI (SHV) Scenario 10	VI (SHV) Scenario 11	VI (SHV) Scenario 12	VI (SHV) Scenario 13	VI (SHV) Scenario 14	VI (SHV) Scenario 15
Day	motorway	33,300 (12.9%)	28,404 (16.2%)	34,315 (11.8%)	35,481 (7.7%)	20,400 (15%)	20,400 (20%)	20,400 (15%)	20,400 (15%)	20,400 (15%)
	trunk	26,928 (5.4%)	21,012 (7.3%)	26,794 (4.4%)	27,705 (2.3%)	8400 (15%)	8400 (20%)	16,800 (15%)	8400 (15%)	33,600 (15%)
	primary	26,928 (5.4%)	21,012 (7.3%)	26,794 (4.4%)	27,705 (2.3%)	4800 (0%)	4800 (5%)	9600 (0%)	4800 (0%)	19,200 (0%)
	secondary	18,192 (10.7%)	14,652 (13.6%)	18,061 (9.8%)	19,562 (5.7%)	6600 (0%)	6600 (5%)	13,200 (0%)	13,200 (0%)	26,400 (0%)
	tertiary	8928 (6.2%)	7476 (7.6%)	9050 (5.1%)	9717 (2.6%)	2100 (0%)	2100 (5%)	2100 (0%)	4200 (0%)	8400 (0%)
	residential	3216 (3.5%)	2400 (4.4%)	3062 (3.1%)	3404 (1.6%)	350 (0%)	350 (5%)	350 (0%)	700 (0%)	1400 (0%)
	service	1098 (2.7%)	768 (3.6%)	1059 (2.4%)	1110 (1.4%)	175 (0%)	175 (5%)	175 (0%)	350 (0%)	700 (0%)
Evening	motorway	20,400 (11%)	20,400 (11%)	20,400 (11%)	20,400 (11%)	20,400 (11%)	20,400 (16%)	20,400 (11%)	20,400 (11%)	20,400 (11%)
	trunk	3200 (11%)	3200 (11%)	3200 (11%)	3200 (11%)	1600 (11%)	1600 (16%)	3200 (11%)	1600 (11%)	12,800 (11%)
	primary	2000 (0%)	2000 (0%)	2000 (0%)	2000 (0%)	1000 (0%)	1000 (5%)	2000 (0%)	1000 (0%)	8000 (0%)
	secondary	1200 (0%)	2400 (0%)	2400 (0%)	2400 (0%)	1200 (0%)	1200 (5%)	2400 (0%)	4800 (0%)	9600 (0%)
	tertiary	600 (0%)	600 (0%)	600 (0%)	600 (0%)	600 (0%)	600 (5%)	600 (0%)	1200 (0%)	2400 (0%)
	residential	400 (0%)	100 (0%)	100 (0%)	100 (0%)	100 (0%)	100 (5%)	100 (0%)	200 (0%)	400 (0%)
	service	50 (0%)	50 (0%)	50 (0%)	50 (0%)	50 (0%)	50 (5%)	50 (0%)	100 (0%)	200 (0%)
Night	motorway	20,400 (32%)	20,400 (32%)	20,400 (32%)	20,400 (32%)	20,400 (32%)	20,400 (37%)	20,400 (32%)	20,400 (32%)	20,400 (32%)
	trunk	1600 (32%)	1600 (32%)	1600 (32%)	1600 (32%)	800 (32%)	800 (37%)	1600 (32%)	800 (32%)	6400 (32%)
	primary	1280 (0%)	1280 (0%)	1280 (0%)	1280 (0%)	640 (0%)	640 (5%)	1280 (0%)	640 (0%)	5120 (0%)
	secondary	1440 (0%)	1440 (0%)	1440 (0%)	1440 (0%)	720 (0%)	720 (5%)	1440 (0%)	2880 (0%)	5760 (0%)
	tertiary	300 (0%)	300 (0%)	300 (0%)	300 (0%)	300 (0%)	300 (5%)	300 (0%)	600 (0%)	1200 (0%)
	residential	50 (0%)	50 (0%)	50 (0%)	50 (0%)	50 (0%)	50 (5%)	50 (0%)	100 (0%)	200 (0%)
	service	25 (0%)	25 (0%)	25 (0%)	25 (0%)	25 (0%)	25 (5%)	25 (0%)	50 (0%)	100 (0%)

References

1. European Environmental Agency. *Environmental Noise in Europe 2020*; EEA report No 22/2019; Publications Office of the European Union: Copenhagen, Denmark, 2020.
2. Licitra, G. *Noise Mapping in the EU: Models and Procedures*; CRC Press: Boca Raton, FL, USA; Taylor and Francis Group: Germantown, NY, USA, 2013.
3. Kessels, F. EURO Advanced Tutorials on Operational Research. In *Traffic Flow Modelling: Introduction to Traffic Flow Theory Through a Genealogy of Models*; Springer: Berlin/Heidelberg, Germany, 2018.
4. END. *Directive 2002/49/EC of the European Parliament and of the Council of 25 June 2002 Relating to the Assessment and Management of Environmental Noise*; European Commission: Brussels, Belgium, 2002.
5. Rickenbacker, H.; Brown, F.; Bilec, M. Creating environmental consciousness in underserved communities: Implementation and outcomes of community-based environmental justice and air pollution research. *Sust. Cities Soc.* **2019**, *47*, 101473. [CrossRef]
6. Barrigón Morillas, J.M.; Gómez Escobar, V.; Méndez Sierra, J.; Vílchez-Gómez, R.; Vaquero Martínez, J.; Trujillo Carmona, J. A categorization method applied to the study of urban road traffic noise. *J. Acoust. Soc. Am.* **2005**, *117*, 2844–2852. [CrossRef]

7. Rey Gozalo, G.; Barrigón Morillas, J.M.; Gómez Escobar, V. Urban streets functionality as a tool for urban pollution management. *Sci. Total Environ.* **2013**, *461–462*, 453–461. [CrossRef] [PubMed]
8. Zambon, G.; Benocci, R.; Brambilla, G. Statistical Road Classification Applied to Stratified Spatial Sampling of Road Traffic Noise in Urban Areas. *Int. J. Environ. Res.* **2016**, *10*, 411–420.
9. Zambon, G.; Benocci, R.; Brambilla, G. Cluster categorization of urban roads to optimize their noise monitoring. *Environ. Mon. Assess.* **2016**, *188*, 26. [CrossRef] [PubMed]
10. Barrigón Morillas, J.M.; Montes González, D.; Gómez Escobar, V.; Rey Gozalo, G.; Vílchez-Gómez, R. A proposal for producing calculated noise mapping defining the sound power levels of roads by street stratification. *Environ. Pollut.* **2021**, *270*, 116080. [CrossRef] [PubMed]
11. Staab, J.; Schady, A.; Weigand, M.; Lakes, T.; Taubenböck, H. Predicting traffic noise using land-use regression—A scalable approach. *J. Exp. Sci. Environ. Epidem.* **2022**, *32*, 232–243. [CrossRef]
12. Can, A.; Van Renterghem, T.; Rademaker, M.; Dauwe, S.; Thomas, P.; De Baets, B.; Botteldooren, D. Sampling approaches to predict urban street noise levels using fixed and temporary microphones. *J. Environ. Monit.* **2011**, *13*, 2710–2719. [CrossRef]
13. Can, A.; Dekoninck, L.; Rademaker, M.; Van Renterghem, T.; De Baets, B.; Botteldooren, D. Noise measurements as proxies for traffic parameters in monitoring networks. *Sci. Total Environ.* **2011**, *410*, 198–204. [CrossRef]
14. Brink, M. A Review of Explained Variance in Exposure-Annoyance Relationships in Noise Annoyance Surveys. In Proceedings of the International Commission on Biological Effects of Noise (ICBEN), Nara, Japan, 18–22 June 2014.
15. WHO. *Environmental Noise Guidelines for the European Region*; WHO Regional Office for Europe: Geneva, Switzerland, 2018.
16. Spence, C.; Zampini, M. Auditory contributions to multisensory product perception. *Act. Acust. Acust.* **2006**, *92*, 1009–1025.
17. Kang, J.; Aletta, F.; Gjestland, T.; Brown, L.; Botteldooren, D.; Schulte-Fortkamp, B.; Lercher, P.; van Kamp, I.; Genuit, K.; Fiebig, A.; et al. Ten questions on the soundscapes of the built environment. *Build. Environ.* **2016**, *108*, 284–294. [CrossRef]
18. Lionello, M.; Aletta, F.; Kang, J. A systematic review of prediction models for the experience of urban soundscapes. *Appl. Acoust.* **2020**, *170*, 107479. [CrossRef]
19. Can, A.; Gauvreau, B. Describing and classifying urban sound environments with a relevant set of physical indicators. *J. Acoust. Soc. Am.* **2015**, *137*, 208–218. [CrossRef]
20. Aumond, P.; Can, A.; De Coensel, B.; Botteldooren, D.; Ribeiro, C.; Lavandier, C. Modeling Soundscape Pleasantness Using perceptual Assessments and Acoustic Measurements Along Paths in Urban Context. *Act. Acust. Acust.* **2017**, *103*, 430–443. [CrossRef]
21. Van Renterghem, T.; Thomas, P.; Dekoninck, L.; Botteldooren, D. Getting insight in the performance of noise interventions by mobile sound level measurements. *Appl. Acoust.* **2022**, *185*, 108385. [CrossRef]
22. De Coensel, B.; Brown, A.L.; Tomerini, D. A road traffic noise pattern simulation model that includes distributions of vehicle sound power levels. *Appl. Acoust.* **2016**, *111*, 170–178. [CrossRef]
23. Kephalopoulos, S.; Paviotti, M.; Anfosso-Lédée, F. *Common Noise Assessment Methods in Europe (CNOSSOS-EU)*; Publications Office of the European Union: Luxembourg, 2012; 180p.
24. Wei, W.; Botteldooren, D.; Van Renterghem, T.; Hornikx, M.; Forssén, J.; Salomons, E.; Ögren, M. Urban background noise mapping: The general model. *Act. Acust. Acust.* **2014**, *100*, 1098–1111. [CrossRef]
25. Aumond, P.; Fortin, N.; Can, A. Overview of the NoiseModelling Open-Source Software Version 3 and its Applications. In Proceedings of the NOISE-CON Congress (261, 4, 2005–2011), Seoul, Republic of Korea, 12 October 2012.
26. Bocher, E.; Guillaume, G.; Picaut, J.; Petit, G.; Fortin, N. NoiseModelling: An Open Source GIS Based Tool to Produce Environmental Noise Maps. *ISPRS Int. J. Geo. Inform.* **2019**, *8*, 130. [CrossRef]
27. Le Bescond, V.; Can, A.; Aumond, P.; Gastineau, P. Open-source modeling chain for the dynamic assessment of road traffic noise exposure. *Transp. Res. Part D Transp. Environ.* **2021**, *94*, 102793. [CrossRef]
28. Forssén, J.; Hornikx, M.; Botteldooren, D.; Wei, W.; Van Renterghem, T.; Ögren, M. A model of sound scattering by atmospheric turbulence for use in noise mapping calculations. *Act. Acust. Acust.* **2014**, *100*, 810–815. [CrossRef]
29. Van Renterghem, T.; Horoshenkov, K.; Parry, J.; Williams, D. Statistical analysis of sound level predictions in refracting and turbulent atmospheres. *Appl. Acoust.* **2022**, *185*, 108426. [CrossRef]
30. Matlab. *The MathWorks Inc., version: 9.13.0 (R2022b)*; The MathWorks Inc.: Natick, MA, USA, 2022. Available online: https://www.mathworks.com (accessed on 1 December 2022).
31. Hagan, M.; Demuth, H.; Beale, M.; De Jesus, O. *Neural Network Design*, 2nd ed.; Martin Hagan: Stillwater, OK, USA, 2014.
32. Wunderli, J.-M.; Pieren, R.; Habermacher, M.; Vienneau, D.; Cajochen, C.; Probst-Hensch, N.; Röösli, M.; Brink, M. Intermittency ratio: A metric reflecting short-term temporal variations of transportation noise exposure. *J. Exp. Sci. Environ. Epidemiol.* **2016**, *26*, 575–585. [CrossRef] [PubMed]
33. Hall, F.L.; Papakyriakou, M.J.; Quirt, J.D. Comparison of outdoor microphone locations for measuring sound insulation of building facades. *J. Sound Vib.* **1984**, *92*, 559–567. [CrossRef]
34. Memoli, G.; Paviotti, M.; Kephalopoulos, S.; Licitra, G. Testing the acoustical corrections for reflections on a facade. *Appl. Acoust.* **2008**, *69*, 479–495. [CrossRef]
35. Mateus, M.; Carrilho, J.D.; Da Silva, M.G. An experimental analysis of the correction factors adopted on environmental noise measurements performed with window mounted microphones. *Appl. Acoust.* **2015**, *87*, 212–218. [CrossRef]

36. Barrigón Morillas, J.M.; Montes González, D.; Rey Gozalo, G. A review of the measurement procedure of the ISO 1996 standard. Relationship with the European Noise Directive. *Sci. Total Environ.* **2016**, *565*, 595–606. [CrossRef] [PubMed]
37. Heutschi, K. A simple method to evaluate the increase of traffic noise emission level due to buildings for a long straight street. *Appl. Acoust.* **1995**, *44*, 259–274. [CrossRef]
38. Montes González, D.; Barrigón Morillas, J.M.; Rey Gozalo, G. The influence of microphone location on the results of urban noise measurements. *Appl. Acoust.* **2015**, *90*, 64–73. [CrossRef]
39. *ISO 9613-2*; Acoustics-Attenuation of Sound Propagation Outdoors, Part 2: General Method of Calculation. International Organization for Standardization: Geneva, Switzerland, 1996; revised in 2017.
40. Salomons, E.; Polinder, H.; Lohman, W.; Zhou, H.; Borst, H.; Miedema, H. Engineering modeling of traffic noise in shielded areas in cities. *J. Acoust. Soc. Am.* **2009**, *126*, 2340–2349. [CrossRef]
41. Thomas, P.; Van Renterghem, T.; De Boeck, E.; Dragonetti, L.; Botteldooren, D. Reverberation-based urban street sound level prediction. *J. Acoust. Soc. Am.* **2013**, *133*, 3929–3939. [CrossRef]
42. Jonasson, H. Acoustical Source Modelling of Road Vehicles. *Act. Acust. Acust.* **2007**, *93*, 173–184.
43. Hadden, W.; Pierce, A. Sound diffraction around screens and wedges for arbitrary point source locations. *J. Acoust. Soc. Am.* **1981**, *69*, 1266–1276. [CrossRef]
44. Öhrström, E.; Skånberg, A.; Svensson, H.; Gidlöf-Gunnarsson, A. Effects of road traffic noise and the benefit of access to quietness. *J. Sound Vib.* **2006**, *295*, 40–59. [CrossRef]
45. Forssén, J.; Hornikx, M. Statistics of A-weighted road traffic noise levels in shielded urban areas. *Act. Acust. Acust.* **2006**, *92*, 998–1008.
46. Farres, J.C. Barcelona Noise Monitoring Network. In Proceedings of the Euronoise 2015, Maastricht, The Netherlands, 31 May–3 June 2015; pp. 2315–2320.
47. Mydlarz, C.; Sharma, M.; Lockerman, Y.; Steers, B.; Silva, C.; Bello, J.P. The Life of a New York City Noise Sensor Network. *Sensors* **2019**, *19*, 1415. [CrossRef]
48. Mietlicki, F.; Mietlicki, C.; Sineau, M. An Innovative Approach for Long Term Environmental Noise Measurement: RUMEUR Network in the Paris Region. In Proceedings of the Euronoise 2015, Maastricht, The Netherlands, 31 May–3 June 2015; pp. 2315–2320.
49. Van Renterghem, T.; Thomas, P.; Dominguez, F.; Dauwe, S.; Touhafi, A.; Dhoedt, B.; Botteldooren, D. On the ability of consumer electronics microphones for environmental noise monitoring. *J. Environ. Mon.* **2011**, *13*, 544–552. [CrossRef]
50. Mydlarz, C.; Salamon, J.; Bello, J.P. The implementation of low-cost urban acoustic monitoring devices. *Appl. Acoust.* **2017**, *117*, 207–218. [CrossRef]
51. Quintero, G.; Balastegui, A.; Romeu, J. A low-cost noise measurement device for noise mapping based on mobile sampling. *Measurement* **2019**, *148*, 106894. [CrossRef]
52. Yang, D.; Zhao, J. Acoustic Wake-Up Technology for Microsystems: A Review. *Micromachines* **2023**, *14*, 129. [CrossRef]
53. Buch, N.; Velastin, S.; Orwell, J. A Review of Computer Vision Techniques for the Analysis of Urban Traffic. *IEEE Trans. Intell. Transport. Syst.* **2011**, *12*, 920–939. [CrossRef]
54. Fredianelli, L.; Carpita, S.; Bernardini, M.; Del Pizzo, L.; Brocchi, F.; Bianco, F.; Licitra, G. Traffic Flow Detection Using Camera Images and Machine Learning Methods in ITS for Noise Map and Action Plan Optimization. *Sensors* **2022**, *22*, 1929. [CrossRef] [PubMed]

sensors

MDPI

Article

Sons al Balcó: A Comparative Analysis of WASN-Based L_{Aeq} Measured Values with Perceptual Questionnaires in Barcelona during the COVID-19 Lockdown

Daniel Bonet-Solà, Pau Bergadà, Enric Dorca, Carme Martínez-Suquía and Rosa Ma Alsina-Pagès *

HER—Human-Environment Research, La Salle-Universitat Ramon Llull, Sant Joan de la Salle, 42, 08022 Barcelona, Spain; daniel.bonet@students.salle.url.edu (D.B.-S.)
* Correspondence: rosamaria.alsina@salle.url.edu

Abstract: The mobility and activity restrictions imposed in Spain due to the COVID-19 pandemic caused a significant improvement in the urban noise pollution that could be objectively measured in those cities with acoustic sensor networks deployed. This significant change in the urban soundscapes was also perceived by citizens who positively appraised this new acoustic scenario. In this work, authors present a comparative analysis between different noise indices provided by 70 sound sensors deployed in Barcelona, both during and before the lockdown, and the results of a perceptual test conducted in the framework of the project *Sons al Balcó* during the lockdown, which received more than one hundred contributions in Barcelona alone. The analysis has been performed by clustering the objective and subjective data according to the predominant noise sources in the location of the sensors and differentiating road traffic in heavy, moderate and low-traffic areas. The study brings out strong alignments between a decline in noise indices, acoustic satisfaction improvement and changes in the predominant noise sources, supporting the idea that objective calibrated data can be useful to make a qualitative approximation to the subjective perception of urban soundscapes when further information is not available.

Keywords: lockdown; soundscape; L_{Aeq}; annoyance; perception; WASN; Barcelona

Citation: Bonet-Solà, D.; Bergadà, P.; Dorca, E.; Martínez-Suquía, C.; Alsina-Pagès, R.M. Sons al Balcó: A Comparative Analysis of WASN-Based L_{Aeq} Measured Values with Perceptual Questionnaires in Barcelona during the COVID-19 Lockdown. *Sensors* **2024**, *24*, 1650. https://doi.org/10.3390/s24051650

Academic Editor: Hector Eduardo Roman

Received: 19 December 2023
Revised: 24 February 2024
Accepted: 24 February 2024
Published: 3 March 2024

1. Introduction

The lockdown period during the COVID-19 outbreak at the beginning of 2020 allowed a new possibility not recorded yet: to be able to dismiss the human effect on many situations and confront unexplored scenarios, which could give clues to better understand human behavior. One of these situations was the reduction in acoustic noise level in urban and suburban areas since many people were confined at home or at least were subjected to severe mobility restrictions.

The lockdown gave the scientific community a unique scenario to assess how the reduction in noise levels affects the human perception of the acoustic environment. It is expected that the lower noise levels measured during the confinement translate to higher levels of acoustic satisfaction by the population. However, noise levels alone do not explain the complexity of the subjective acoustic comfort appraisal. In fact, the type of predominant noise source should be taken into account [1]. Therefore, a combined analysis of objective noise indices and the subjective assessment of the soundscapes performed during the lockdown, taking into account the predominant noise sources present in different areas, could give a very useful insight into which regulations should have a greater impact on acoustic comfort (e.g., reducing traffic density in quieter residential areas or limiting night-time leisure activities). As these kind of regulations normally face a strong rejection by part of the citizenship (such as private vehicle owners or people related to the leisure sector), assessing and comparing their expected effectiveness beforehand is essential.

In a previous research, ref. [2] data gathered from acoustic sensors scattered in diverse acoustic points in a medium-sized city (i.e., Girona, Spain) were compared with the answers reported in a poll and conducted around the same points. The authors were aware of the very small set of subjective data analyzed. However, it satisfied the goal of opening the door to the possibility of evaluating the soundscape by means of both objective and calibrated measurements and perceived appraisal by citizens. The answers reported in the survey, despite being a preliminary analysis with limited data, were able to distinguish the sound sources present around a certain sensor and match those noise sources detected in sensors data analysis. The results presented in that work also aligned with the objective measurements related to noise levels and the perception of the neighbors. There was a remarkable coherence with a prior analysis [3] when the authors also analyzed the environmental sound scenario before and during the lockdown for each of the sensors. In the current work, we make a step forward and move to a much bigger city (i.e., Barcelona, Spain) and we correlate data from 70 acoustic sensors with the answers of the 119 volunteers who assessed soundscapes from different locations within the city.

In this study, the authors explore the possibility of deriving some human perceptions by means of objective data provided by a number of noise sensors in the city of Barcelona [4]. Objective data are compared with subjective data and possible correlations are pursued between sensors data and subjective responses gathered by a survey in the same city. All sensor data and subjective surveys were collected during the COVID-19 lockdown in 2020. Sensors are assembled by common main source of noise with the aim to check whether there is a clear correlation with the survey outcomes. Clusters of sensors are build and both the type and level of the noise are taken into account when comparing subjective soundscape assessment with objective acoustic data. Authors will also present an analysis on the acoustic satisfaction assessment prior and during the lockdown per type of main noise source, as well as an assessment of perceptual constructs for soundscapes for each cluster of contributions.

This paper is structured as follows. Section 2 shows the state-of-the-art of the sensor data collection, with special emphasis on the lockdown period. Section 3 details the lockdown restrictions in Barcelona, the data used in this work, the description of the studied areas and the data treatment performed. Next, Section 4 details several clusters of sensors in Barcelona with the comparison of the objective measurements and the subjective results. Finally, Section 5 offers a discussion on the results and on the strengths and limitations of the study. Section 6 details the conclusions of this work.

2. State-of-the-Art

The mobility and activity restrictions imposed by the COVID-19 pandemic provided, unintentionally, a unique scenario to evaluate how these restrictions affect the noise pollution levels and their perception by citizens. In the past three years, many studies have been conducted to assess the effects of the lockdown on different soundscapes. Most of the literature has focused on one of two approaches. On the one hand, some authors have chosen to analyze objective data collected by a sensor or a network of sensors during the lockdown in order to compare the results with normal pre-pandemic data. On the other hand, other researchers opted to conduct a survey about the perception of people during these same periods.

Most of the published articles on data gathered by sensors during the lockdown are from urban or sub-urban areas. However, several studies have also been conducted in other environments such as underwater soundscapes [5–8], marine soundscapes [9], offshore human activities [10] and airport surroundings [11]. Furthermore, most of the research conducted in populated areas focused mainly on determining the L_{Aeq} reduction caused by the restrictions. Nonetheless, other changes in the soundscape have also been highlighted, e.g., in the San Francisco Bay Area, where a shift in the song frequency in some birds has been detected during the COVID-19 lockdown [12]. Apparently, the more favorable conditions caused by the drastic reduction in human activity and anthropogenic noise not only affected aspects like activity schedules, movement dynamics or exploratory behavior of different species [13] but also the vocalizations used in their songs in order to maximize communication distance in this new acoustic environment [12]. Another study [14] proved that the distribution of anomalous noise events and the intermittency ratio showed statistically significant differences in urban and suburban areas in Milan and Rome during the COVID-19 lockdown translating to a noticeable decrease in the negative impact of noise pollution in the population of both areas.

The geographical extent of these analyses varies from a single location in a city, e.g., Stockholm [15], to the combined contribution of seven of the major conurbations in India [16,17]. Lately, sensor networks consisting of sound meters have been deployed in many cities. Most of the recent publications that have studied changes in noise levels in 2020 have taken advantage of data collected from them. However, the scope of these networks differs significantly from one city to another ranging from 3 sound meters (as of the date of the study) in Montreal [18] to the impressive 70 sensors in Barcelona [4]. It is also worth noting that, in some cases, portable monitoring stations have been used [19,20] to gather noise data in a number of different locations when a permanent network of sensors was not available.

As expected, most of the literature verified a noteworthy reduction in the mean L_{Aeq} level of noise pollution during the lockdown. Nonetheless, some notable exceptions have also been spotted. In a quiet residential area in the city of Kobe [21]. Noise levels were higher during the state of emergency declared, apropos of the COVID-19 disease. According to the author, this area experiences seasonal changes in noise levels, making it more difficult to correctly set target values of the acoustic environment planning by referring to the measured noise level during the shutdown. Also, in Boston [22], in one of the three protected areas assessed, located near a highway, sound levels were between 4 and 6 dB higher during the lockdown. The probable explanation provided by the authors is that in a scenario with reduced traffic, vehicles could travel faster, thus creating more noise. On the contrary, the two other protected areas, which were closer to the city center, experimented a decrease in 1–3 dB during the same period.

Several acoustic metrics have been chosen in the different studies, with L_{Aeq}, L_d (stands for daytime equivalent noise level), L_n (stands for night-time equivalent noise level) and L_{den} (stands for day-night equivalent noise level) being the most widely used, which are consistent with some of the minimum indicators proposed by Asensio et al. [23]. Some authors offered the global average reduction of L_{Aeq} during the lockdown in the city or region studied. The decrease in L_{Aeq} is widespread but some differences are spotted according to the current restrictions present in each situation. Some of these changes in the mean L_{Aeq} documented in the literature are 5.4 dBA in London [24] (ranging from 1.2 to 10.7 dBA), 5.1 dBA in the Ruhr Area (Germany) [25], 6–7 dBA in Montreal [18], 6–10 dBA in Monza (Italy) [26], a daily average peak drop of more than 4 dBA in Stockholm [15], 7 dBA in Rome and Milan [27] or 5.2–5.9 dBA during the peak of the restrictions in Barcelona [4].

This average reduction in the noise level is not necessarily consistent in all the locations where data were gathered. Some authors documented differences according to land use categories. In Rio [28], there was a noise reduction between 10 and 15 dBA in those areas with a predominance of human activities whereas there was no major reduction near major arteries. In Granada [29], the L_{Aeq} variation ranged between 13.3 and 30.5 dBA depending on the location. Also, in seven Indian cities [16] the noise reduction ranged between 4 and 14 dBA for residential, industrial and commercial areas. Studies conducted in Madrid [30], the Ruhr Area [25] and Barcelona [4] among others also spotted differences according to the type of location.

Beyond the average L_{Aeq} reduction, a comparison between day and night variations of the noise levels during the pandemic has also carried out done, e.g., in Buenos Aires [20], with decreases of 1.4–4.7 dBA during the day and 2.7–6.9 dBA during the night or in the Île-de-France region [31], with decreases in the road traffic noise of 4.6 dBA during the daytime and 7 dBA in the night. Differences in daily noise indicators have also been spotted in Barcelona [4], Girona [3], Dublin [32] or Madrid [30], among others, showing that time patterns were also affected during the lockdown. Moreover, other acoustic and even psychoacoustic metrics such as loudness and sharpness have also been calculated by some researchers, e.g., in London [24] where clustering of the 11 locations has been applied.

Regarding the people's perception of the changes in the soundscape during the COVID-19 shutdown, there are two main approaches in the literature. On the one hand, some studies have surveyed people directly, asking about their perception of the soundscape with a guided set of questions. This is the case of a study set in Italy [33] where an 18-questions survey was answered by 323 participants and the results confirmed the expectation of a decrease in the noise pollution levels. Another similar work was conducted in France [31] where residents of the Île-de-France region perceived a significant reduction in the level of noise from a set of sources such as human activities, road traffic noise or airborne noise. In Argentina [34], a survey among 1371 social network users detected that people preferred the new acoustic environment caused by the COVID-19 lockdown. Also, in London [35,36] a mixed-method approach consisting of triangulating data from surveys and spontaneous descriptions offered by the participants (home workers) is being performed with the goal of finding associations between perception of the indoor soundscape and psychological well-being. Not only changes in residential soundscapes have been analyzed but also the impact on a historic soundscape such as the Berlin Wall Memorial [37] by means of soundwalks and informal interviews with staff members and tourists on the site. In Milan, ref. [38] the authors present a wide analysis of the changes in L_{Aeq} in the city of Milan, evaluating the impact of the Anomalous Noise Events [39] in the different periods of the lockdown. Also, in [40] the situation was faced worldwide with questionnaires, including both indoors and outdoors, finding a clear improvement in the perception of the citizens facing the unpredictable situation of the pandemic. Another study conducted in Madrid on noise perception and related health effects during the lockdown presented a cross-sectional study by noise sources based on data collected from 582 participants who answered a questionnaire [41].

On the other hand, in some cases where people could not be surveyed, other approaches were selected. Mitchell et al. [42] developed a model to predict the soundscape pleasantness and eventfulness during the lockdown in London and Venice based on a database of previous binaural recordings and soundscape questionnaires and new recordings made during the pandemic. In the Basque Country [43], experts in soundscape and architecture listened to recordings taken between March and May 2020 and made two perceptual analyses, i.e., they annotated perceived sound events and assessed the pleasantness and eventfulness.

There are some precedents that combined an objective and a subjective approach in order to analyze the impact of the lockdown using a multidimensional approach such as in the city of Lorient, France [44], where data gathered from a network of sensors were used in addition to the citizen's perception of the soundscape during 2019 and 2020, collected by two questionnaires, to improve the accuracy in describing changes in a sound environment. Finally, a systematic review on 119 studies about the perceptual change or the noise level change during the COVID-19 pandemic lockdown can be found in [45].

3. Methods and Data Gathered

The research presented in this work combines data of a diverse nature: (1) A-weighted equivalent sound pressure levels (L_{Aeq}) and other noise indices (L_d), (L_e) and (L_n) provided by a network of calibrated sound sensors deployed in Barcelona and (2) questionnaires answered by participants in the *Sons al Balcó* project including their perception of the soundscape around their dwellings both before and during the lockdown and details about the most annoying sounds spotted.

3.1. Lockdown Restrictions

In this subsection, the exact restrictions that were enforced in Barcelona during the lockdown and de-escalation stages are described to give context to the causes behind the decrease in noise levels and improvement of the acoustic comfort of citizens.

The lockdown period in Spain started on 14 March 2020, initially affecting only students and other professionals working in the education sector and ended on 3 May 2020. Within this period, there were weeks with a stricter confinement affecting all non-essential workers. However, the final days allowed children to go out for a walk and adults to practice sports outdoors in different time spots. In addition, shops, bars, restaurants museums, libraries, sport facilities and leisure establishments were all closed. Public and private transport was highly reduced and only used to commute to the work place for the population still working.

After the lockdown began, the de-escalation process extended from 4 May 2020 to 17 June 2020. During these weeks, shops and other business started opening by appointment and with limited capacity. In the final stages of the de-escalation process, even the mobility to second residences was finally restored.

3.2. Questionnaire Design

The questionnaire was designed following some of the previous works of the team about perception and sound [46] related to health, and also other tests in the framework of former projects as LIFE-DYNAMAP [47], and analyzed and used to model the annoyance as in [48]. All the citizens answering the questionnaire were informed of the use that the research team would have from their answers and videos in terms of ethics and publication, and they signed a consent to use their information.

The main questions asked to citizens were related to the comparison of their soundscape before the lockdown and during the lockdown, as can be found in [49,50]. Some of the more relevant are the following:

- How do you describe the soundscape of your home, before the lockdown and during the lockdown?
- How do the following adjectives describe the soundscape you recorded? Loud, shrill, noisy, disturbing, sharp, exciting, calming, pleasant?
- Which sounds are present in the soundscape you recorded? Road traffic, plane, train, industry works, commercial activities, leisure activities, neighbors, pets, birds, water, vegetation?
- Please, indicate how much the former different sounds disturb you
- Compare the annoyance related to those sounds before and during the lockdown.

The researchers had the support of the X (formerly Twitter) accounts of both institutions involved in the *Sons al Balcó* project (ISGlobal and La Salle Campus Barcelona), and the dissemination of its own project X (formerly Twitter) account (@SonsalBalco). The researchers chose to map the soundscape of the lockdown in Catalonia due to the size, the potential population and the possibility of having contributions from both big cities like Barcelona, but also from small villages where the soundscape may not have changed so much during the lockdown. The citizens contributing were only asked to give a nickname, and did not require to register or log in to any platform, as all was conducted via web, which probably increased the participation but did not allow the team to contact the contributors after the data collection.

3.3. Data Collection Campaign

Several campaigns have been conducted to collect data in the *Sons al Balcó* project [49,51]. This present work is focused on data obtained from the first campaign performed in 2020 during the final stages of the lockdown caused by the COVID-19 pandemic and the initial stages of the de-escalation process. A socio-acoustic online questionnaire was designed to obtain perceptive data representative of the soundscapes across Catalonia. Some requirements were considered before launching the digital survey. First, LimeSurvey [52] was chosen as the web service platform to implement the online question-and-answer survey. The setting included different response formats and video uploading capacities. One of the main advantages of LimeSurvey compared to other survey applications is that it is an open-source solution that can be deployed to any server that supports it. Therefore, it is not constrained to the servers of the survey application provider. The specific implementation consisted on an Amazon EC2 cloud computing instance running a Bitnami Stack for LimeSurvey 4.2.3-0 on Ubuntu 16.04.6 LTS. Furthermore, for the purpose of reducing traffic, an Amazon S3 bucket was also applied to upload the recorded videos directly from the smartphones of the participants. Lastly, a Fine Uploader library running on EC2 was installed to manage and sign the requests allowing access to the aforementioned S3 bucket.

The questionnaire included different topics such as sociodemographic data, soundscape location and perceived quality, both before and during the confinement. Additionally, participants had to report the presence of different noise sources and their respective annoyance. These sound categories included different kinds of motorized traffic (automobiles, trains or planes), industry, construction works, commercial and recreational activities, neighborhood noise, pets, birds, water and vegetation. Further details on the survey can be found in [49].

3.4. Sensors Data

For this study, noise levels were obtained from the Wireless Acoustic Sensor Network (WASN) of Barcelona, which has already been used to conduct, in a previous work by the authors [4], a thorough analysis of several noise indices during the COVID-19 lockdown in the city.

The WASN deployed in Barcelona (also named Barcelona Noise Monitoring Network) consists of 112 devices, 86 of which are sensors and 26 are sound level meters. These 86 sensors are placed for long-term analysis in several pre-analyzed places around the city. Since not all sensors worked properly during the lockdown, only 70 sensors out of the 86 deployed to conduct this study were analyzed. All the used sensors are CESVA's TA120 Class 1 sound level meters. The location of these 70 sensors is depicted in Figure 1 along with the location of the assessed soundscapes. Furthermore, if we look at the distribution of noise sensors in Barcelona, there is a higher number of devices deployed in the city center and in leisure areas. This means that this network mainly monitors road traffic, commerce and leisure activities. A detailed depiction of the locations and maps can be found in [4]. Moreover, see [53,54] for more information about the Barcelona Noise Monitoring Network.

Sensors were active both in a normal pre-pandemic scenario and during the lockdown. Data from the first semester of 2018 and 2019 has been used as baseline levels for comparison with noise levels obtained during the lockdown. Most of the sensors were working 24 h a day during the studied periods and provided A-weighed equivalent sound pressure levels at one minute time resolution.

3.5. Description of the Studied Areas

As one of the main purposes of this study is to analyze the specific relationships between the decrease in noise levels and the improvement of the perceived acoustic comfort in different areas according to the main type of noise source, sensors have been manually grouped in different clusters. The first three groupings correspond to the majority of sensors deployed in areas where road traffic noise is the main source of noise exposure. They are divided into Heavy-Traffic Areas (sensors that measured mean values above 67.5 dBA during the baseline time-frame, i.e., the first semesters of 2018 and 2019), Moderate-Traffic Areas (mean values between 64.5 and 67.5 dBA) and Low-Traffic Areas (mean values below 64.5 dBA).

Figure 1. Location of the sensors and the assessed soundscapes during the 2020 *Sons al Balcó* campaign.

There are 16 sensors located in heavy-traffic areas. They are located in some of the main arteries of the city, which communicate different urban districts and connect with the entry and exit points to other cities in the metropolitan area. These streets are mainly used by people going and returning from their jobs and by delivery services and public transportation. Thus, they were less affected by the mobility restrictions, especially during the final stages of the lockdown. For this case, 22 respondents, both men and women, were considered to be in the scope of the selected sensors. The ages of the volunteers were between 19 and 72 years.

A total of 12 sensors are located in Moderate-Traffic Areas. These sensors are usually in mid-sized streets, often with multiple lanes. They are also occupied by people commuting using both private and public transportation. Even though road traffic is the predominant noise source present, these areas combine residential and office buildings with commerce and restaurants or even with some recreational facilities. For this cluster, 17 people were considered for the following results. They were both men and women from 32 to 69 years old.

There are sixteen sensors located in Low-Traffic Areas. They are located in quieter and smaller streets usually in the middle of residential buildings, with the occasional bar, grocery store or supermarket. The main sound source is traffic coming from the mobility of the neighbors. However, there is also commerce related noise and neighborhood noise. During weekends and especially Sundays, these locations are even quieter. In these sensor' areas, there are 26 respondents whose age is between 28 and 71 years and they are both men and women.

The fourth grouping corresponds to sensors placed in SuperBlock areas which consist mostly in pacified streets. Barcelona's SuperBlock project was conceived to reduce the road traffic in residential areas, mostly traffic derived from private vehicles [55]. The main project's goal is to create greener and healthier urban public space. However, some of them are still under construction. In fact, the survey's results reported that in 57.14% of the videos collected inside the area of influence of the SuperBlock sensors there were construction works spotted in the audios sent. A total of six sensors are located in SuperBlocks. In this case, seven participants evaluated soundscapes considered to be in the scope of SuperBlocks' sensors. The age of the volunteers was between 23 and 50 years old and they were both men and women.

The next two clusters contain the sensors placed in areas were leisure activities, commercial activities and restaurants are the main source of noise exposure (both daytime and night-time leisure activities). Particularly, six sensors are located in areas where the main sound sources are related to Daytime Leisure activities and restaurants. Most of these sensors are located in small squares in the middle of residential areas in different city districts. Road traffic is limited or nonexistent. The main source of noise comes from the passersby, pedestrians, tourists, playgrounds and restaurants. Occasionally, there is also some commercial and nightlife activity in these areas. For this cluster of sensors, 24 people answered the questionnaire.

Next, there are nine sensors located in areas where the main noise source comes from Night-Time Leisure activities. They are in neighborhoods full of bars, pubs and restaurants and it is especially active on afternoons, evenings and nights all along the week with lots of interaction from tourists and students. Trading activity also produces a moderate activity during the day. In this area, there were 6 volunteers that fit the profile, who were both men and women from 32 to 50 years old.

From the 70 sensors studied, 65 are distributed among these 6 types of locations. The other 5 sensors are placed in industrial areas or parks and 5 videos were collected inside the area of influence of those 5 sensors. However, they are too few to extract any meaningful aggregated information from them and they have not been included in the clustered study.

3.6. Data Treatment

While data from different sensors were gathered during the 2020 lockdown period, a socio-acoustic digital participatory survey was also performed. It aimed to gather the positive and negative perception of noise experienced from home before and during the lockdown. In this survey, the respondents were told to record a video of their sound environment (representative of a typical daily soundscape during the lockdown) and answer a questionnaire about their perception. In total, 366 volunteers from 132 different locations completed the questionnaire and uploaded their videos. One of these contributions had to be discarded because the location was ambiguous. Almost 40% out of the 365 accepted participants came from cities with a WASN deployed. Twenty-two were from Girona and preliminary results on their analysis have already been published by authors [2]. This present work will be anchored in the largest portion of volunteers (119) that described soundscapes from Barcelona in the survey. The contributors' profile was both men and women between 29 and 86 years old.

To ensure the reliability of the responses, the videos recorded were manually labeled and the sound events reported by volunteers were compared with the sound events spotted by annotators. In addition, only the questionnaires with the essential data correctly provided were accepted for this study, i.e., the exact location of the soundscape and the general assessment of the soundscape both during and before the lockdown.

Authors opted for a clustered approach to perform the study. As stated in Section 3.5, both sensors and assessed soundscapes are being manually grouped in clusters according to the predominant noise source and density of traffic. This approach aims to offer more insight in which kind of noise sources regulation have a greater impact in the acoustic satisfaction of citizenship.

Contributions from Barcelona were assigned to the area of influence of their nearest sensor. Several preliminary experiments were performed with accepted radii from 400 m to 1 km with comparable results. Finally, contributions from a maximum of 1 km distant to the sensor were taken into account for the analysis, assuming that farther contributions may have a very different soundscape and even different urban sound environment (leisure, schools, traffic, etc.). Schafer [56] already explains that focus listening with its implication of distance separating the listener from the sound event is disintegrating before the sound walls, so lo-fi soundscapes do not have perspective, but mask the listener with a constant presence. Distances between the sensors and the location of the assessed soundscapes were evaluated using the Haversine formula [57], an easy to implement method for obtaining optimal approximations of distances over a circle when the longitude and latitude are known.

Each one of the studied sensors provided L_{Aeq} noise levels at a one-minute time resolution. The mean values for the different noise indices (L_{Aeq}, L_d, L_e and L_n) are computed for each sensor both for the lockdown period (from 14 March 2020 to 3 May 2020) and for the baseline time-frame (first semesters of 2018 and 2019). After that, the mean noise indices corresponding to sensors from the same cluster are also averaged to obtain the final mean noise indices for the given cluster. Variances for the measured L_{Aeq} decreases will also be calculated.

Volunteers assessed the acoustic comfort around their dwellings both for the lockdown period and for the pre-lockdown period using a Likert Scale [58] (Very Negative, Negative, Neutral, Positive and Very Positive). This Likert Scale is converted to a 5-point scale (1 to 5). After that, both ratings are compared to obtain the soundscape rating improvement for each dwelling. The average and the variance for all the soundscape rating improvements of the dwellings included in the same cluster are subsequently computed.

Subsequently, the L_{Aeq} decrease for each one of the studied clusters will be compared with the reported improvement of the subjective acoustic comfort (soundscape rating by volunteers). The hypothesis is that a general alignment will be found between both objective and subjective data but that there will be significant differences between some of the clusters, especially related to the different predominant noise sources, e.g., traffic or leisure.

The previous statistical results will be complemented with a more nuanced study, both quantitative and qualitative, of the daily noise indices and other answers provided in the survey (for each cluster). Specifically, distinct intra-day pattern variations in the mean sound pressure level for each cluster will be described. Also, the main types of actual noise sources spotted in the area will be commented. The specific subjective assessment both before and during the lockdown (using a Likert scale) for the overall quality of the soundscape and the assessment of several perceptual constructs will also help give more insight into the appraisal of the soundscapes and the possible outlier opinions included.

4. Results

This section offers a detailed description of the comparison between the objective acoustic data gathered by the sensors in Barcelona and the subjective data collected from the citizen science campaign. An individual analysis and interpretation has been conducted for each one of the six clusters described in Section 3.5.

4.1. Barcelona General Results

In this section, we describe the most relevant sensors results in Barcelona together with the subjective contributions of the citizens. The goal is to show whether the conclusions reached on previous research studies, ref. [4] and summarized below might show some qualitative coincidence with the answers to the poll.

Barcelona experimented a significant drop in noise level during the lockdown stages in 2020. The decrease was especially steep during the stages with stricter mobility and activity constraints and during the night hours. This decline has to be put in context as Barcelona was already showing a mild but steady noise reduction trend in most of its sensors from 2018 to the first months of 2020, probably due to the pacifying efforts being implemented in recent years by the local administration.

Although all areas of the city produced lower noise levels during the lockdown stages, they were not equally affected. Areas with heavy traffic experienced lower noise reduction than areas with moderate traffic, mainly during the day. In residential and low-traffic areas, the reduction was more restrained because the pre-lockdown noise levels were also lower. As for the other sources of studied noise, they also showed differences. On the one hand, daytime leisure, restaurant areas and nightlife areas were among the most affected, with distinct intraday noise variation. Nightlife areas took a huge plunge during the evening and night-time frames, whereas daytime leisure and restaurant areas were more affected during afternoons and evenings. Furthermore, Superblocks and shopping areas presented a similar drop irrespective of the hour of the day. On the other hand, industrial and services areas were among the less affected by the restrictions, basically during the morning hours.

Table 1 compares the mean improvement in the L_{Aeq} level during the lockdown with the subjective improvement perceived by participants in the *Sons al Balcó* project during the same time-frame. Data collected came from 70 sensors deployed in Barcelona during the lockdown and from 119 surveys answered by Barcelona citizens. Each row in the table corresponds to the different clusters described in Section 3.5. In the second column, the total number of sensors included in each investigated area is shown.

Table 1. Comparative of the mean improvement in the L_{Aeq} level during the lockdown and the mean subjective improvement in the soundscape rating (acoustic satisfaction) reported by the participants in the *Sons al Balcó* project during the same time-frame.

Type of Location	Num. of Sensors	Num. of Soundscapes	Mean Distance Sensor-Video [km]	Mean L_{Aeq} Decrease	Variance L_{Aeq} Decrease	Mean Soundscape Rating Improvement	Variance Soundscape Rating Improvement
Heavy Traffic	16	22	0.39	−4.84	2.14	+1.77 (62.77%)	1.42
Moderate Traffic	12	17	0.31	−5.54	1.21	+2.12 (92.58%)	1.24
Low Traffic	16	26	0.47	−4.63	1.07	+1.5 (58.14%)	3.06
SuperBlocks	6	7	0.4	−7.2	15.98	+1.14 (34.65%)	1.81
Daytime Leisure	6	24	0.44	−7.34	5.45	+1.04 (32.81%)	1.17
Night-Time Leisure	9	6	0.32	−8.25	1.54	+1.17 (35.45%)	0.97
Aggregated data	70	119	0.55	−5.72	5.0	+1.47 (51.63%)	1.7

The third column in Table 1 reports the number of videos and questionnaires collected inside the area of influence of the corresponding sensors. The videos have been assigned to the area of influence of a single sensor, the nearest one. Only videos collected within a 1 km radius from the nearest sensor have been considered. The mean distance in km from the locations where the videos were recorded to the nearest sensors are shown in Column 4. Column 5 shows the mean dip in the L_{Aeq} level during the lockdown in 2020 compared to the same levels measured during the same weeks in 2018 and 2019. Column 6 contains the variance of the L_{Aeq} decrease for the different sensors. Column 7 shows the perceived mean improvement in the soundscape rating (and the percentage of improvement over the original rating), after converting the original Likert scale used by participants to numerical values from 1 to 5. Finally, Column 8 includes the variance of the soundscape rating improvement according to the volunteers.

Even though the mean L_{Aeq} improvement is higher for the SuperBlocks and Leisure areas than for the areas mostly affected by road traffic noise, the perceived subjective improvement is higher for those areas where road traffic noise is predominant. That hints at the fact that road traffic noise is especially annoying for most of the participants. Reducing road traffic noise exposure has a greater impact on the general subjective assessment of the improvement than reducing other types of noises.

Focusing on the three different traffic areas (heavy, moderate and low), there is a correlation between the objective mean improvement in the L_{Aeq} levels and the subjective mean improvement reported by contributors. Moderate-Traffic Areas are where the improvement is higher, followed by Heavy-Traffic Areas and, finally, Low-Traffic Areas.

SuperBlocks experienced a high drop in the L_{Aeq} level that was translated to a more modest improvement in the subjective assessment. One of the reasons is that construction works were already resumed by the time the videos were collected and they were especially abundant inside the areas of influence of the SuperBlock sensors.

Leisure areas were especially affected by the activity restrictions conducting to mean L_{Aeq} drops of more than 7.3 dB during the lockdown. However, the subjective assessment only improved by 33 to 35% (which pales in comparison to the improvement in road-traffic areas). The main reason for that is that the original (before the lockdown) subjective assessment of the soundscapes collected in those areas was higher than the subjective assessment of the road-traffic-exposed areas. Leisure areas had an original mean rating of 3.2 points. Therefore, it was virtually impossible to achieve the improvements reported in the road-traffic-exposed areas.

The improvement for both the objective measurement and the subjective perception was slightly higher for the nightlife areas than for the Day-time Leisure Areas showing, again, a correlation between the subjective perception and the objective reduction in sound levels.

Aggregated data in the last row of Table 1 include all the sensors active in Barcelona during the lockdown and all the valid contributions received during the 2020 campaign of *Sons al Balcó* from the city of Barcelona, even the ones beyond the 1 km threshold.

Figure 2 shows the presence of several kinds of noise sources and other sound events in the received videos sorted by the type of area in which the nearest sensor is circumscribed. The percentage depicted in the figure is related to the number of contributions where participants spotted each of the sound events compared with the total of contributions for each cluster. The absolute mean L_{Aeq} improvement during the lockdown compared to the mean noise level for the two previous years is also represented in the figure for comparison purposes. Noise sources have been grouped in five categories: (a) Traffic, which includes all types of motorized traffic; (b) Industry/Construction, which includes noises from both industrial sources and construction sites; (c) Commerce/Leisure, which includes noises from recreational activities, restaurants and shopping areas; (d) Neighbors, which includes neighborhood noise and pets and (e) Nature, which includes sound events related to nature elements and wildlife (mainly water, vegetation and birds).

The most prevalent sound in most of the clusters is traffic noise, which appears in 50% or more of the videos independently of the type of area where the nearest sensor is located. Road traffic noise prevalence is higher in those areas where the L_{Aeq} improvement caused by the restrictions of the lockdown was less significant. That hints at the fact that road traffic noise is one of the main contributors to the global L_{Aeq} in urban locations. On the contrary, nature sounds appear more frequently in areas where the L_{Aeq} took a steeper dip. That is caused because some sound sources such as birds that are easily squelched in noisy soundscapes became apparent when the louder noise sources decreased.

It is also noteworthy that in the lockdown context, the third sound event most prevalent after traffic noise and nature related sounds is neighborhood noise, which was spotted in above 50% of the videos for all the studied groups except for SuperBlocks. In contrast, industry and construction noise and, especially, leisure and commerce noise was greatly decreased. These three categories (neighbors, industry/construction and commerce/leisure) appear to be independent of the measured L_{Aeq} improvement.

A more detailed analysis of each type of urban area is conducted in the subsequent subsections.

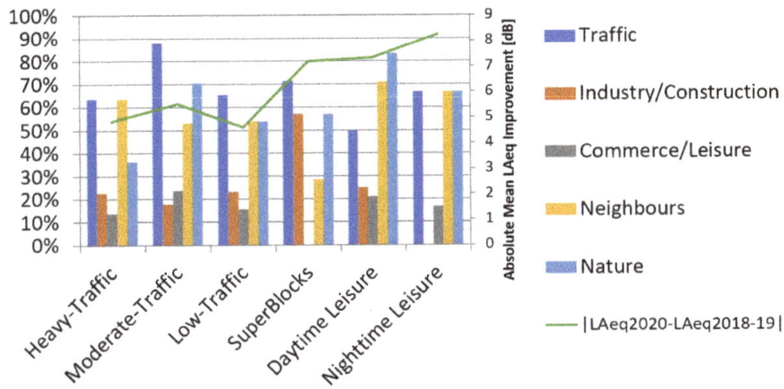

Figure 2. Percentage of appearance of sound events reported by type of area compared with the L_{Aeq} improvement during the lockdown.

4.1.1. Sensors in Heavy-Traffic Areas

As seen in Table 2, mean pre-lockdown daily noise indices for theses sensors were significantly higher than those from sensors of other clusters, with more than 70 dB during the day and almost 65 dB during the night. Even though indices were clearly improved during the lockdown, they remained notably high when compared to sensors in other areas. Furthermore, the improvement caused by the restrictions in this group of sensors was inferior to all the other studied groupings during days and evenings and the second-to-smallest during nights.

Table 2. Mean daily noise indices during the lockdown (2020) and the same time-frame for previous years (2018–2019) for the sensors in Heavy-Traffic Areas.

Period	L_d [dB]	L_e [dB]	L_n [dB]
2018–2019	70.86	69.84	64.97
2020	66.63	64.21	58.11

The soundscape evaluation before the lockdown was generally poor, with 50% of participants rating it as "Negative" or "Very Negative" (Table 3). In general, contributions located in areas where road traffic is the main noise source reported more deteriorated soundscapes (before the lockdown) than areas where the predominant noise source is different (always according to the opinions reported in the surveys). In contrast, during the lockdown the subjective evaluation of the quality of the soundscape experienced one of the higher boosts compared to other clusters. As observed in Table 3, all the 50% negative assessments pre-lockdown changed to neutral or positive assessments. In addition, 63.64% of dwellings considered to have "Very Positive" soundscapes during the lockdown is the second-to-highest figure for the studied clusters only after SuperBlocks.

Comparing the huge improvement in the subjective acoustic satisfaction assessment (Table 3) with the rather modest reduction in the levels in noise indices (Table 2), it seems that the original pre-lockdown noise levels being as high was a critical cause of dissatisfaction and that the modest reduction they experienced during the lockdown was enough to significantly change the perception of the soundscape.

Table 3. Subjective acoustic satisfaction assessment before and during the lockdown for dwellings near sensors in Heavy-Traffic Areas (Likert Scale).

Period	Very Negative	Negative	Neutral	Positive	Very Positive
Pre-Lockdown	9.09%	40.91%	13.64%	31.82%	4.55%
Lockdown	0%	0%	4.55%	31.82%	63.64%

The perceptual constructs' assessment is quite heterogeneous. However, there is general consensus in all the studied clusters that positive perceptual constructs (with the exception of excitement) are more representative of the studied soundscapes than negative perceptual constructs. That being said, 36% of respondents in this cluster find their soundscape noisy and 23% find it disturbing (Figure 3). These percentages are generally higher than in other clusters (with the exception of Low Traffic) which is consistent with the higher noise indices in Heavy-Traffic Areas. The main noise sources reported in this group are road traffic and neighborhood noise and the most annoying, according to survey results, is road traffic, which is coherent with sensors located in Heavy-Traffic Areas.

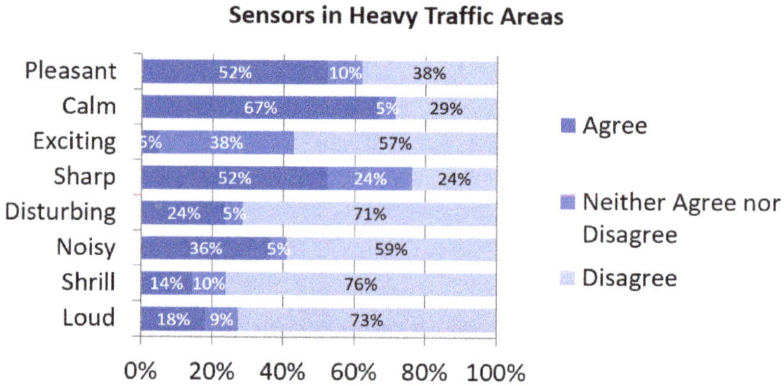

Figure 3. Assessment of perceptual constructs for soundscapes near sensors in Heavy-Traffic Areas.

4.1.2. Sensors in Moderate-Traffic Areas

Before the lockdown, the mean noise indices in this cluster (Table 4) were approximately 3 dB lower than in the Heavy Traffic cluster (Table 2) but significantly higher than in Low-Traffic Areas (see Section 4.1.3). However, from the three traffic focused clusters, they are the ones that showed a larger decrease during the lockdown.

Table 4. mean daily noise indices during the lockdown (2020) and the same time-frame for previous years (2018–2019) for the sensors in Moderate-Traffic Areas.

Period	L_d [dB]	L_e [dB]	L_n [dB]
2018–2019	67.62	66.23	61.48
2020	62.56	59.77	54.53

According to the answers, there was a huge improvement in the sound environment of these sensors when comparing before and after the lockdown periods. In fact, it is the cluster that showed a bigger amelioration of the global subjective rating of the soundscape during the lockdown (Table 1). Before the lockdown, more than 50% of the respondents rated their dwelling's soundscape as "Negative" or "Very Negative" and not a single participant considered it to be "Very Positive" (Table 5). However, during the lockdown, almost 95% of them considered that the soundscape was "Positive" or "Very Positive" with no reported cases of negative assessments. Objective and subjective data are clearly aligned for the traffic clusters. Moderate-Traffic Areas show both the steeper dip in the noise levels during the lockdown (compared to Low and Heavy-Traffic Zones) and also show the bigger improvement in the acoustic satisfaction degree of its inhabitants.

Table 5. Subjective acoustic satisfaction assessment before and during the lockdown for dwellings near sensors in Moderate-Traffic Areas (Likert Scale).

Period	Very Negative	Negative	Neutral	Positive	Very Positive
Pre-Lockdown	35.29%	17.65%	29.41%	17.65%	0%
Lockdown	0%	0%	5.88%	47.06%	47.06%

This vastly positive evaluation of the soundscapes could be explained by the fact that road traffic was drastically reduced and although it progressively increased in the de-escalation process, it never recovered the former density, as mentioned in [4]. In fact, even though road traffic is the most spotted sound event in Moderate-Traffic Areas, appearing in almost 90% of the recordings (Figure 2), the reported annoyance caused by road traffic noise is lower than the annoyance caused by the less prevalent leisure and commerce activities and construction works in the area. Again, the fact that the most predominant sound source is not considered especially annoying in the surveys is consistent with the reduced noise indices during the lockdown.

In general terms, perceptual constructs' assessment in this cluster is similar to the other groups. Participants agree more regarding the representation of positive constructs such as calmness or pleasantness than in negative constructs (Figure 4). It is to be noted, though, that the exceeding percentage of dissent for disturbance, noisiness and shrillness is considerably higher than in the other traffic-related clusters.

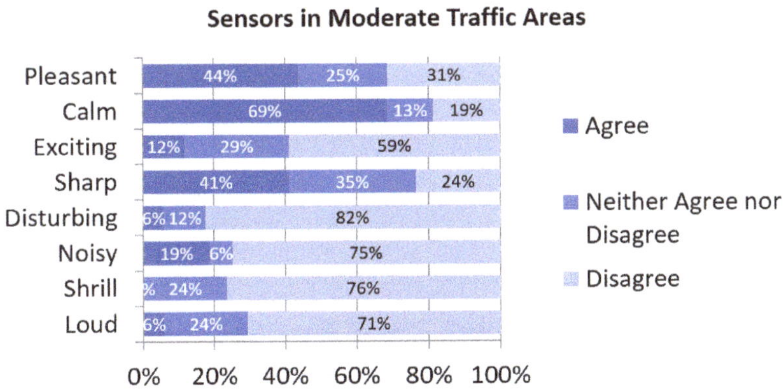

Figure 4. Assessment of perceptual constructs for soundscapes near sensors in Moderate-Traffic Areas.

4.1.3. Sensors in Low-Traffic Areas

Noise indices for this group of sensors were the lowest during 2018 and 2019 (Table 6). However, they experienced a milder decline during the lockdown and, in fact, they were no longer the lowest during the restrictions, surpassed by Day-time Leisure Areas.

Table 6. Mean daily noise indices during the lockdown (2020) and the same time-frame for previous years (2018–2019) for the sensors in Low-Traffic Areas.

Period	L_d [dB]	L_e [dB]	L_n [dB]
2018–2019	62.18	60.26	54.35
2020	57.86	54.08	49

The initial pre-lockdown assessment of the soundscapes in the surveys amounts to fewer negative evaluations overall, with around 42% of "Negative" or "Very Negative" appraisals compared to the 50% or more of the other clusters related to traffic noise (Table 7). They also achieve a substantial improvement during the lockdown. However, it is the only cluster where some soundscapes rated "Very Negative" remain. Results seem a little inconsistent with other clusters but a more detailed analysis detected some outlier opinions among the respondents which is consistent with the higher than normal variance in the soundscape rating improvement reported in Table 1. Surprisingly, about 10% of the respondents considered that the quality of the soundscape had deteriorated during the lockdown. Analyzing the responses, construction works were reported near the locations of two of the dissenting citizens, which is the most probable cause for this exceptional dissatisfaction with the acoustic environment. In fact, in a similar way to Moderate-Traffic Areas, traffic was the most reported noise source among contributors but it was also significantly less annoying than construction and commerce activities (much less prevalent in the studied soundscapes).

Table 7. Subjective acoustic satisfaction assessment before and during the lockdown for dwellings near sensors in Low-Traffic Areas (Likert Scale).

Period	Very Negative	Negative	Neutral	Positive	Very Positive
Pre-Lockdown	34.62%	7.69%	34.62%	11.54%	11.54%
Lockdown	7.69%	3.85%	3.85%	42.31%	42.31%

Perceptual constructs' assessment in Low-Traffic Areas shows a wide range of dissenting opinions among citizens (Figure 5). It is the only cluster where positive and negative perceptual constructs are similarly rated. Surprisingly, in these quieter parts of the city where noise indices are lower, *noisy* and *loud* are more often depicted as adequate adjectives than *pleasant*. Again, an explanation can be found in many construction works resuming their activity in the final stages of the lockdown after a long period of virtually no noise in the surroundings. Furthermore, inhabitants of these quieter areas are less used to high levels of noise pollution and their expectations may be more demanding.

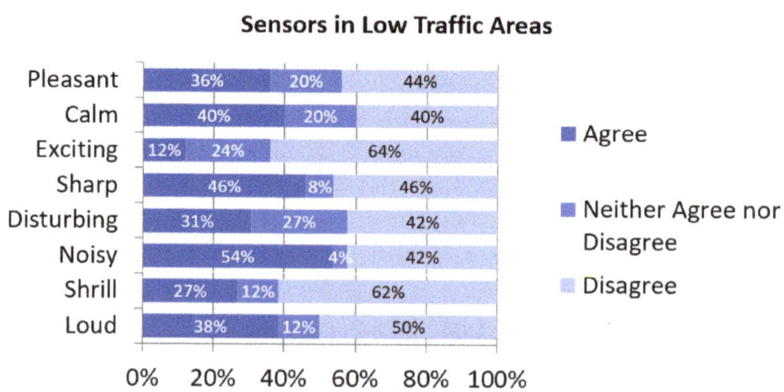

Figure 5. Assessment of perceptual constructs for soundscapes near sensors in Low-Traffic Areas.

4.1.4. Sensors in SuperBlock Areas

One would expect that sensors in this cluster would provide similar noise indices to those of the Low-Traffic Areas as they are both mainly located in residential quieter zones. However, noise levels in Table 8 are significantly higher, similar to the indices of Moderate-Traffic Areas. There are two causes that explain these figures. First, some of the SuperBlock sensors are really located near the bordering streets of the pacified area where there is an exceeding traffic density caused by the mobility restrictions inside the SuperBlock and do not really represent the noise levels present in the inner buildings. Second, and even more relevant, intensive construction works were being executed in SuperBlocks (see Figure 2) in order to convert spaces previously dedicated to motor traffic to pedestrian areas. As a consequence, there are significant differences between the reported decrease in the noise indices during the lockdown among the sensors included in SuperBlocks as it can be observed by the overly high value of the variance reported in Table 1 which is accentuated by the reduced number of sensors available in the cluster.

Table 8. mean daily noise indices during the lockdown (2020) and the same time-frame for previous years (2018–2019) for the sensors in SuperBlocks areas.

Period	L_d [dB]	L_e [dB]	L_n [dB]
2018–2019	68.91	66.24	62.23
2020	61.69	58.91	55.19

The analysis of the subjective results for this cluster has to be taken as a qualitative approximation with limited reliability due to the few contributors who reported soundscapes in SuperBlocks areas and the higher variance also detected in the subjective aprraisals. First, it should be noted that a single participant rated the soundscape around his dwelling as "Negative" during the lockdown (Table 9). In fact, this citizen considered that the quality of the soundscape had worsened compared to the pre-lockdown scenario. Again, construction works are the probable cause for his outlier opinion according to his answers to the survey. The other six participants agree that the soundscape quality vastly improved in 2020 giving, in fact, the highest percentage of "Very Positive" ratings among all the studied clusters. This overly elevated degree of acoustic satisfaction is aligned with the L_d variation during the lockdown which, at −7.22 dB, is significantly higher than in the other clusters.

Table 9. Subjective acoustic satisfaction assessment before and during the lockdown for dwellings near sensors in SuperBlocks areas (Likert Scale).

Period	Very Negative	Negative	Neutral	Positive	Very Positive
Pre-Lockdown	0%	28.57%	28.57%	28.57%	14.29%
Lockdown	0%	14.29%	0%	14.29%	71.43%

The evaluation of perceptual constructs by inhabitants of SuperBlocks' surroundings is similar to the other clusters. However, it is noteworthy that all of the volunteers disagreed with their soundscape being *disturbing* (Figure 6). It may be related to the total absence of commerce or leisure-related sounds reported in this subset of questionnaires, as shown in Figure 2.

4.1.5. Sensors in Day-Time Leisure Areas

Intraday patterns for this group of sensors are significantly different from road-traffic-dominated domains. Noise levels increase during the afternoon after class and after finishing the working day and remain high through the evening, taking into account that in Spain dinner time can easily extend to 23 h. Therefore, in a normal scenario, L_e is especially elevated, usually surpassing L_d, as seen in Table 10. As leisure activities and restaurants were severely affected by the restrictions, this cluster presents the lower noise indices during the lockdown, including the steeper drop in L_e. Owing to that, the noise indices pattern during the confinement followed the usual trend in other areas where L_d towers over both L_e and L_n.

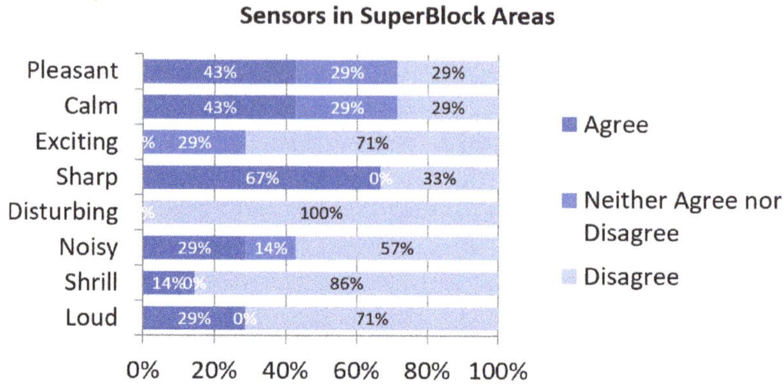

Figure 6. Assessment of perceptual constructs for soundscapes near sensors in SuperBlocks areas.

Table 10. Mean daily noise indices during the lockdown (2020) and the same time-frame for previous years (2018–2019) for the sensors in Day-time Leisure Areas.

Period	L_d [dB]	L_e [dB]	L_n [dB]
2018–2019	63.72	64.3	57.03
2020	57.28	54.02	47.35

The acoustic satisfaction of the respondents for the pre-pandemic soundscape was fair-to-middling with a majority of people considering it neither positive nor negative (Table 11). On the contrary, a vast percentage of people changed their appraisal to "Positive" or "Very Positive" when the restrictions remained in force. This improvement completely aligned with the drastic decrease in the L_e index, which is the most representative of the noise caused by daytime leisure activities and restaurants as it has been already stated.

Table 11. Subjective acoustic satisfaction assessment before and during the lockdown for dwellings near sensors in Day-time Leisure Areas (Likert Scale).

Period	Very Negative	Negative	Neutral	Positive	Very Positive
Pre-Lockdown	8.33%	4.17%	58.33%	20.83%	8.33%
Lockdown	0%	0%	12.5%	54.17%	33.33%

Sound events spotted in Day-time Leisure Areas when the campaign was performed are significantly different from other areas. Specifically, it is the only grouping where nature-originated sounds were reported in more than 80% of the contributions, surpassing road traffic noise. In fact, it is the only cluster where nature sound categories are more prevalent than any of the other sound classes, followed by neighborhood noise (Figure 2). Again, this is consistent with the fact that noise indices were lower than in other areas in the period affected by the restrictions. These exceedingly low indices also contribute to a sharper soundscape, which is the perceptual construct more people agree on in the survey (Figure 7). As for the other perceptual constructs, their assessment is along the lines of most of the other clusters.

4.1.6. Sensors in Night-Time Leisure Areas

Noise indices in Night-Time Leisure Areas also follow intraday patterns significantly different from road traffic areas with elevated L_e and L_n which are consistent with the peak hours of night-time activity. In fact, L_n in this cluster is especially high, only second to the levels of Heavy Traffic sensors (Table 12). These areas were especially affected during the lockdown due to the curfew and mobility restrictions, which cancelled most of the activity. Therefore, it is not surprising that the L_n collapsed from 62.74 dB to 49.49 dB, the most impressive drop among all indices. Also noticeable is the decrease for the L_e, only surpassed by the drop in the Daytime Leisure cluster.

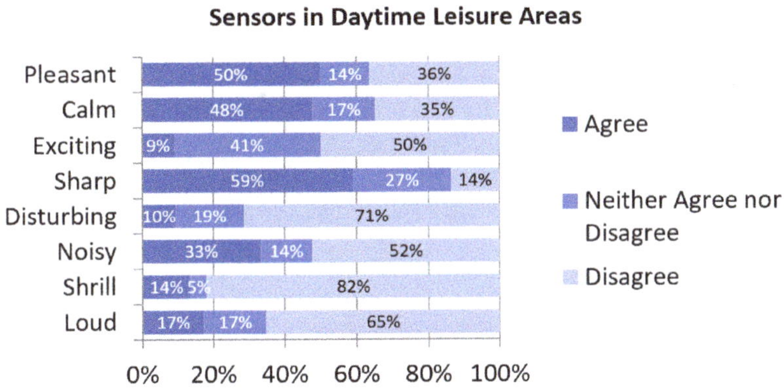

Figure 7. Assessment of perceptual constructs for soundscapes near sensors in Day-time Leisure Areas.

Table 12. mean daily noise indices during the lockdown (2020) and the same time-frame for previous years (2018–2019) for the sensors in Night-Time Leisure Areas.

Period	L_d [dB]	L_e [dB]	L_n [dB]
2018–2019	63.97	64.54	62.74
2020	57.6	54.81	49.49

As in the case for SuperBlocks, the number of contributions is limited and results of this cluster should be complemented with additional data when it is available. It is the only cluster of contributions where there was a 100% consensus regarding the fact that all the soundscapes where deemed "Positive" or "Very Positive" during the lockdown (Table 13). It is also noteworthy that there were not negative assessments of the soundscapes in the pre-lockdown scenario; the majority of participants considered their acoustic environment neither positive nor negative, which is aligned with the average noise indices detected in the 2018–19 period.

Table 13. Subjective acoustic satisfaction assessment before and during the lockdown for dwellings near sensors in Night-Time Leisure Areas (Likert Scale).

Period	Very Negative	Negative	Neutral	Positive	Very Positive
Pre-Lockdown	0%	0%	66.67%	33.33%	0%
Lockdown	0%	0%	0%	50%	50%

Typical noise sources almost disappeared during the lockdown, with very few instances of leisure or commerce-related noise reported (Figure 2). In addition, there were not any industry or construction work noises in the surroundings of the participants. The predominant sound events reported were traffic noise, neighborhood noise and nature sounds with the same prevalence. These changes in the soundscape elements combined with very low noise levels are correlated with a higher consensus on some of the positive perceptual constructs (Figure 8). Inhabitants of these areas are the ones that mostly agree with their surroundings being *calm* and they also majorly agree with it being *pleasant*. Also, in most negative perceptual constructs there is not a single respondent that agrees with them as valid describers of their acoustic environment.

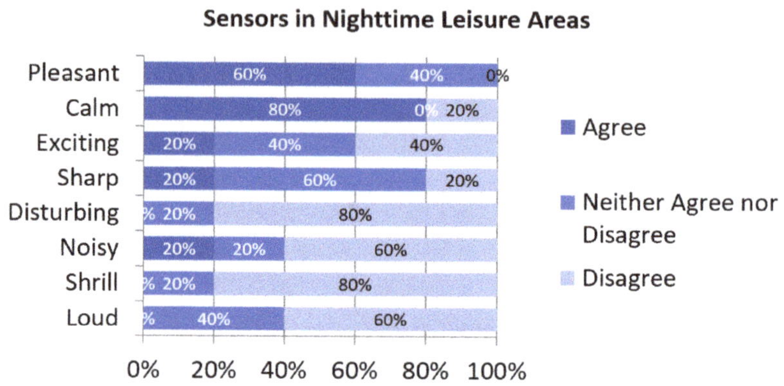

Figure 8. Assessment of perceptual constructs for soundscapes near sensors in Night-Time Leisure Areas.

5. Discussion

Results in Table 1 showed a strong alignment between the mean L_{Aeq} improvement during the lockdown for each cluster and the correspondent mean soundscape rating improvement for the same period. In fact, correlation between Columns 5 and 6 of Table 1 are 97.43% for the first three clusters dedicated to road-traffic noise and 58.32% for the next three clusters with different predominant noise sources. Therefore, the clustered approach seems to be appropriate for the purpose of comparing objective and subjective data.

On the contrary, comparing individual improvements for each one of the videos' subjective assessment with the individual drops in the L_{Aeq} observed in the nearest sensor provided a very low correlation of barely 4%. This very modest correlation rises significantly when the predominant noise source in the area is taken into account. If this individual comparison is performed only in areas where road-traffic is predominant (which include 65 of the contributors) the correlation is more than six times higher (26.44%). Also, the individual comparison performed only in areas where leisure noise is predominant (30 participants in total) gave an even higher correlation of 31.72%.

These figures show that with the clustered point of view the relationship between both sensor data and perception is clearly stronger than when the individual soundscapes are evaluated. In addition, they also highlight the importance of taking into account the predominant noise source present in the surroundings of a given soundscape along with the noise levels to maximize the alignment with its perceived quality.

The clustered approach offers a global perspective of the relationship between the perception of the soundscapes during the lockdown and sensor data. However, there are some dissenting opinions that do not align with the general appraisal of the soundscapes during the lockdown. Three of the participants rated their degree of acoustic satisfaction during the confinement lower than in the pre-lockdown period and another one kept the "Very Negative" assessment both before and during the lockdown. Three of these dissenting opinions are explained by the presence of construction works which resumed exactly when the survey took place. The other outlier assessment seems less coherent after analyzing all the answers provided in the survey by this specific contributor. Even though there are only four dissenting opinions, they have a significant impact in the correlation between the decreasing of the L_{Aeq} level and the improvement of the subjective perception due to the limited total number of contributions available.

In areas where road-traffic noise is predominant, daily noise indices follow a clearly defined pattern where L_d is higher than L_e and L_e is higher than L_n. That does not happen in areas where the primary noise sources are leisure activities. In Day-time Leisure Areas, the higher level usually corresponds to L_e, followed by L_d and L_n. In Night-time Leisure Areas, the pattern is similar but night levels are significantly higher and can be very close to L_d and L_e levels. Therefore, lockdown regulations affected differently the noise indices depending on the type of area. For areas dominated by road traffic, daily noise indices were similarly decreased and correlated with the L_{Aeq} drop. For that reason, an analysis based only on L_{Aeq} may be sufficient. However, for areas dominated by leisure noise, L_e and L_n experienced especially higher drops compared to L_d. In consequence; it is convenient to take into account daily noise indices to obtain a clearer picture of the type of noise sources more affected and the correspondent changes in citizens' appraisal of their acoustic environment.

There are some strengths in this study that are worth mentioning. Available objective data are especially sound because the Barcelona Noise Monitoring Network is an exceptionally large WASN with up to 70 working sound sensors during the lockdown. This vast amount of sensors facilitated that most assessed soundscapes were located relatively near a sensor (normally within 300 to 500 m depending on the cluster).

Most of the surveys in this project were answered in May 2020 during the initial stages of the de-escalation process (the last one accepted was from 12 June). Therefore, the time elapsed from the stricter lockdown in April was relatively short and there were still activity and mobility restrictions in force. In contrast, other studies in the same field found in the literature collected data in the later stages of the de-escalation process, which can have a significant effect on the subjective appraisal of the acoustic environment.

There are also limitations in this work that must be stated. Available subjective data are the main constraint of this work. It is true that more than one hundred of surveys were included in the comparison (which is a number consistent with other similar questionnaire-based studies published [59–61]). However, for attaining higher representation, a larger amount of opinions should be taken into account.

Not all clusters include the same number of sensors. The clusters with fewer sensors usually have a higher variance in the reported decrease in noise levels that have to be considered. In fact, results for the SuperBlocks cluster are not significant because they are based both in a very limited number of sensors and in a reduced set of assessed soundscapes. In addition, the reported noise and soundscape rating improvements have a very high variance affecting the reliability of the comparison.

There are no significant biases in gender and age between the respondents. However, there is a clear bias in their educational level as 87.72% of the participants had a university degree, which is an usual bias in this type of research projects [41]. Also, the study is focused in Barcelona, which is a big metropolis. Results may not be representative of smaller and less populated urban areas.

6. Conclusions

In this research, authors have compared objective data from acoustic sensors deployed in several districts in a big city (i.e., Barcelona, Spain) with the answers obtained from a questionnaire assessing the soundscapes in the surroundings of the sensors as a continuation of the preliminary research conducted in Girona, Spain, by the same researchers [2]. The comparison clearly manifests a coincidence between the reported soundscape quality during the lockdown and the improved level of noise indices collected by the sensors.

In a number of locations, especially in areas with higher levels of road-traffic noise, the reduction in the overall noise level highlighted other noise sources not perceived before the lockdown, such as Birds and Neighbors. This phenomenon is proved by both the gathered data from the sensors and the answers given by the participants. However, the new noise sources were not perceived as a nuisance but as pleasant. Therefore, according to the respondents, this change in the sound environment after confinement was for the better. That implies that a decrease in the road traffic noise has two direct benefits. Not only does it provides lower noise indices but it also changes the soundscape constitution from a scenario where only annoying noise sources are spotted (usually related to motorized traffic) to a scenario where both annoying and pleasant sound events are equally represented (both traffic and nature related).

The clustering of the contributions according to different kinds of predominant noise sources highlighted significant differences among them. Road traffic areas experienced the highest increases in the subjective evaluation by the volunteers even though the corresponding decrease in the noise indices was smaller in the same zones. On the contrary, in the surroundings of spots with predominance of leisure noise, a major drop in the noise indices translated into a milder increase in the subjective appraisal of the soundscape. This hints at the fact that road traffic noise is especially annoying. Therefore, its reduction has a greater effect on the acoustic comfort of citizens than other sound sources. These results have a direct impact both in environmental governance and urban planning. Measures taken by the public administration in order to increase the acoustic comfort of the residents should consider giving priority to the decrease in road traffic noise among other noise sources as they will be more effective. Accordingly, keeping road traffic noise as isolated as possible from dwellings and working places should also be a priority for urban planners.

Also, construction work has an important impact both on the noise levels and in the negative appraisal of the soundscape by those living nearby. On the contrary, other noise sources such as neighborhood noise or commercial noise are not as clearly correlated. For that reason, it is highly advisable to take into account the type of noise sources in addition to the measured noise levels when modeling the soundscape perception for prediction purposes.

The assessment of perceptual constructs by participants is heterogeneous, especially in comparison with the global assessment of the quality of the soundscape. They are generally aligned with the overall improvement reported with the objective data provided with sensors. In fact, there is a general consensus in the predominance of positive perceptual constructs over negative ones. However, differences among the various positive perceptual constructs are less correlated with the noise indices. Likewise, individual percentages for each one of the negative perceptual constructs are not exactly correlated with variations of the objective noise levels. Therefore, to obtain information more aligned with the objective data, a combined analysis of the perceptual constructs is more recommended than an individual approach.

In the future, the authors plan to start the automatic detection of sounds and objects in the videos uploaded by citizens in the framework of the still opened project *Sons al Balcó*. After the lockdown, the project has been conducted into local collecting campaigns, as the ones conducted in Sabadell (https://sonsalbalco.salle.url.edu/sonsdesabadell/, last accessed 24 August 2022) and in Granollers (https://sonsalbalco.salle.url.edu/sonsdegranollers/, last accessed 24 August 2022). The video and questionnaires collection has been wider in smaller cities, and this fact opens the possibility of starting the work in designing the indicators to evaluate both the objective and calibrated measurements (L_{Aeq}, etc.) with the annoyance and pleasantness evaluated subjectively via questionnaires. The automatic detection of objects and sounds in the videos would increase substantially the information for each citizen's contribution and enrich the indicators design. The next steps are focused on using more data coming from citizen participation to start the definition of indicators useful for administration and researchers, combining objective and subjective evaluation of noise and soundscapes, and finally, attempting to predict the subjective perception of a soundscape using several indicators including WASN-based L_{Aeq} levels.

Author Contributions: Conceptualization, R.M.A.-P. and P.B.; methodology, D.B.-S. and E.D.; validation, R.M.A.-P. and C.M.-S.; resources, R.M.A.-P.; data curation, D.B.-S., C.M.-S. and E.D.; writing—original draft preparation, P.B., D.B.-S. and E.D.; writing—review and editing, C.M.-S. and R.M.A.-P.; supervision, R.M.A.-P.; project administration, R.M.A.-P.; funding acquisition, R.M.A.-P. All authors have read and agreed to the published version of the manuscript.

Funding: The research that led to this contribution has been conducted thanks to funding from Secretaria d'Universitats i Recerca from the Departament d'Empresa i Coneixement (Generalitat de Catalunya) and Universitat Ramon Llull, under the grant 2020-URL-Proj-054 (Rosa Ma Alsina-Pagès).

Institutional Review Board Statement: Not applicable.

Informed Consent Statement: Informed consent was obtained from all subjects involved in the study.

Data Availability Statement: No new data were created or analyzed in this study. Data sharing is not applicable to this article.

Acknowledgments: The authors would like to thank all the participants of this first data collection campaign of project *Sons al Balcó*. The authors would also like to thank *Ajuntament de Barcelona* for the access to the data coming from the sensors in Barcelona Noise Monitoring Network.

Conflicts of Interest: The authors declare no conflicts of interest.

Abbreviations

The following abbreviations are used in this manuscript:

ANR	Active Noise Reduction
EU	European Union
L_{Aeq}	A-weighted equivalent sound level
L_d	Day Noise Level. A-weighted L_{eq} , over the 7 h to 21 h period
L_e	Evening Noise Level. A-weighted L_{eq} , over the 21 h to 23 h period
L_n	Night Noise Level. A-weighted L_{eq} , over the 23 h to 7 h period
L_{den}	Day-Evening-Night noise level. A-weighted L_{eq}, over a whole day, but with a penalty of 10 dBA for night-time noise (23 to 7) and 5 dBA for evening noise (21 to 23)
SPL	Sound Pressure Level
WASN	Wireless Acoustic Sensor Network
WHO	World Health Organization

References

1. Bonet-Solà, D.; Vidaña-Vila, E.; Alsina-Pagès, R.M. Prediction of the acoustic comfort of a dwelling based on automatic sound event detection. *Noise Mapp.* **2023**, *10*, 20220177. [CrossRef]
2. Dorca, E.; Bonet-Solà, D.; Bergadà, P.; Martínez-Suquía, C.; Alsina-Pagès, R.M. Sons al balcó: A subjective approach to the WASN-based LAeq measured values during the COVID-19 lockdown. In Proceedings of the 10th International Electronic Conference on Sensors and Applications, Online, 15–30 November 2023.
3. Alsina-Pagès, R.; Bergadà, P.; Martínez-Suquía, C. Sounds in Girona during the COVID Lockdown. *J. Acoust. Soc. Am.* **2021**, *149*, 3416–3423. [CrossRef]
4. Bonet-Solà, D.; Martínez-Suquía, C.; Alsina-Pagès, R.; Bergadà, P. The Soundscape of the COVID-19 Lockdown: Barcelona Noise Monitoring Network Case Study. *Int. J. Environ. Res. Public Health* **2021**, *18*, 5799. [CrossRef] [PubMed]
5. Bertucci, F.; Lecchini, D.; Greeven, C.; Brooker, R.M.; Minier, L.; Cordonnier, S.; René-Trouillefou, M.; Parmentier, E. Changes to an urban marina soundscape associated with COVID-19 lockdown in Guadeloupe. *Environ. Pollut.* **2021**, *289*, 117898. [CrossRef] [PubMed]
6. Basan, F.; Fischer, J.G.; Kühnel, D. Soundscapes in the German Baltic Sea Before and During the COVID-19 Pandemic. *Front. Mar. Sci.* **2021**, *8*, 689860. [CrossRef]
7. Smith, K.B.; Leary, P.; Deal, T.; Joseph, J.; Ryan, J.; Miller, C.; Dawe, C.; Cray, B. Acoustic vector sensor analysis of the Monterey Bay region soundscape and the impact of COVID-19. *J. Acoust. Soc. Am.* **2022**, *151*, 2507–2520. [CrossRef] [PubMed]
8. Duane, D.; Dolan, K.; Iafrate, J.; Freeman, L. The Effect of COVID-19 Shutdowns on Hawaiian Coral Reef Soundscapes. In Proceedings of the OCEANS 2023-Limerick, Limerick, Ireland, 5–8 June 2023; pp. 1–6.
9. Longden, E.G.; Gillespie, D.; Mann, D.; McHugh, K.A.; Rycyk, A.M.; Wells, R.; Tyack, P.L. Comparison of the marine soundscape before and during the COVID-19 pandemic in dolphin habitat in Sarasota Bay, FL. *J. Acoust. Soc. Am.* **2022**, *152*, 3170–3185. [CrossRef]
10. Sertlek, H. Hindcasting Soundscapes before and during the COVID-19 Pandemic in Selected Areas of the North Sea and the Adriatic Sea. *J. Mar. Sci. Eng.* **2021**, *9*, 702. [CrossRef]
11. Montano, W.; Gushiken, E. Lima soundscape before confinement and during curfew. Airplane flights suppressions because of Peruvian lockdown. *J. Acoust. Soc. Am.* **2020**, *148*, 1824–1830. [CrossRef]
12. Derryberry, E.P.; Phillips, J.N.; Derryberry, G.E.; Blum, M.J.; Luther, D. Singing in a silent spring: Birds respond to a half-century soundscape reversion during the COVID-19 shutdown. *Science* **2020**, *370*, 575–579. [CrossRef]
13. Montgomery, R.A.; Raupp, J.; Parkhurst, M. Animal Behavioral Responses to the COVID-19 Quietus. *Trends Ecol. Evol.* **2021**, *36*, 184–186. [CrossRef]
14. Alías, F.; Alsina-Pagès, R.M. Effects of COVID-19 lockdown in Milan urban and Rome suburban acoustic environments: Anomalous noise events and intermittency ratio. *J. Acoust. Soc. Am.* **2022**, *151*, 1676–1683. [CrossRef]
15. Rumpler, R.; Venkataraman, S.; Göransson, P. An observation of the impact of CoViD-19 recommendation measures monitored through urban noise levels in central Stockholm, Sweden. *Sustain. Cities Soc.* **2020**, *63*, 102469. [CrossRef]

16. Mimani, A.; Singh, R. Anthropogenic noise variation in Indian cities due to the COVID-19 lockdown during March-to-May 2020 *J. Acoust. Soc. Am.* **2021**, *150*, 3216–3227. [CrossRef] [PubMed]
17. Garg, N.; Gandhi, V.; Gupta, N. Impact of COVID-19 lockdown on ambient noise levels in seven metropolitan cities of India *Appl. Acoust.* **2022**, *188*, 108582. [CrossRef] [PubMed]
18. Steele, D.; Guastavino, C. Quieted City Sounds during the COVID-19 Pandemic in Montreal. *Int. J. Environ. Res. Public Health* **2021**, *18*, 5877. [CrossRef] [PubMed]
19. Mishra, A.; Das, S.; Singh, D.; Maurya, A.K. Effect of COVID-19 lockdown on noise pollution levels in an Indian city: A case study of Kanpur. *Environ. Sci. Pollut. Res.* **2021**, *28*, 46007–46019. [CrossRef] [PubMed]
20. Said, G.; Arias, A.; Carilli, L.; Stasi, A. Urban noise measurements in the City of Buenos Aires during the mandatory quarantine *J. Acoust. Soc. Am.* **2020**, *148*, 3149–3152. [CrossRef] [PubMed]
21. Sakagami, K. A Note on Variation of the Acoustic Environment in a Quiet Residential Area in Kobe (Japan): Seasonal Changes in Noise Levels Including COVID-Related Variation. *Urban Sci.* **2020**, *4*, 63. [CrossRef]
22. Terry, C.; Rothendler, M.; Zipf, L.; Dietze, M.C.; Primack, R.B. Effects of the COVID-19 pandemic on noise pollution in three protected areas in metropolitan Boston (USA). *Biol. Conserv.* **2021**, *256*, 109039. [CrossRef]
23. Asensio, C.; Aumond, P.; Can, A.; Gascó, L.; Lercher, P.; Wunderli, J.M.; Lavandier, C.; de Arcas, G.; Ribeiro, C.; Muñoz, P.; et al. A Taxonomy Proposal for the Assessment of the Changes in Soundscape Resulting from the COVID-19 Lockdown. *Int. J. Environ. Res. Public Health* **2020**, *17*, 4205. [CrossRef]
24. Aletta, F.; Oberman, T.; Mitchell, A.; Tong, H.; Kang, J. Assessing the changing urban sound environment during the COVID-19 lockdown period using short-term acoustic measurements. *Noise Mapp.* **2020**, *7*, 123–134. [CrossRef]
25. Hornberg, J.; Haselhoff, T.; Lawrence, B.T.; Fischer, J.L.; Ahmed, S.; Gruehn, D.; Moebus, S. Impact of the COVID-19 Lockdown Measures on Noise Levels in Urban Areas—A Pre/during Comparison of Long-Term Sound Pressure Measurements in the Ruhr Area, Germany. *Int. J. Environ. Res. Public Health* **2021**, *18*, 4653. [CrossRef]
26. Bartalucci, C.; Borchi, F.; Carfagni, M. Noise monitoring in Monza (Italy) during COVID-19 pandemic by means of the smart network of sensors developed in the LIFE MONZA project. *Noise Mapp.* **2020**, *7*, 199–211. [CrossRef]
27. Alsina-Pagès, R.M.; Alías, F.; Bellucci, P.; Cartolano, P.P.; Coppa, I.; Peruzzi, L.; Bisceglie, A.; Zambon, G. Noise at the time of COVID-19: The impact in some areas in Rome and Milan, Italy. *Noise Mapp.* **2020**, *7*, 248–264. [CrossRef]
28. Gevú, N.; Carvalho, B.; Fagerlande, G.C.; Niemeyer, M.L.; Cortês, M.M.; Torres, J.C.B. Rio de Janeiro noise mapping during the COVID-19 pandemic period. *Noise Mapp.* **2021**, *8*, 162–171. [CrossRef]
29. Manzano, J.V.; Pastor, J.A.A.; Quesada, R.G.; Aletta, F.; Oberman, T.; Mitchell, A.; Kang, J. The "sound of silence" in Granada during the COVID-19 lockdown. *Noise Mapp.* **2021**, *8*, 16–31. [CrossRef]
30. Asensio, C.; Pavón, I.; de Arcas, G. Changes in noise levels in the city of Madrid during COVID-19 lockdown in 2020. *J. Acoust. Soc. Am.* **2020**, *148*, 1748–1755. [CrossRef] [PubMed]
31. BruitParif. Effets du Confinement puis du Déconfinement sur le Bruit en ÎLE-de-FRANCE. Available online: https://www.bruitparif.fr (accessed on 12 May 2021).
32. Basu, B.; Murphy, E.; Molter, A.; Sarkar Basu, A.; Sannigrahi, S.; Belmonte, M.; Pilla, F. Investigating changes in noise pollution due to the COVID-19 lockdown: The case of Dublin, Ireland. *Sustain. Cities Soc.* **2021**, *65*, 102597. [CrossRef]
33. Bartalucci, C.; Bellomini, R.; Luzzi, S.; Pulella, P.; Torelli, G. A survey on the soundscape perception before and during the COVID-19 pandemic in Italy. *Noise Mapp.* **2021**, *8*, 65–88. [CrossRef]
34. Maggi, A.L.; Muratore, J.; Gaetán, S.; Zalazar-Jaime, M.F.; Evin, D.; Pérez Villalobo, J.; Hinalaf, M. Perception of the acoustic environment during COVID-19 lockdown in Argentina. *J. Acoust. Soc. Am.* **2021**, *149*, 3902–3909. [CrossRef] [PubMed]
35. Torresin, S.; Albatici, R.; Aletta, F.; Babich, F.; Oberman, T.; Elzbieta Stawinoga, A.; Kang, J. Indoor soundscapes at home during the COVID-19 lockdown in London—Part I: Associations between the perception of the acoustic environment, occupantś activity and well-being. *Appl. Acoust.* **2021**, *183*, 108305. [CrossRef] [PubMed]
36. Torresin, S.; Albatici, R.; Aletta, F.; Babich, F.; Oberman, T.; Stawinoga, A.E.; Kang, J. Indoor soundscapes at home during the COVID-19 lockdown in London—Part II: A structural equation model for comfort, content, and well-being. *Appl. Acoust.* **2022**, *185*, 108379. [CrossRef] [PubMed]
37. Jordan, P.; Fiebig, A. COVID-19 Impacts on Historic Soundscape Perception and Site Usage. *Acoustics* **2021**, *3*, 594–610. [CrossRef]
38. Zambon, G.; Confalonieri, C.; Angelini, F.; Benocci, R. Effects of COVID-19 outbreak on the sound environment of the city of Milan, Italy. *Noise Mapp.* **2021**, *8*, 116–128. [CrossRef]
39. Socoró, J.C.; Alías, F.; Alsina-Pagès, R.M. An anomalous noise events detector for dynamic road traffic noise mapping in real-life urban and suburban environments. *Sensors* **2017**, *17*, 2323. [CrossRef] [PubMed]
40. Caniato, M.; Bettarello, F.; Gasparella, A. Indoor and outdoor noise changes due to the COVID-19 lockdown and their effects on individuals' expectations and preferences. *Sci. Rep.* **2021**, *11*, 1–17. [CrossRef]
41. Casla-Herguedas, B.; Romero-Fernández, A.; Carrascal, T.; Navas-Martín, M.Á.; Cuerdo-Vilches, T. Noise Perception and Health Effects on Population: A Cross-Sectional Study on COVID-19 Lockdown by Noise Sources for Spanish Dwellings. *Buildings* **2023**, *13*, 2224. [CrossRef]
42. Mitchell, A.; Oberman, T.; Aletta, F.; Kachlicka, M.; Lionello, M.; Erfanian, M.; Kang, J. Investigating urban soundscapes of the COVID-19 lockdown: A predictive soundscape modeling approach. *J. Acoust. Soc. Am.* **2021**, *150*, 4474–4488. [CrossRef]

43. Lenzi, S.; Sádaba, J.; Lindborg, P. Soundscape in Times of Change: Case Study of a City Neighbourhood During the COVID-19 Lockdown. *Front. Psychol.* **2021**, *12*, 570741. [CrossRef]
44. Aumond, P.; Can, A.; Lagrange, M.; Gontier, F.; Lavandier, C. Multidimensional analyses of the noise impacts of COVID-19 lockdown. *J. Acoust. Soc. Am.* **2021**, *151*, 911–923. [CrossRef]
45. Hasegawa, Y.; Lau, S.K. A qualitative and quantitative synthesis of the impacts of COVID-19 on soundscapes: A systematic review and meta-analysis. *Sci. Total. Environ.* **2022**, *844*, 157223. [CrossRef] [PubMed]
46. Alsina-Pagès, R.M.; Freixes, M.; Orga, F.; Foraster, M.; Labairu-Trenchs, A. Perceptual evaluation of the citizen's acoustic environment from classic noise monitoring. *Cities Health* **2021**, *5*, 145–149. [CrossRef]
47. Sevillano, X.; Socoró, J.C.; Alías, F.; Bellucci, P.; Peruzzi, L.; Radaelli, S.; Coppi, P.; Nencini, L.; Cerniglia, A.; Bisceglie, A.; et al. DYNAMAP–Development of low cost sensors networks for real time noise mapping. *Noise Mapp.* **2016**, *3*. [CrossRef]
48. Orga, F.; Mitchell, A.; Freixes, M.; Aletta, F.; Alsina-Pagès, R.M.; Foraster, M. Multilevel annoyance modelling of short environmental sound recordings. *Sustainability* **2021**, *13*, 5779. [CrossRef]
49. Alsina-Pagès, R.M.; Orga, F.; Mallol, R.; Freixes, M.; Baño, X.; Foraster, M. Sons al balcó: Soundscape Map of the Confinement in Catalonia. In Proceedings of the Engineering Proceedings, Online, 15–30 November 2020; Volume 2, p. 77.
50. Alsina-Pagès, R.M.; Orga, F.; Mallol, R.; Freixes, M.; Baño, X.; Foraster, M. Soundscape of Catalonia during the first COVID-19 lockdown: Preliminary results from the Sons al Balcó project. *Eng. Proc.* **2021**, *2*, 77.
51. Baño, X.; Bergadà, P.; Bonet-Solà, D.; Egea, A.; Foraster, M.; Freixes, M.; Ginovart-Panisello, G.J.; Mallol, R.; Martín, X.; Martínez, A.; et al. Sons al Balcó, a Citizen Science Approach to Map the Soundscape of Catalonia. *Eng. Proc.* **2021**, *10*, 54.
52. LimeSurvey Online Survey Tool. Available online: https://www.limesurvey.org/en/ (accessed on 12 October 2020).
53. Camps, J. Barcelona noise monitoring network. In Proceedings of the EuroNoise, Maastricht, The Netherland, 31 May–3 June 2015.
54. Farrés, J.C.; Novas, J.C. Issues and challenges to improve the Barcelona Noise Monitoring Network. In Proceedings of the 11th European Congress and Exposition on Noise Control Engineering, Heraklion, Crete, Greece, 27–31 May 2018; pp. 27–31.
55. Alsina-Pagès, R.M.; Ginovart-Panisello, G.J.; Freixes, M.; Radicchi, A. A Soundwalk in the heart of Poblenou superblock in Barcelona: Preliminary study of the acoustic events. *Noise Mapp.* **2021**, *8*, 207–216. [CrossRef]
56. Schafer, R.M. *The Soundscape: Our Sonic Environment and the Tuning of the World*; Simon and Schuster: Toronto, ON, Canada 1993.
57. Purnomo, R.; Putra, T.D.; Kusmara, H.; Priatna, W.; Mukharom, F. Haversine Formula to Find The Nearest PetShop. *JATISI (J. Tek. Inform. Dan Sist. Inf.)* **2022**, *9*, 2205–2221. [CrossRef]
58. Likert, R. A technique for the measurement of attitudes. *Arch. Psychol.* **1932.**
59. Aletta, F.; Van Renterghem, T. Associations between personal attitudes towards COVID-19 and public space soundscape assessment: An example from Antwerp, Belgium. *Int. J. Environ. Res. Public Health* **2021**, *18*, 11774. [CrossRef] [PubMed]
60. Paunovic, K.; Jakovljevic, B.; Mircic, R.; Pajic, D.; Konatarevic, M. New soundscape after the Covid-19 lockdown. In Proceedings of the INTER-NOISE and NOISE-CON Congress and Conference Proceedings, Institute of Noise Control Engineering, Grand Rapids, MI, USA, 15–18 May 2023; Volume 265, pp. 1374–1382.
61. Qu, F.; Li, Z.; Zhang, T.; Huang, W. Soundscape and subjective factors affecting residents' evaluation of aircraft noise in the communities under flight routes. *Front. Psychol.* **2023**, *14*, 1197820. [CrossRef] [PubMed]

sensors

Article

Blind Calibration of Environmental Acoustics Measurements Using Smartphones

Ayoub Boumchich [1], Judicaël Picaut [1,*], Pierre Aumond [1], Arnaud Can [1] and Erwan Bocher [2]

1 UMRAE, CEREMA, Univ Gustave Eiffel, F-44344 Bouguenais, France; ayoub.boumchich@univ-eiffel.fr (A.B.); pierre.aumond@univ-eiffel.fr (P.A.); arnaud.can@univ-eiffel.fr (A.C.)
2 Lab-STICC, UMR 6285, CNRS, Université Bretagne Sud, F-56000 Vannes, France; erwan.bocher@cnrs.fr
* Correspondence: judicael.picaut@univ-eiffel.fr

Abstract: Environmental noise control is a major health and social issue. Numerous environmental policies require local authorities to draw up noise maps to establish an inventory of the noise environment and then propose action plans to improve its quality. In general, these maps are produced using numerical simulations, which may not be sufficiently representative, for example, concerning the temporal dynamics of noise levels. Acoustic sensor measurements are also insufficient in terms of spatial coverage. More recently, an alternative approach has been proposed, consisting of using citizens as data producers by using smartphones as tools of geo-localized acoustic measurement. However, a lack of calibration of smartphones can generate a significant bias in the results obtained. Against the classical metrological principle that would aim to calibrate any sensor beforehand for physical measurement, some have proposed mass calibration procedures called "blind calibration". The method is based on the crossing of sensors in the same area at the same time, which are therefore supposed to observe the same phenomenon (i.e., measure the same value). The multiple crossings of a large number of sensors at the scale of a territory and the analysis of the relationships between sensors allow for the calibration of the set of sensors. In this article, we propose to adapt a blind calibration method to data from the NoiseCapture smartphone application. The method's behavior is then tested on NoiseCapture datasets for which information on the calibration values of some smartphones is already available.

Keywords: environmental noise; noise mapping; smartphone application; calibration

Citation: Boumchich, A.; Picaut, J.; Aumond, P.; Can, A.; Bocher, E. Blind Calibration of Environmental Acoustics Measurements Using Smartphones. *Sensors* **2024**, *24*, 1255. https://doi.org/10.3390/s24041255

Academic Editor: Hector Eduardo Roman

Received: 18 July 2023
Revised: 25 January 2024
Accepted: 5 February 2024
Published: 16 February 2024

1. Introduction

Managing environmental noise, particularly in urban areas, is a major health and social issue. Numerous environmental policies encourage local authorities to produce noise maps of their territory with the aim of establishing an inventory of the noise environment and then proposing action plans to improve its quality. This is the case, for example, with the European directive 2002/49/EC [1] relating to the assessment and management of environmental noise.

The production of noise maps remains the most widely used tool when considering environmental policies. In general, these maps are produced using simulations based on calculation models requiring traffic data for the calculation of acoustic emission and spatial data for the modeling of acoustic propagation. Because access to these data is sometimes complicated, and their quality is sometimes questionable, the result of the simulations only partially reflects the existing state of the sound environment. Conversely, the use of acoustic sensors arranged within noise observatories gives a more detailed and realistic image of the noise environment of an area, but the insufficient number of sensors available does not allow for covering the whole territory and producing a detailed noise map [2].

The densification of sensors through the deployment of low-cost sensor networks is an interesting alternative, but the network thus produced may prove difficult to maintain in the long term. Although several experiments have already taken place, to our knowledge, there is no functional network of this type that can produce noise maps.

Another alternative is for citizens to become data producers themselves, using smartphones as measuring instruments, as part of a participative or crowd-sourcing approach. On this subject, since the pioneering work in the early 2010s [3–5], many studies have been conducted [6,7], notably on the quality of acoustic measurements produced with a smartphone, as well as on the implementation of a participatory approach to collect data on a large scale and over the long term. Among these approaches, the one based on the NoiseCapture application, which was developed in our laboratory, is the most advanced today [8]. Since the application was released in 2017 (for Android smartphones only), a considerable amount of data has been collected worldwide [9]. Analysis of the data revealed a wide range in the quality of the noise indicators collected due to the measurement protocol and, in particular, the lack of acoustic calibration of the smartphones in most cases. A lack of calibration, or even a bad calibration, can indeed generate a significant bias in the measurement results. The realization of a calibration in the state of the art, from a reference device (for example, an acoustic calibrator), would normally constitute a prerequisite for the realization of measurements, but the access to such reference devices by any citizen makes this procedure difficult to apply in practice. The proportion of calibrated smartphones in the totality of collected data is then very low, making its use for the production of noise maps more difficult.

In contrast to the classical metrological principle of calibrating any sensor for physical measurement, others have proposed so-called "blind" mass calibration procedures. The method is based on the crossing of sensors in the same area, at the same time, which are therefore supposed to observe the same phenomenon (i.e., to measure the same value). The repetition of these crossings of a large number of sensors at the scale of a territory, and the analysis of the relations between sensors allow, in theory, to calibrate all the sensors. This type of blind calibration seems particularly interesting for data such as those collected by NoiseCapture, especially in urban areas, where several sensors can cross each other in the same area at equivalent time periods.

In this paper, we propose to implement a blind calibration method for uncalibrated mobile noise measurements. The approach itself is not novel, since blind calibration has already been applied in other fields, but its application to a database consisting of geo-localized acoustic measurements is, in our opinion, a major step forward, calling into question the need to calibrate each smartphone individually. In the present work, this approach is applied on NoiseCapture data, but it could be generalized for any equivalent dataset. In the present case, we have exploited the NoiseCapture dataset for the 2017–2020 period available for download [10] as well as more recent additional data obtained by connecting to the online database [11]. It is important to mention at this point that all data collected with the NoiseCapture application are totally anonymous. The application fully respects users' privacy [12].

The method, described in Section 2, is based on modeling the relationships between sensors, which can be written in matrix form, and which can then be solved as a linear algebra problem. The behavior of the method, as well as a modified model, is then tested on NoiseCapture datasets for which information on the calibration values of some smartphones is available (Section 3). Finally, as an experiment, the method is applied to the dataset of the City of Rezé in France, allowing the production of a "calibrated" noise map based on the collected raw data (Section 3.5). Section 4 concludes on the next challenges to deploy this method on a large variety of territories.

2. Methodology

2.1. The Problem of the Acoustic Calibration of Smartphones on a Large Scale

2.1.1. General Considerations about Smartphone Acoustic Calibration

The principle of involving citizens in a participative science approach in the acoustical context is to collect massively geo-localized objective and subjective acoustic data. These data can then be used to produce noise maps for the benefit of local authorities, for example in the context of establishing action plans to reduce noise pollution. The project can also be part of an educational [13,14] or citizen approach to raising awareness and the co-construction of public policies [15,16]. Whatever the purpose of the collected data the calibration of smartphones is an issue that is often discussed.

Several works have shown that different acoustic measurement applications installed on the same smartphone or the same application installed on different smartphones can generate differences in the measured acoustic indicators [17–19] that can reach up to nearly 30 dB compared to a reference device [20]. It can be explained in particular by the different coding of the applications as well as by hardware differences between the smartphones. In this context, particular attention was paid to the development of the NoiseCapture application to ensure compliance with the acoustic acquisition protocol on Android smartphones. One can expect that the dispersion of measured noise values within the NoiseCapture application is lower. However, the calibration of the application/smartphone pairs is still required to obtain acoustic results with a minimum of bias [18,20,21].

In this paper, acoustic calibration is seen as the correction of a measured sound pressure signal so that this measurement coincides with a reference signal (i.e., an acoustic calibrator most of the time). This correction allows for a systematic error between the device to be calibrated and the reference device. In the simplest case, if X is the temporal sound pressure signal measured by the smartphone, then the true value Y of the observable is related to the measured measurement X via a calibration coefficient k such that:

$$Y = k \times X. \quad (1)$$

Within a smartphone application for noise measurement, the calibration consists of estimating this coefficient k, which normally takes into account all the elements of the analog–digital conversion chain, such as the correction linked to the sensitivity of the microphone and the effects of the digital discretization of the signal. Considering sound level in decibels (dB) instead of acoustic pressure, the estimated sound level L_Y can be calculated using the measured sound level L_X by the smartphone with the following relation:

$$L_Y = L_X + 20 \log k = L_X + \Delta. \quad (2)$$

Without the correction, the smartphone will produce a systematic offset (in dB) of a value equal to Δ.

In most experiments, the calibration procedure consists of evaluating the difference Δ in measurement between a smartphone and a reference device (e.g., a class 1 sound level meter) and then proceeding to a correction in the overall sound level, possibly A-weighted, by using an acoustic correction factor [22]. Most of the time, this correction is assumed to be a constant compared to the reference device; however, linearity problems can occur at low and high levels and in frequencies, which could justify a more adapted calibration [20,21], such as proposed by [23] for example. Instead of using reference devices, some alternative calibration methods have also been proposed, based, for example, on the measurement of a quiet sound level [18] or on the in situ measurement of road traffic noise [24]. In addition, if the calibration corrections are collected for different smartphone models and integrated in a reference database, the calibration of a smartphone can also be performed indirectly by searching for the corresponding calibration value in this database [18]. Nevertheless, some works have also shown possible differences between two identical models of smartphones, depending on different versions of the operating system or due to hardware changes on two generations of the same model [20]. Note also that the use of an external microphone instead of the smartphone's internal microphone can improve the accuracy of the measurement, but it still requires microphone calibration [25–28].

2.1.2. Smartphone Calibration with NoiseCapture

Like other similar applications, NoiseCapture allows for defining a calibration value Δ either directly by manually entering a calibration value in the application parameters or automatically using one of the calibration methods proposed by the application: using either an acoustic calibrator, a reference measurement device, a smartphone already calibrated with the NoiseCapture application, or based on a road traffic measurement. This calibration can be carried out several times, possibly giving rise to different calibration values, but for each measurement carried out, the calibration value applied is systematically associated with the measurement track (i.e., for all the measurement points in the corresponding track). In the first public release of the application, only the calibration value was collected. Since NoiseCapture release 51 (1.2.15) at the end of 2020, the method used for calibration is also part of the information collected.

In practice, over the period 2017–2020 (before release 1.2.15), 24% of measurement points have a calibration value different from the default value (0 dB), which suggests that the corresponding smartphones may have undergone acoustic calibration. However, even though it represents a very large mass of data, the observed calibration values may call into question the quality of the calibration: 61.12% of the calibrated smartphones, for example, have calibration values higher than $\pm$15 dB, which does not seem realistic even considering the low metrological quality of some smartphones. Finally, only specific events organized by specialists, for example with the objective of raising awareness among citizens or for research purposes, can ensure a high quality of data by considering a state-of-the-art calibration and a training of the users [22,29,30]. It is particularly the case for NoiseCapture Party events, which aim at collecting data during a specific event, supervised by qualified persons, generally over a short period of time and a limited spatial extent. However, such data represent only 0.6% of the data collected over the 2017–2020 period [9].

2.1.3. Mass Calibration vs. Individual Calibration of Smartphones

Relevant exploitation of mobile data at a large scale is therefore hampered by the heterogeneity of the collected data, which is mainly due to the lack or misapplication of a calibration protocol. To solve this problem, a relevant solution consists of simultaneously calibrating a posteriori all the collected data, including those that would have given rise to a calibration, in order to ensure total coherence between the data. In the literature, this mass calibration of data measured with mobile sensors, instead of considering the individual calibration of sensors, has led to the development of specific methodologies referred to as blind calibration, self-calibration, or re-calibration. In [31], the authors propose, for example, to take advantage of the multiple rendezvous between an uncalibrated smartphone and several calibrated smartphones to estimate its bias; a consensus is then found to calibrate all the smartphones simultaneously by solving a discrete average consensus problem. Here again, the fact of having only a few reference data points limits the use of the method. On the contrary, in [32,33], the Moments-Based Calibration approach does not require reference data but considers that all mobile sensors move in the same way in the whole study domain with the same probability. The ergodicity property then simplifies the mathematical analysis of the problem; in practice, as in our case, it is however not verified, since at the scale of a large territory, it is admitted that two smartphones will never meet. In [34], the calibration method does not rely on any such assumption and formulates the mutual calibration problem as a linear algebra problem whose solution relies on the resolution of a Laplacian matrix. This last method seems particularly well-suited to NoiseCapture data, and we have therefore chosen to adapt it to the present problem.

2.2. *NoiseCapture Application and Database*

2.2.1. Application Principle

The principle of mobile noise measurements is to collect geo-referenced acoustic data in a spatial area (Figure 1). A given user starts a measurement, moves along a path, and then stops the measurement. At each time step of 1 s, several acoustic indicators are calculated on the fly, recorded on the smartphone, and sent anonymously to a remote server. The transmitted data are verified and archived, and then they are processed in a simplified way in order to represent them in a cartographic representation. This representation takes the form of a noise map, where some acoustic indicators are aggregated on a hexagonal elementary spatial extent, the network of hexagons covering the entire globe.

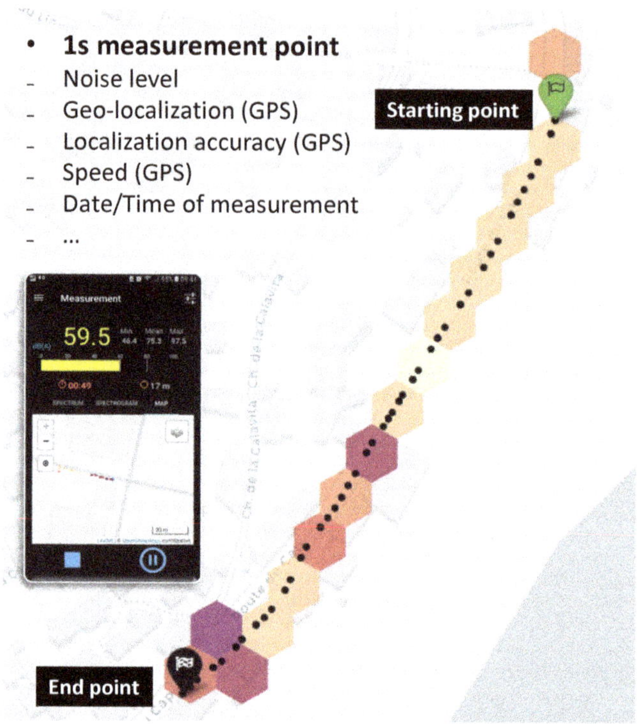

Figure 1. NoiseCapture approach. Using the NoiseCapture application, a user moves along a path (i.e., a track in the NoiseCapture vocabulary); each second (i.e., a measurement point), several noise indicators (sound level, spectrum) and other information (date/time, localization, speed, etc.) are calculated. When the user stops the measurements, the data are stored within the smartphone, and, if authorized by the user, uploaded to the NoiseCapture remote server. Raw data collected by the entire NoiseCapture community is preprocessed and averaged in hexagonal spatial zones (15 m radius); then, they are displayed in the form of noise maps.

2.2.2. Database and Privacy Policy

All data collected by the application are detailed in reference [35]. Particular attention has been paid to strict respect for privacy as well as the use of these data by the contributor and by third parties. The application's data confidentiality policy clearly explains the purpose of the application, namely to meet the needs of scientific research, in a context of participatory science and open science, carried out by French public research establishments (Université Gustave Eiffel—formerly Ifsttar—and CNRS). It is also specified that no personal data are collected. The development of this application and the redaction of the privacy notice was performed in consultation with the departments in charge of open science and legal aspects at Université Gustave Eiffel—formerly Ifsttar—in 2016–2017.

In particular, it is clearly specified from the very first run of the application that its use does not require registration, does not collect any personal data, does not record any audio data (acoustic indicators are calculated on the fly), and does not perform any background tasks. Furthermore, it is specified that the user can choose whether or not to contribute to the community database, can stop data collection at any time (by no longer using the application or by uninstalling it from the smartphone), and can access the application's source code and all data collected by the community. This short version of the privacy policy refers to a detailed online version on the NoiseCapture application website.

Each installation of the smartphone application creates a unique universal identifier (UUID), which is then associated with each set of measurements carried out with the smartphone. This enables all the data collected by a single smartphone to be postprocessed under a relevant process (for example, in this case, for the blind calibration of the smartphone measurements). This UUID is not linked to any other smartphone identifier. In addition, a user can generate a new UUID by uninstalling and reinstalling the application. If a user contacts the NoiseCapture project administrators to have their measurement data removed from the community server, they will be asked to provide this UUID.

From an application functional point of view, and like any application developed on the Android platform and deposited on the official repository (Google Play), the person installing and using the application must validate authorizations. In our case, only two authorizations are required: device location authorization and audio recording authorization. When the application is run for the first time, the short version of the NoiseCapture privacy policy is displayed and detailed [36], and the user must validate and accept these conditions in order to use the application. Also on first execution, a text is displayed to justify the application's activation of device localization (NoiseCapture localization policy [37]). Users are also regularly warned that this application can never replace a calibrated professional sound-level meter.

The data collected by the application and uploaded to the community server are accessible in their entirety, free of charge, and distributed under the Open Data Commons Open Database License (ODbL) [38], meaning that it is possible to share (to copy, distribute and use the database), to create (to produce works from the database) and to adapt (to modify, transform and build upon the database) as long as the database user attributes, shares and keeps open the database. This seemed to us to be a fair return for the community contributing to the development of this database and also in the context of Open Science.

2.3. Blind Calibration Model

2.3.1. Natural Graph Model

Among the solutions proposed in the literature for blind calibration, as a first attempt, the Natural Graph Model (NGM)-based blind calibration scheme proposed in [34] seems adapted to the mobile noise measurements, such as those collected using the NoiseCapture application. This method consists of exploiting the multiple appointments of sensors at positions close in time and space (i.e., in the same hexagon at a nearby time period) in order to establish mutual calibrations between sensors (Figure 2). In other words, if two smartphones simultaneously measure the same acoustic phenomenon, they should produce the same indicators (in the next development, we will say that there is a link between the two smartphones). This approach assumes that the sound level is homogeneous in a hexagonal zone whatever the position of the smartphones in that zone. There is no interaction between the zones; they are independent.

Due to differences in calibration for both smartphones, this rendezvous leads to the establishment of a correction factor between the two smartphones, i.e., a relative calibration, which can be generalized to the scale of a network of smartphones to establish relative calibrations between devices. For a very dense sensor network, the multiple appointments create a redundancy of information, which can also be exploited to improve the quality of the calibration.

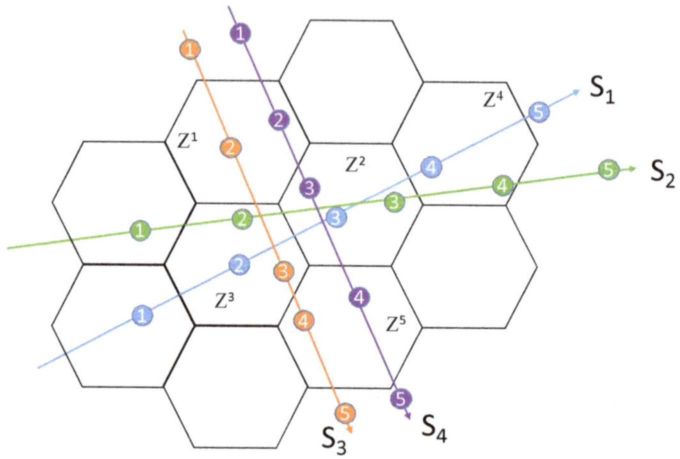

Figure 2. Principle of the blind calibration methodology applied to mobile noise measurements During the procedure, several sensors noted that S_1, S_2, S_3, and S_4 crossed the same spatial area (i.e., an hexagonal zone, noted Z^i) at the same time (t_1, t_2, t_3, t_4, t_5). In theory, these sensors should measure exactly the same acoustic event and therefore produce the same noise indicators. The path of a user is symbolized by a colored arrow; at each time step, the user is localized at a given position, which is symbolized by the colored circle with the time increment inside.

In the following, the original NGM methodology [34] is detailed and applied to the mobile noise measurement, using the same notations. However, we do not repeat all of the original developments so as not to make this article too long. Readers are invited to consult the original article.

Let us consider, for example, four sensors (S_1, S_2, S_3, S_4) traveling a path passing indifferently several hexagons covering a spatial extent at different times t. All the users numbered i present at the same time t in the same area define a zone Z of sensors that measure the value x of the same observable y of the event. Using the relation (2), we have [34]:

$$y = x_i + \Delta_i = x_i + d_i + n_i. \tag{3}$$

In the relation, it is assumed that the offset Δ that is estimated for a sensor is the sum of the exact drift d related to the calibration, assumed to be systematic and stationary over time, and an error n is associated with a non-predictable external effect and non-systematic, which is assumed to be white noise.

Thus, we can define the zone Z^α containing N^α co-located sensors (Table 1), performing the measurement x^α of the same observable y^α (i.e., the true value) such that [34]:

$$\{y^\alpha = x_i^\alpha + d_i + n_i^\alpha\}_{S_i \in Z^\alpha}. \tag{4}$$

For each sensor $S_i \in Z^\alpha$, the corresponding drift d_i can thus be expressed by the other smartphone drifts d_j ($S_j \in Z^\alpha$, $S_j \neq S_i$) using the following relation [34]:

$$d_i = \frac{1}{N^\alpha - 1} \sum_{(S_j \in Z^\alpha, S_j \neq S_i)} \left(d_j + \Delta x_{ji}^\alpha + \Delta n_{ji}^\alpha \right), \tag{5}$$

with $\Delta x_{ji}^\alpha = x_j^\alpha - x_i^\alpha$ and $\Delta n_{ji}^\alpha = n_j^\alpha - n_i^\alpha$.

Since the sensor i moves along other zones and the drift d_i is stationary over time, one can derive a set of linear equations. Considering the whole set of sensors, the linear equations can be written following a matrix form [34]:

$$\mathbf{L}\vec{d} = \Delta\vec{x} + \Delta\vec{n}, \tag{6}$$

where $\mathbf{L}$ is the calibration matrix, $\vec{d}$ is the drift vector, $\Delta\vec{x}$ is the differential vector, and $\Delta\vec{n}$ is the differential white noise vector. Due to the properties of $\mathbf{L}$, the calibration matrix is the Laplacian matrix. Lastly, the authors consider two more hypotheses: (1) the differential white noise vector $\Delta\vec{n}$ is negligible when considering a large number of sensors, meaning that $\mathbf{L}\vec{d} \approx \Delta\vec{x}$; (2) the mean value of all smartphone drifts is nearly zero, which leads to the equivalent constraint $\mathbf{M_1}\vec{d} = 0$ where the elements of $\mathbf{M_1}$ are all equal to 1. Finally, the authors show that the drift vector can be obtained by resolving the following matrix inversion [34]:

$$\vec{d} = (\mathbf{L} + \mathbf{M_1})^{-1}\Delta\vec{x}. \tag{7}$$

Once the drift vector is obtained, the estimated true value in a zone can be calculated using relation (4).

Table 1. Co-location sensor measurements based on the scenario of Figure 2.

Smartphone\Zone	Z^1	Z^2	Z^3	Z^4	Z^5
S_1		x_1^2	x_1^3	x_1^4	
S_2		x_2^2	x_2^3	x_2^4	
S_3	x_3^1		x_3^3		x_3^5
S_4	x_4^1	x_4^2			x_4^5

2.3.2. Simple Mean Model

Instead of using the NGM methodology, one can consider a very simplified approach, the Simple Mean Model [39], which is also considered in the reference works for predicting the gain calibration value for each smartphone. First, we take the average of the measurement values in each column of Table 1 to estimate the true input value of a zone. The SMM assumes a large number of sensors and estimates the true input value y^α using:

$$\hat{y}^\alpha = \frac{1}{N^\alpha} \sum_{(S_i \in Z^\alpha)} (x_i^\alpha) = \frac{1}{N^\alpha} \sum_{(S_i \in Z^\alpha)} (y^\alpha - d_i - n_i^\alpha). \tag{8}$$

Next, the drift value of a sensor can be estimated by calculating:

$$d_i = \hat{y}^\alpha - x_i^\alpha. \tag{9}$$

The linear Equation (5) for the NGM model, plus the constraint $\sum_i (d_i + n_i^\alpha) \approx 0$, is then equivalent to the SMM. In other words, the NGM is a generalized extension of the SMM.

2.3.3. Validation of the NGM Implementation

The NGM implementation was validated by direct comparison with the results published in the reference article [34] for a test dataset. This dataset is based on $S = 100$ simulated measurements located in $G = 100$ zones. Each measurement is simulated as the sum of the true value of the measurement y (a random number between 0 and 100 according to a uniform distribution), of a drift d (a random number according to a Gaussian distribution of variance Δ_{drift}) and of a noise n (a random number according to a Gaussian distribution of variance Δ_{noise}). The membership of a measurement in a zone is obtained randomly. Note that at this step, this dataset has no relation to sound levels and is only used for evaluating the NGM behavior.

On the basis of this dataset, a network graph can be generated to (Figure 3). The system (6) is then solved in order to determine the estimated value of the drift according to the relation (7) as well as the estimated value of the measurement in the corresponding zone according to the relation (4). The Mean Square Error (MSE) between the true value y and the estimated value $\hat{y}$ can then be computed in order to evaluate the model efficiency. In the reference article, the authors choose to represent the results through the link density metric $l_d \equiv 2L/[S(S-1)]$, which represents, on average, the number of times a given smartphone encounters other smartphones, with L designating the number of links. In addition to the application of the present NGM, the results obtained by the Simple Mean Model defined in Section 2.3.2 are also represented. The results are presented in the two following Figures 4 and 5, and they are very similar to Figures 3 and 4 of the reference article [34]. This simple comparison validates our implementation of the NGM.

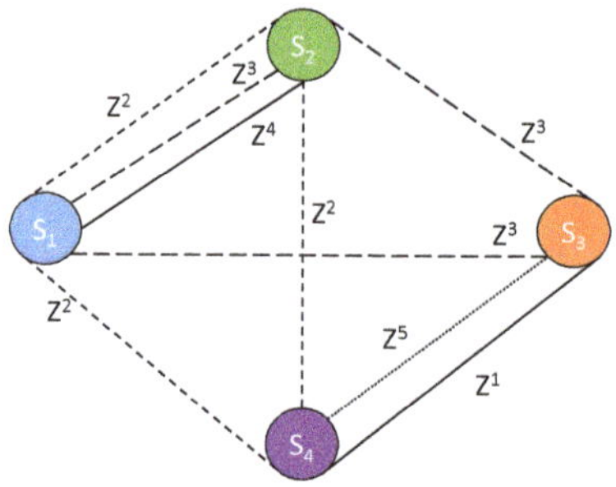

Figure 3. Network graph based on the scenario of Figure 2 and Table 1.

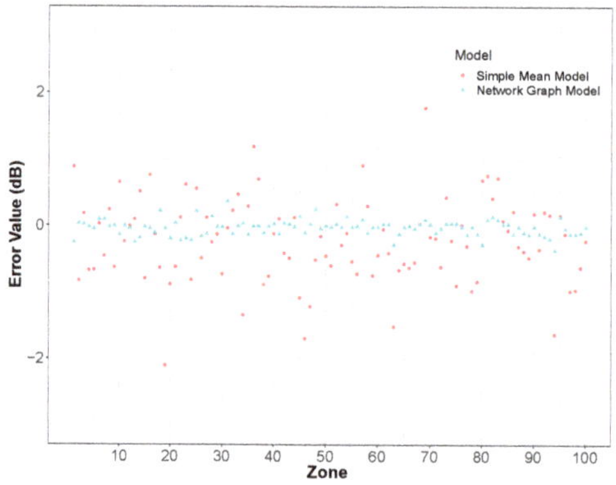

(**a**) $l_d = 1.0$

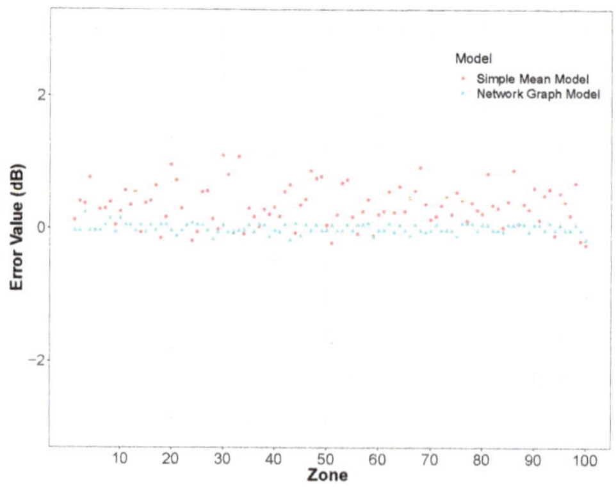

(**b**) $l_d = 10.0$

Figure 4. Comparison between the NGM and the SMM: error between the true value and the estimated value, for a link density (**a**) $l_d = 1.0$ and (**b**) $l_d = 10.0$, with $N = 100$ smartphones in G zones.

Figure 4 illustrates the error between the estimated value and the true value of the measurement for two values of the link density ($l_d = 1.0$ and $l_d = 10.0$). As expected, when the number of links between sensors increases (when l_d increases), the estimation error decreases. Moreover, this figure shows very clearly that the NGM gives a better estimation than the SMM.

Figure 5 generalizes this conclusion by summarizing the results for several values of the link density l_d. The NGM converges quickly to the true values, even for low link densities, while the SMM requires a larger number of links to reach an equivalent level of performance.

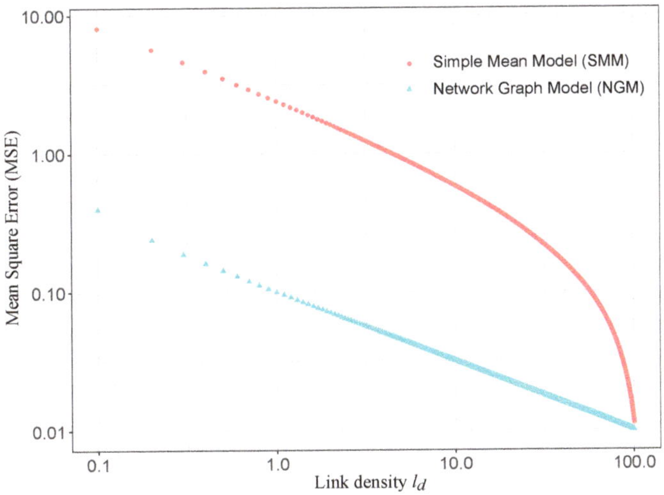

Figure 5. Comparison between the NGM and the SMM: mean square error in function of the link density l_d, with $N = 100$ sensors in G zones.

From a practical point of view, the optimization of the results of the model requires both an increase in the number of sensors and in the link density. Understandably, the more links there are between different sensors and the higher the number of sensors, the better the results.

3. Application of the NGM to a Mobile Acoustic Dataset

3.1. Discussion of NGM Application Assumptions

The development of the NGM is based on several assumptions that need to be discussed regarding its applicability to a mobile acoustic data dataset. Overall, the reliability of all these assumptions, although questionable, is also supported by the results that will be presented later.

3.1.1. NGM Mathematical Assumptions

Regarding the mathematical assumptions of the model, one can consider the following discussion:

- First assumption: the drift d of a given sensor is stationary over time. In principle, the variation of drift over time of a professional microphone is small, especially with respect to its impact on measured noise indicators. A smartphone microphone, on the other hand, is exposed to numerous constraints that may partially modify its acoustic characteristics over time. To our knowledge, there is no published study on the acoustic monitoring of smartphones over time, at least for environmental acoustics applications, but our experience within the NoiseCapture project has not revealed any anomalies on this subject. Moreover, considering the rapid change in

the smartphone fleet, the assumption of stationarity over a short or medium time period seems quite acceptable. In the event of a full deterioration of the smartphone microphone, following an accident, for example, the smartphone will become unusable for its primary function, and it is likely that it will no longer be used to collect data.
- Second assumption: the average value of drifts d on all sensors is null. The average value of all known calibration values in the NoiseCapture database, if we exclude calibration values at zero (default value in the absence of calibration), is of the order of -0.43 dB, i.e., close to zero. This hypothesis, therefore, seems globally acceptable. It is important to note first of all that this assumption is introduced by the authors to ensure the uniqueness condition of the solution of Equation (6) [34]. The assumption can therefore be discussed but is, in any case, required in the approach.
- Third assumption: the noise vector $\vec{n}$ is small in front of $\vec{x}$ for a large number of sensors It is difficult to quantify the error introduced by external conditions or insufficient control of the measurement protocol (noise generated by the operator, bad holding of the smartphone, effect of the wind on the microphone, etc.). However, one can consider that this noise is negligible in comparison with the measurement, and that it can be assimilated to a white noise.

3.1.2. Sensor Definition in the Context of a Mobile Acoustic Measurement

It is also important to consider the definition of a sensor in the context of a mobile acoustic measurement. Indeed, in the present application, we consider a sensor as a (smartphone model, NoiseCapture user) pair (noted later as a (smartphone, user) pair), even if several users can use the same smartphone model. It allows us to consider a specific calibration for each pair: it enables taking into account the fact that two users can, for example, use the same smartphone model with a different measurement protocol, or that the same smartphone model can give rise to several technically different generations and then different calibration corrections. In the NoiseCapture approach, a given user is defined by an Universally Unique Identifier (UUID) that is associated to the corresponding smartphone.

3.1.3. Assumption of Simultaneous Measurements between Two Sensors

The major assumption of the NGM model, which requires matching data that were measured at the same time and at the same place, is very crucial and raises the question of the choice of "homogeneous" time periods for the collected data in the context of a mobile acoustic dataset. In reference [40], the authors consider, for example, that a measurement of 10 min duration can be sufficient to characterize the sound environment equivalent to a period of one hour and that "homogeneous" periods of the same day can be discriminated by measurements of 10 to 20 min. For the moment, the temporal distribution of the collected data with NoiseCapture is not controllable, and only the accumulation of a large number of data with time will be able to ensure, in the future, a sufficient number of data for all temporal and homogeneous reference periods of a day. At this stage, within the framework of the present work, we will consider a larger time period of 1 h or more with the hypothesis of homogeneous sound environments.

3.2. *Comparison with Reference Datasets: NoiseCapture Parties*

In the NoiseCapture approach, specific events can be specifically organized in order to collect acoustic data over a defined spatial extent and over a given period. These events, called NoiseCapture Parties, are organized, for example, by researchers to collect data on a specific territory as part of their research into exposure to noise pollution [41–43], by teachers to train school and university students in environmental noise issues [14], or by local authorities wishing to raise awareness on the subject of noise environments [16]. In general, these events are run by professionals who are very familiar with the practice of acoustic measurement in the environment. For such events, smartphone calibration is systematically provided and the measurement protocol is detailed. Therefore, on these reference datasets, some calibration data are available for a large number of smartphones (i.e., the initial calibration value).

In this section, and as a preliminary step, we propose to apply the NGM to several reference datasets (Table 2). Each dataset is defined by an identifier 'pk_party' that identifies the corresponding data in the reference database [9]. The total number of 1 s measurement points, the number of tracks (consisting of all 1 s measurement points during the same track), the measurement time period, as well as the total number of (calibrated) smartphones, are also indicated. In addition, in the framework of the application of the NGM model to these datasets, the number of links and the value of the link density l_d are also given.

Table 2. Application of the NGM on NoiseCapture Parties datasets (reference data).

'pk_party'	Country	Tracks	Points	Time Period (24-Hour Format)	Nb of Sensors	Nb of Cal. Sensors	Zones	Links	l_d
10	Italy	149	15,912	11:00–12:00	12	11	479	357	5.4
13	France	100	21,470	10:00–11:00	11	11	817	508	9.2
22	France	192	17,309	12:00–19:00	23	23	403	1902	7.5
26	Italy	332	23,220	10:00–12:00	20	20	619	2526	13.3

Each event allows for the collection of data on a spatial extent defined by a set of contiguous hexagonal areas, as illustrated for example in Figure 1. The rayon of the hexagons is set to 15 m by default in the NoiseCapture approach, but the influence of this size on the behavior of the model will be discussed later in Section 3.4.

By construction, it is expected that the NGM performance will increase as the number of links between sensors increases and, therefore, as link density increases, too. In view of the l_d values in Table 2 and by looking at Figure 6, this hypothesis does not appear so clearly, even if the trend is globally respected.

Beyond a high l_d value, it is important that all smartphones are linked together. For example, in the case of NoiseCapture Parties N°13 and 26, one can observe that there are several groups of smartphones, with many links within each of these groups but not between smartphones from different groups (see, for example, Figure 7a for the NoiseCapture Party N°26). Conversely, the NoiseCapture Party event N°22 yields satisfactory results because most of the sensors are linked together (see Figure 7b for the NoiseCapture Party N°22). However, the variance is greater for NoiseCapture Party N°22, but this can be explained by a longer measurement period (7 h) than for the other NoiseCapture Parties, possibly generating a greater variability in sound levels.

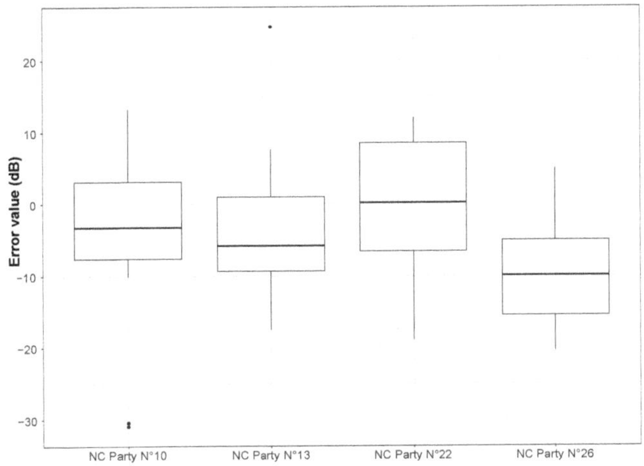

Figure 6. Error value between the estimated drift value and the initial calibration value for each calibrated smartphone used in the NoiseCapture (NC) Parties N°10, 13, 22 and 26.

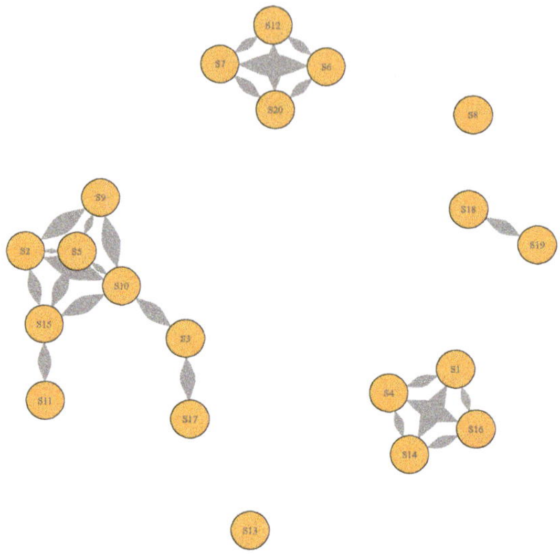

(**a**) NoiseCapture Party N°26

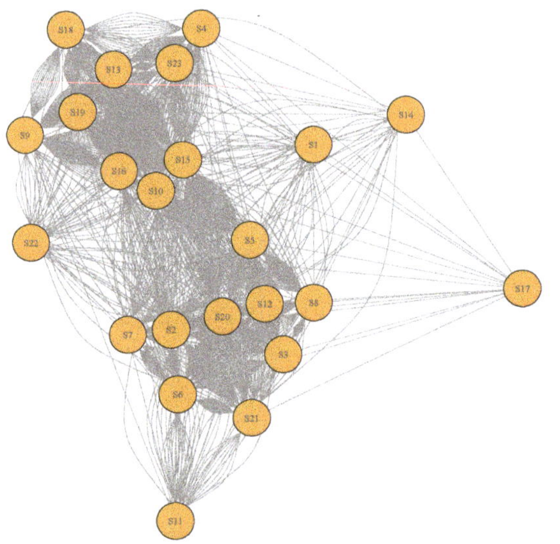

(**b**) NoiseCapture Party N°22

Figure 7. Smartphone network graph for the NoiseCapture Party (**a**) N°26 with 20 linked smartphones within 6 distinct subsets of data and (**b**) N°22 with 23 linked smartphones within the same subset of data.

3.3. *Hybrid NGM-SMM*

As discussed in the last paragraph, the improvement of the NGM method relies on the increase in the number of links between smartphones and, thus, the increase in link density. Obviously, if there are too many smartphones with few links with other smartphones, then the link density will decrease and the model efficiency will also decrease. An alternative to the original approach consists of applying the NGM to the pairs (smartphone and user) with the most links and then using the corresponding calibrated pairs to determine the drift of the other pairs by using SMM. This methodology, which can be qualified as a hybrid

NGM-SMM method, makes it possible to "focus" the NGM efficiency on the most relevant pairs by optimizing the link density and to determine the calibration values for the other pairs more easily with the SMM.

This methodology has been first tested on the dataset of the NoiseCapture Party N°22, but the conclusion is similar for the other NoiseCapture Parties. Several values of the minimal number of links per pair (smartphone, user) to be considered as a cut-off between NGM and SMM in the hybrid method were tested: from more than 1 link (this corresponds to the full NGM, with 23 (smartphone, user) pairs) to more than 140 links (12 remaining pairs), in order to evaluate the hybrid model efficiency. As expected, when the minimum number of links increases, the number of remaining (smartphone, user) pairs naturally decreases.

Figure 8 illustrates the results of this hybrid method through the mean error between the estimated drift values and the initial smartphone calibration values. In these results, all smartphones are concerned, whether they have been calibrated by the NGM method or by the SMM method. Compared to the NGM reference, we observe a better behavior of the hybrid approach (the variance decreases), and this is more so as the minimum number of links increases. This result clearly shows the contribution of the hybrid NGM-SMM method compared to the NGM method alone.

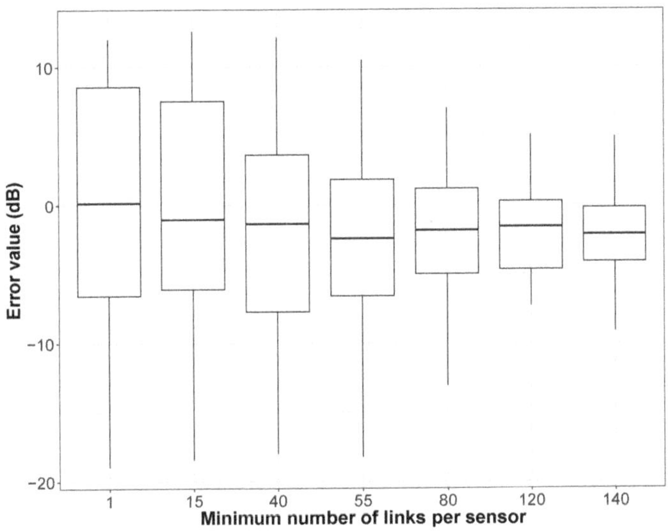

Figure 8. Application of hybrid NGM-SMM methodology on the NoiseCapture Party N°22 dataset. Error (in dB) between the estimated drift and the initial calibration of smartphones as a function of the number of links between (smartphone, user) pairs from 1 (this corresponds to the full NGM, i.e., the reference using the initial 23 smartphones) to 140 (12 remaining smartphones).

3.4. *Effect of the Size of the Spatial Area on the Hybrid Method*

As mentioned below, the size of the spatial area may have an effect on the method's efficiency. In this paragraph, we compare the effect of the size of the hexagon on the result of the hybrid model using the NoiseCapture Party N°22 dataset. Results are detailed in Table 3 in terms of mean error (in dB) between the estimated drift and the initial calibration value and in terms of uncertainty (i.e., the interval between the 75 and 25 quantiles after correcting with the bias value). It should be noted that the larger the area, the fewer the links between smartphones; this explains why some of the rows in the Table 3 do not give any results. Whether for the mean error or for the uncertainties, the results in Table 3 show that for the corresponding dataset, the best compromise is obtained for a hexagonal size of

15 m. These results confirm the initial hypothesis of the NoiseCapture approach, suggesting that the sound environment may be considered as homogeneous in an area of 15 m size.

Table 3. Effect of the size of the hexagon on the hybrid method as a function of the minimum number of links per smartphone. When the minimum number of links is equal to 1, it corresponds to the reference NGM. Mean error and uncertainty are given in dB.

Minimum Number of Links		Hexagon Size 10 m	15 m	30 m	50 m
1 (NGM)	Mean error	−2.33	0.36	−0.97	−1.28
	Uncertainty	±7	±8	±7	±6.5
15	Mean error	−2.33	−0.04	−0.97	−1.21
	Uncertainty	±7	±7.5	±7	±6.5
40	Mean error	−2.77	−2.13	−3.12	−3.69
	Uncertainty	±6.5	±5.5	±7.5	±7
55	Mean error	−3.86	−2.77	−3.71	−2.54
	Uncertainty	±6	±5	±5	±5
80	Mean error	−3.37	−2.34	−3.72	
	Uncertainty	±5	±4	±4.5	
120	Mean error	−3.54	−1.88		
	Uncertainty	±4	±3.2		
140	Mean error	−3.48	−1.92		
	Uncertainty	±4	±2.5		
190	Mean error	−4.76			
	Uncertainty	±3.5			

3.5. Comparison with Large Realistic Dataset: City of Rezé (France)

3.5.1. Description of the Dataset

The previous analysis is now extended to the City of Rezé, part of the Nantes metropolitan area, in France (Figure 9), for which a very large amount of data has been collected, both in the context of NoiseCapture Party events (NoiseCapture Party N°2, N°9, and N°52) and by "independent" contributors. In this area, additional data have also been collected similarly to a NoiseCapture event, in the framework of the Sonorezé research project [16], but are not a part of NoiseCapture Parties. The involved area represents a surface of 13,780,000 m^2, gathering a total of 450,335 of 1 s measurement points and 2336 tracks on 10,365 hexagons (Figure 10), collected by 331 (smartphone, user) pairs with 163 different smartphone models. Reference data (NoiseCapture Parties) represents 1877 of 1 s measurement points (0.4% of the whole dataset) and 16 tracks (0.7%), collected by 4 (smartphone, user) pairs (1.2%) and 3 different smartphone models (1.8%). Of the 331 pairs, only 134 smartphones were calibrated by users, which corresponds to 278,561 (61.9%) of 1 s calibrated measurement points and 1529 (65.5%) calibrated tracks. The map shown in Figure 10 is obtained by averaging the sound levels at all the measurement points in each hexagon over the entire data collection period [35].

(**a**) (**b**)

Figure 9. Localization of the City of Rezé in France. (**a**) Localization of the City of Rezé (France); (**b**) Boundaries of the City of Rezé (France).

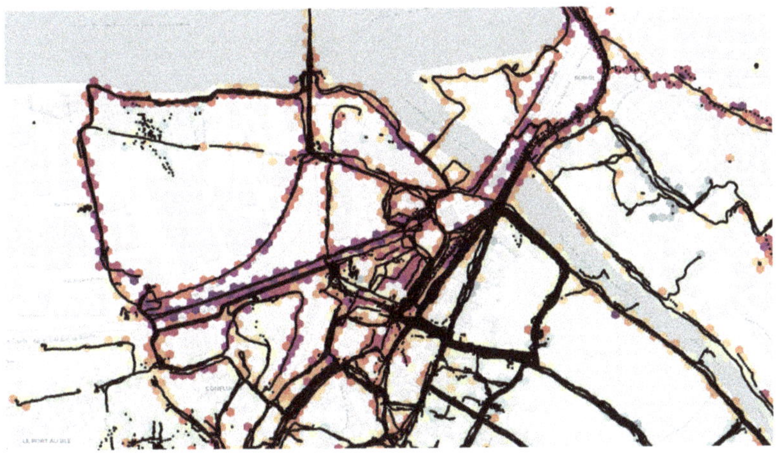

Figure 10. NoiseCapture data collected on a small part of the City of Rezé in France: measurement points and noise map (in dBA) built with raw data.

This dataset was collected over 6 years (2017–2023) at different times of the day and on different weekdays and weekends. In the present work, we have chosen to limit the application of the hybrid method to 08:00–20:00 (as a unique time period), for which a large number of datasets are available, considering that the long-term sound environment would be homogeneous during these periods. It corresponds to 315,598 of 1 s measurement points (i.e., 70.1% of the initial dataset) and 1712 tracks (73.3%). Moreover, to avoid the high variation when it comes to short measurements, a more 'homogeneous' approach was considered. This approach was to consider (smartphone/user) measures that stay less than 30 s (it corresponds to 65.3% of the initial dataset in terms of measurement points), 20 s (71.6% of the initial dataset) or 10 s (91.8% of the initial dataset) in each hexagon. These sub-datasets will be referred to in the next paragraph to as 'filtered data' (Table 4).

Table 4. Errors between the gain calibration value of smartphone and the obtained drift value, as a function of the number of links, in terms of mean error, median error and interquartile range (IQR), for the full dataset and the filtered data.

Minimum Number of Links per Sensor	1	5	10	20	50	100	200	500
Full dataset								
IQR	19.4	18.6	18	16.2	11.4	6.8	21.1	26.6
Mean	−6	−6	−6	−5.4	−2.7	−3.4	−11.8	−12
Median	−6.5	−6.5	−6.5	−6	−2.5	−3.4	−12.2	−11.8
Number of (smartphone, user) pairs	201	169	155	145	94	72	37	30
Filtered dataset—10 s								
IQR	18.9	17.9	17.1	15.7	11.1	19.6	22.6	
Mean	−4.2	−5.1	−2.1	−4	−2.9	−3.6	−9.9	
Median	−3.7	2.8	−1.3	−4.9	−1.8	−4.7	−12.6	
Number of (smartphone, user) pairs	163	131	108	85	57	26	19	
Filtered dataset—20 s								
IQR	17	16.9	20.4					
Mean	−3.8	−3.6	−4.8					
Median	−2.9	−2.8	−5.5					
Number of (smartphone, user) pairs	101	45	18					
Filtered dataset—30 s								
IQR	16.3	18.9	19					
Mean	−3.1	−0.7	−3.6					
Median	−3.9	−1.5	−0.8					
Number of (smartphone, user) pairs	63	20	12					

3.5.2. Time Slot Variability for a Rendezvous

Similarly to Figure 8, Figure 11 illustrates the mean error and uncertainty of the hybrid method, applied to data collected for the City of Rezé, as a function of the minimum number of links between (smartphone, user) pairs. Overall, we can already see that the variance is greater with this Rezé dataset than for the results shown in Figure 8 for the NoiseCapture Party N°22. It is due to the fact that this dataset contains a large amount of data produced outside the NoiseCapture Parties, some of which is of lower quality.

The approach is also applied on the sub-dataset with a minimum presence time of 30 s, 20 s and 10 s in a hexagon. Here again, the hybrid approach seems to give better results as the number of minimum links increases (the mean error decreases, as does the uncertainties). For the full dataset, the limit of improvement is reached a priori when the number of remaining (smartphone, user) pairs becomes insufficient. In the present case, this limit seems to appear for a number of links between 50 (94 remaining pairs) and 200 (37 remaining pairs), and it is visible for a number of links equal to 100. In this case, the average error is −3.4 dB between the smartphone calibration values and the drift values obtained using the hybrid method. The uncertainty is also much lower in this situation.

When considering a minimum time of presence in an hexagon area, we observe that the mean error decreases in comparison with the full data (results for a minimum number of links of 5 and 10), while the uncertainty is quite similar and constant. For a larger number of links, there are no more remaining (smartphone, user) pairs, and the hybrid method cannot give a result. When comparing the results for the full dataset with the results for a time of presence of 10 s, we observe that the optimum minimum number of links is reached earlier for the filtered data. It is difficult to conclude, since there are not enough data for 20 and 30 s, but one could expect that increasing the temporal filter duration will increase the quality of the results of the hybrid method.

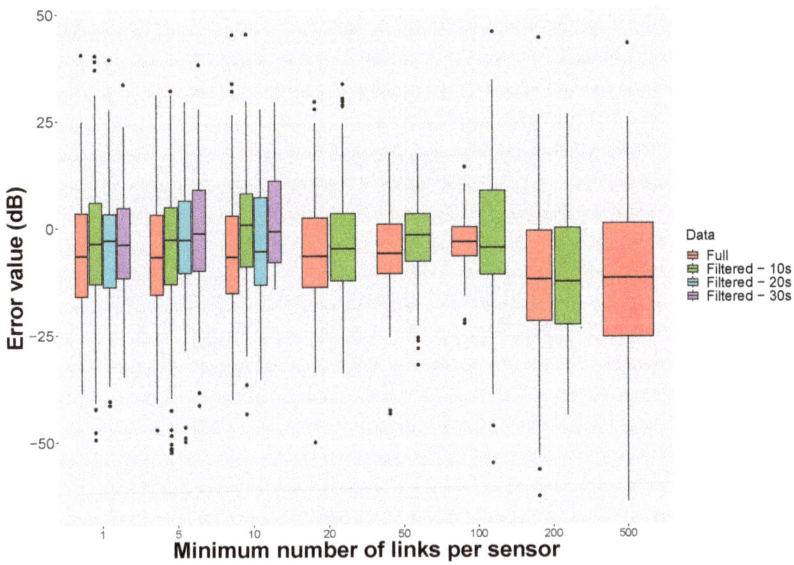

Figure 11. Application of the hybrid NGM-SMM methodology on the City of Rezé. Mean error (in dB) between the estimated drift and the initial calibration of smartphones as a function of the number of links between couples (smartphone, user) from 1 (i.e., the NGM reference) to 500, for each filter duration. The hybrid method is applied to both the 'full' dataset and the sub-dataset ('filtered') that correspond to a presence time of at least 30 s, 20 s and 10 s in a hexagon area.

3.5.3. Qualitative Results

In addition, we now consider the application of the hybrid method on the City of Rezé, with a minimum number of links of 100, which corresponds to the best configuration for the full dataset. As an illustration, Figure 12 shows the comparison between calibrated noise maps, either by considering the individual smartphone calibration values (as measured on the smartphone), or by considering the calibration values obtained using the hybrid blind calibration method, for a small part of the City of Rezé:

- The noise map (in dBA) produced with the initial calibration values ('Initial' noise map, Figure 12a). It considers only data for smartphones with an initial calibration (134 pairs).
- The noise map (in dBA) obtained by applying the blind calibration, using the hybrid method with a minimum threshold of 100 links per smartphone, but only for the smartphones that were initially calibrated ('Blind calibrated' noise map, Figure 12b). In this case, 52.7% of smartphones were calibrated (54 using the NGM method and 53 using the SMM method), enabling 71.9% of measurement points to be corrected.
- The difference map (in dBA) between the Initial and the Blind calibrated noise maps (Figure 12c); this difference map is calculated on the basis of the differences in the sound level in each hexagon. This map is completed in Figure 13 by a representation of the distribution of sound level differences as a percentage of the total number of corresponding hexagons in the whole City of Rezé (8464 hexagons contain data on all 10,365 hexagons).

(**a**)

(**b**)

(**c**)

Figure 12. Noise maps of a part of the City of Rezé: (**a**) data with initial calibration (134 calibrated (smartphone, user) pairs); (**b**) data after applying the blind calibration on the initially calibrated smartphones only; (**c**) difference noise map (**a**,**b**), see also details of the differences in Figure 13.

A qualitative comparison of the map produced using the calibration values initially entered by users and the map produced after blind calibration provides some first insights into the method. The initial map (Figure 12a) cannot completely serve as a reference, because there may be errors in the calibration values entered by users. On the other hand, the blind calibration method also allows a calibration value to be estimated for smartphones that have not been calibrated, which is an asset that is not evaluated here.

The findings are as follows. The blind calibration method tends to result in a noise map with higher noise values in this case study. It is probably due to a bias linked to the assumption that the calibration values are centered on zero, which is not necessarily the case for a small number of smartphones. In fact, of the 134 smartphones, 10 correspond to almost half of the measurements, and in this particular case, the average gain calibration given for these smartphones is negative and slightly overestimated by the method. This is visually accentuated in the neighbourhoods where few measurements contribute to the estimated value for each hexagon. This is the case for instance in the northwest where the density of measurements is small (see Figure 6 of Reference [16]).

That said, Figure 13 shows that the dispersion of the differences between the two maps is fairly small with a large part of the points concentrated between −8 and +2 dB, which confirms the validity of the method (this distribution would be probably centered for an input dataset whose calibration values are centered on zero). The average value of sound level differences in the hexagons is overall very close to the average value of differences between the initial calibration values and after the blind calibration mentioned in the previous Section 3.5.2 (−3.4 dB), which seems consistent.

Particular behaviors can be observed on this distribution, such as a peak at +12 dB. A detailed analysis of each of the blind calibration values obtained for each (smartphone, user) pair and an assessment of the individual contribution of each pair in each of the hexagonal zones would be required in order to make any assumptions about these peaks. This question could be the subject of a future investigation.

It will also be interesting, in a further study, to test the behavior of the method as a function of the input datasets in order to adapt it to the study areas; this point is discussed in the following section.

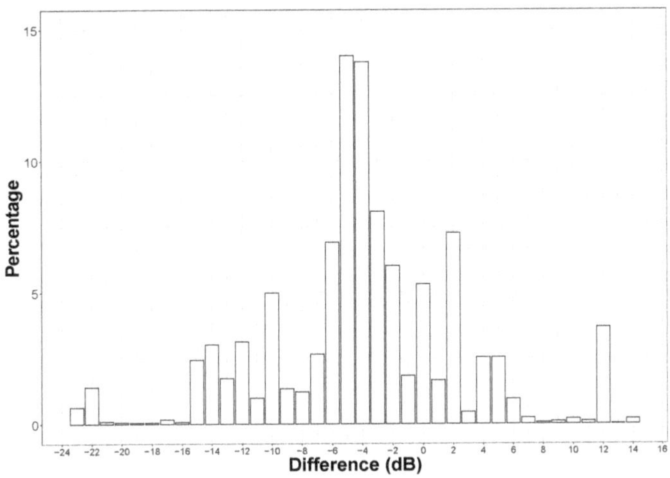

Figure 13. Distribution of the differences (in dB) of the noise level measured in each hexagonal zone, between the initial calibrated noise map and the noise map after blind calibration with the hybrid model, for the whole City of Rezé (for smartphones that were initially calibrated only). Differences are calculated for each hexagonal area (15 m) that composed the City of Rezé. *Y*-axis is given in terms of a percentage (%) of hexagonal area characterized by a given difference in dB.

4. Conclusions

Mobile noise measurements offer an alternative way of producing noise maps and collecting data on the noise environment through a participatory approach in which every citizen can become a data producer. Over and above the interest in contributing to the evaluation and development of public policies, this project raises real research questions, particularly in relation to the quality of the data produced and its use in an operational or regulatory context. Recent work on NoiseCapture data has shown a certain heterogeneity in the data collected, for example, in the absence of acoustic calibration of smartphones, a lack of expertise in the field of environmental acoustics by the contributors, or difficulties in implementing a measurement protocol that could be shared by all contributors. Data cleaning and quality control are therefore essential stages in the relevant use of the information collected. The work presented in this article is part of this approach and was aimed more specifically at implementing a generic calibration method for all data simultaneously. It is now accepted that it will never be possible to calibrate each smartphone individually and that a mass calibration should therefore be considered instead.

Among the solutions envisaged, those based on blind calibration approaches, already tested on other studies such as for air quality measurements, are an interesting perspective. In the present article, we have exploited a method that takes advantage of the multiple rendezvous of several smartphones "at the same place" and "at the same time", measuring the same acoustic event. Written as a network graph model, the resolution of the associated matrix system can then be used to determine a mean drift for each smartphone, which is similar to a calibration correction in acoustics. The method relies on certain constraints, which are discussed in the paper, such as the temporal distribution of the data at our disposal to verify the "at the same time" condition, or more accurately at similar periods in the day, as well as the size of the spatial area to verify the "at the same place" condition. In addition, the number of rendezvous a smartphone can have with others is an important factor for the quality of results. In particular, we proposed a hybrid approach to address this critical point, enabling us firstly to improve the quality of the calibration on a limited number of smartphones by using the Network Graph Model, then, secondly, using these calibrated smartphones to calibrate the other ones using a simpler approach. With regard to the first limitation, the progressive accumulation of new data over time should make it possible to obtain a more relevant temporal distribution of data. We have also observed that considering only smartphones with a minimum time of presence in each spatial area could be a way to enhance the behavior of the hybrid method. Regarding the second limitation related to the size of the spatial area, the results show that a 15 m radius spatial area was sufficient to verify a relatively homogeneous noise environment in the context of the hybrid method.

The obtained results seem particularly interesting and demonstrate the feasibility of such a blind calibration approach for mobile noise data. The method can also be improved by taking advantage of the simultaneous presence of reference sensors in a given area, such as noise observatories or calibrated smartphones, as suggested in [44].

The behavior of the method could also be studied on the basis of a perfectly controlled virtual mobile noise measurement dataset, as shown in Section 2.3.3. For example, it would be possible to study in more detail the effects of time of presence in hexagons, temporal and spatial variability, minimum number of links, or the presence of reference sensors. It could be useful to identify with more confidence the best conditions for applying the hybrid blind calibration method and to adapt its parameter values to the characteristics of the dataset. A virtual mobile noise measurement dataset will also enable testing other spatial and temporal grids, replacing for instance hexagons by streets with similar traffic behavior or refining the "at the same time" condition relying on temporal periods with similar sound levels. It will be of interest finally to test the sensibility of the method to datasets with different levels of heterogeneity in the participatory contributions, as this first analysis suggests that some main contributors might have an influence on the method if they collect a large proportion of the data and have calibration values not centered on zero.

More generally, to improve the method, it might also be useful to improve the quality of the data collected. It could be envisaged at the source, by improving the mobile application to ensure better control of the measurement procedure but also a better understanding of the measurement context. One example is the possibility offered by specific libraries for the development of smartphone applications to obtain information on the user mode of travel (on foot, by bike, on public transport), or even the location of the smartphone itself (in the hand, in the pocket, in a bag). Improving the quality of the database can also be achieved a posteriori by searching for and then removing any data collected that could be assimilated to anomalies. It can be taken into account, for example, by considering methods such as the Local Outlier Factor (LOF) [45] or the Isolation Forest [46] methods.

To conclude, the blind calibration approach, possibly considering improvements, is a very interesting way of tackling the difficulty of calibrating each individual smartphone. With this methodology and as part of a participative approach to noise map production, it would no longer be necessary to ask users to calibrate their smartphones, as this can be completed a posteriori. It would also be interesting if each user could know the calibration value in return and have it automatically integrated into the application parameters.

Author Contributions: Conceptualization, A.B. and J.P.; Formal analysis, A.B., J.P., P.A. and A.C.; Funding acquisition, J.P.; Investigation, A.B., J.P., P.A., A.C. and E.B.; Methodology, A.B. and J.P.; Project administration, J.P. and E.B.; Software, A.B.; Supervision, J.P. and E.B.; Validation, A.B., J.P., P.A. and A.C.; Visualization, A.B.; Writing—original draft, A.B. and J.P.; Writing—review and editing, A.B., J.P., P.A., A.C. and E.B. All authors have read and agreed to the published version of the manuscript.

Funding: This research was initially funded in the framework of the ENERGIC-OD Project (European Network for Redistributing Geospatial Information to user Communities-Open Data) under the ICT Policy Support Programme (ICT PSP) (CIP-ICT-PSP-2013-7) as part of the Competitiveness and Innovation Framework Programme by the European Community. A part of this research is funding by the Région Pays de La Loire grand number 2020_10361.

Institutional Review Board Statement: Not applicable.

Informed Consent Statement: Not applicable.

Data Availability Statement: The data presented in this study for the 2017–2020 period are openly available from the Université Gustave Eiffel Dataverse Repository at https://doi.org/10.25578/J5DG3W (accessed on 13 July 2023).

Acknowledgments: The authors would like to thank Gwendall Petit and Nicolas Fortin, from the Environmental Acoustics Research Laboratory (UMRAE), for their help in manipulating and representing the NoiseCapture data. Map data are from OpenStreetMap licensed under the Open Data Commons Open Database License (ODbL) by the OpenStreetMap Foundation (OSMF) (see https://www.openstreetmap.org/copyright/ (accessed on 13 July 2023)).

Conflicts of Interest: The authors declare no conflicts of interest.

References

1. European Parlamient Directive 2002/49/EC of the European Parliament and of the Council of 25 June 2002 relating to the assessment and management of environmental noise-Declaration by the Commission in the Conciliation Committee on the Directive relating to the assessment and management of environmental noise. *Off. J.* **2022**, *L 189*. Available online: http://data.europa.eu/eli/dir/2002/49/oj/eng (accessed on 13 July 2023).
2. Can, A.; Dekoninck, L.; Botteldooren, D. Measurement network for urban noise assessment: Comparison of mobile measurements and spatial interpolation approaches. *Appl. Acoust.* **2014**, *83*, 32–39. [CrossRef]
3. Maisonneuve, N.; Stevens, M.; Ochab, B. Participatory Noise Pollution Monitoring Using Mobile Phones. *Inf. Polity* **2010**, *15*, 51–71. [CrossRef]
4. Kanjo, E. NoiseSPY: A Real-Time Mobile Phone Platform for Urban Noise Monitoring and Mapping. *Mob. Netw. Appl.* **2010**, *15*, 562–574. [CrossRef]
5. D'Hondt, E.; Stevens, M.; Jacobs, A. Participatory noise mapping works! An evaluation of participatory sensing as an alternative to standard techniques for environmental monitoring. *Pervasive Mob. Comput.* **2013**, *9*, 681–694. [CrossRef]

6. Guillaume, G.; Can, A.; Petit, G.; Fortin, N.; Palominos, S.; Gauvreau, B.; Bocher, E.; Picaut, J. Noise mapping based on participative measurements. *Noise Mapp.* **2016**, *3*, 140–156. [CrossRef]
7. Brambilla, G.; Pedrielli, F. Smartphone-Based Participatory Soundscape Mapping for a More Sustainable Acoustic Environment. *Sustainability* **2020**, *12*, 7899. [CrossRef]
8. Picaut, J.; Fortin, N.; Bocher, E.; Petit, G.; Aumond, P.; Guillaume, G. An open-science crowdsourcing approach for producing community noise maps using smartphones. *Build. Environ.* **2019**, *148*, 20–33. [CrossRef]
9. Picaut, J.; Boumchich, A.; Bocher, E.; Fortin, N.; Petit, G.; Aumond, P. A Smartphone-Based Crowd-Sourced Database for Environmental Noise Assessment. *Int. J. Environ. Res. Public Health* **2021**, *18*, 7777. [CrossRef]
10. Picaut, J.; Fortin, N.; Bocher, E.; Petit, G. Université Gustave Eiffel Online Repository for Research Data. NoiseCapture Data Extraction from August 29, 2017 until August 28, 2020 (3 Years). 2021. Available online: https://data.univ-gustave-eiffel.fr/dataset.xhtml?persistentId=doi:10.25578/J5DG3W (accessed on 13 July 2023).
11. Noise-Planet Website. Exploit NoiseCapture Data. 2023. Available online: https://noise-planet.org/noisecapture_exploit_data.html (accessed on 13 July 2023).
12. Noise-Planet Website. NoiseCapture Privacy Policy. 2022. Available online: https://noise-planet.org/NoiseCapture_privacy_policy.html (accessed on 13 July 2023).
13. Zipf, L.; Primack, R.B.; Rothendler, M. Citizen scientists and university students monitor noise pollution in cities and protected areas with smartphones. *PLoS ONE* **2020**, *15*, e0236785. [CrossRef]
14. Guillaume, G.; Aumond, P.; Bocher, E.; Can, A.; Ecotière, D.; Fortin, N.; Foy, C.; Gauvreau, B.; Petit, G.; Picaut, J. NoiseCapture smartphone application as pedagogical support for education and public awareness. *J. Acoust. Soc. Am.* **2022**, *151*, 3255–3265. [CrossRef]
15. Lefevre, B.; Agarwal, R.; Issarny, V.; Mallet, V. Mobile crowd-sensing as a resource for contextualized urban public policies: A study using three use cases on noise and soundscape monitoring. *Cities Health* **2021**, *5*, 179–197. [CrossRef]
16. Can, A.; Audubert, P.; Aumond, P.; Geisler, E.; Guiu, C.; Lorino, T.; Rossa, E. Framework for urban sound assessment at the city scale based on citizen action, with the smartphone application NoiseCapture as a lever for participation. *Noise Mapp.* **2023**, *10*, 20220166. [CrossRef]
17. Kardous, C.A.; Shaw, P.B. Evaluation of smartphone sound measurement applications. *J. Acoust. Soc. Am.* **2014**, *135*, EL186–EL192. [CrossRef]
18. Zhu, Y.; Li, J.; Liu, L.; Tham, C.K. iCal: Intervention-free Calibration for Measuring Noise with Smartphones. In Proceedings of the 2015 IEEE 21st International Conference on Parallel and Distributed Systems (ICPADS), Melbourne, Australia, 14–17 December 2015; pp. 85–91. [CrossRef]
19. Murphy, E.; King, E.A. Testing the accuracy of smartphones and sound level meter applications for measuring environmental noise. *Appl. Acoust.* **2016**, *106*, 16–22. [CrossRef]
20. Ventura, R.; Mallet, V.; Issarny, V.; Raverdy, P.G.; Rebhi, F. Evaluation and calibration of mobile phones for noise monitoring application. *J. Acoust. Soc. Am.* **2017**, *142*, 3084–3093. [CrossRef] [PubMed]
21. Nast, D.R.; Speer, W.S.; Prell, C.G.L. Sound level measurements using smartphone "apps": Useful or inaccurate? *Noise Health* **2014**, *16*, 251. [CrossRef] [PubMed]
22. Aumond, P.; Lavandier, C.; Ribeiro, C.; Boix, E.G.; Kambona, K.; D'Hondt, E.; Delaitre, P. A study of the accuracy of mobile technology for measuring urban noise pollution in large scale participatory sensing campaigns. *Appl. Acoust.* **2017**, *117*, 219–226. [CrossRef]
23. Garg, S.; Lim, K.M.; Lee, H.P. An averaging method for accurately calibrating smartphone microphones for environmental noise measurement. *Appl. Acoust.* **2019**, *143*, 222–228. [CrossRef]
24. Aumond, P.; Can, A.; Rey Gozalo, G.; Fortin, N.; Suárez, E. Method for in situ acoustic calibration of smartphone-based sound measurement applications. *Appl. Acoust.* **2020**, *166*, 107337. [CrossRef]
25. Kardous, C.A.; Shaw, P.B. Evaluation of smartphone sound measurement applications (apps) using external microphones—A follow-up study. *J. Acoust. Soc. Am.* **2016**, *140*, EL327–EL333. [CrossRef]
26. Roberts, B.; Kardous, C.; Neitzel, R. Improving the accuracy of smart devices to measure noise exposure. *J. Occup. Environ. Hyg.* **2016**, *13*, 840–846. [CrossRef] [PubMed]
27. Celestina, M.; Hrovat, J.; Kardous, C.A. Smartphone-based sound level measurement apps: Evaluation of compliance with international sound level meter standards. *Appl. Acoust.* **2018**, *139*, 119–128. [CrossRef]
28. Celestina, M.; Kardous, C.A.; Trost, A. Smartphone-based sound level measurement apps: Evaluation of directional response. *Appl. Acoust.* **2021**, *171*, 107673. [CrossRef]
29. Can, A.; Guillaume, G.; Picaut, J. Cross-calibration of participatory sensor networks for environmental noise mapping. *Appl. Acoust.* **2016**, *110*, 99–109. [CrossRef]
30. Pődör, A.; Szabó, S. Geo-tagged environmental noise measurement with smartphones: Accuracy and perspectives of crowd-sourced mapping. *Environ. Plan. Urban Anal. City Sci.* **2021**, *48*, 2710–2725. [CrossRef]
31. Miluzzo, E.; Lane, N.D.; Campbell, A.T.; Olfati-Saber, R. CaliBree: A self-calibration system for mobile sensor networks. In *Distributed Computing in Sensor Systems*; Nikoletseas, S.E., Chlebus, B.S., Johnson, D.B., Krishnamachari, B., Eds.; Springer: Berlin/Heidelberg, Germany, 2008; Volume 5067, pp. 314–331.

32. Wang, C.; Ramanathan, P.; Saluja, K.K. Moments based blind calibration in mobile sensor networks. In Proceedings of the 2008 IEEE International Conference on Communications, Beijing, China, 19–23 May 2008; Volume 1–13, pp. 896–900. [CrossRef]
33. Wang, C.; Ramanathan, P.; Saluja, K.K. Blindly Calibrating Mobile Sensors Using Piecewise Linear Functions. In Proceedings of the 2009 6th Annual IEEE Communications Society Conference on Sensor, Mesh and Ad Hoc Communications and Networks (secon 2009), Rome, Italy, 22–26 June 2009; pp. 216–224.
34. Lee, B.T.; Son, S.C.; Kang, K. A Blind Calibration Scheme Exploiting Mutual Calibration Relationships for a Dense Mobile Sensor Network. *IEEE Sens. J.* **2014**, *14*, 1518–1526. [CrossRef]
35. Bocher, E.; Petit, G.; Picaut, J.; Fortin, N.; Guillaume, G. Collaborative noise data collected from smartphones. *Data Brief* **2017**, *14*, 498–503. [CrossRef]
36. NoiseCapture App. NoiseCapture Privacy Policy (Short Version). 2017. Available online: https://github.com/Universite-Gustave-Eiffel/NoiseCapture/blob/master/app/src/main/assets/html/privacy_policy.html (accessed on 13 July 2023).
37. Noise-Planet App. Localization Policy. 2021. Available online: https://github.com/Universite-Gustave-Eiffel/NoiseCapture/blob/master/app/src/main/assets/html/localisation_notice.html (accessed on 13 July 2023).
38. Open Data Commons Open Database License (ODbL). 2011. Available online: https://opendatacommons.org/licenses/odbl/ (accessed on 13 July 2023).
39. Whitehouse, K.; Culler, D. Calibration as Parameter Estimation in Sensor Networks. In Proceedings of the 1st ACM International Workshop on Wireless Sensor Networks and Applications, Atlanta, GA, USA, 28 September 2002; pp. 59–67. [CrossRef]
40. Brocolini, L.; Lavandier, C.; Quoy, M.; Ribeiro, C.F. Measurements of acoustic environments for urban soundscapes: Choice of homogeneous periods, optimization of durations, and selection of indicators. *J. Acoust. Soc. Am.* **2013**, *134*, 813–821. [CrossRef]
41. Graziuso, G.; Grimaldi, M.; Mancini, S.; Quartieri, J.; Guarnaccia, C. Crowdsourcing Data for the Elaboration of Noise Maps: A Methodological Proposal. *J. Phys. Conf. Ser.* **2020**, *1603*, 012030. [CrossRef]
42. Graziuso, G.; Mancini, S.; Francavilla, A.B.; Grimaldi, M.; Guarnaccia, C. Geo-Crowdsourced Sound Level Data in Support of the Community Facilities Planning. A Methodological Proposal. *Sustainability* **2021**, *13*, 5486. [CrossRef]
43. Siliézar, J.; Aumond, P.; Can, A.; Chapron, P.; Péroche, M. Case study on the audibility of siren-driven alert systems. *Noise Mapp.* **2023**, *10*, 20220165. [CrossRef]
44. Dorffer, C.; Puigt, M.; Delmaire, G.; Roussel, G. Informed Nonnegative Matrix Factorization Methods for Mobile Sensor Network Calibration. *IEEE Trans. Signal Inf. Process. Netw.* **2018**, *4*, 667–682. [CrossRef]
45. Breunig, M.M.; Kriegel, H.P.; Ng, R.T.; Sander, J. LOF: Identifying density-based local outliers. *ACM SIGMOD Rec.* **2000**, *29*, 93–104. [CrossRef]
46. Liu, F.T.; Ting, K.M.; Zhou, Z.H. Isolation Forest. In Proceedings of the 2008 Eighth IEEE International Conference on Data Mining, Pisa, Italy, 15–19 December 2008; pp. 413–422. [CrossRef]

sensors

MDPI

Article

Coupling Different Road Traffic Noise Models with a Multilinear Regressive Model: A Measurements-Independent Technique for Urban Road Traffic Noise Prediction

Domenico Rossi [1,*], Antonio Pascale [2,3], Aurora Mascolo [1] and Claudio Guarnaccia [1,*]

1 Department of Civil Engineering, Campus of Fisciano, University of Salerno, Via Giovanni Paolo II, 132, 84084 Fisciano, Italy; amascolo@unisa.it
2 Department of Mechanical Engineering/Centre for Mechanical Technology and Automation (TEMA), Campus Universitário de Santiago, University of Aveiro, 3810-193 Aveiro, Portugal; a.pascale@ua.pt
3 LASI—Intelligent Systems Associate Laboratory, 4800-058 Guimarães, Portugal
* Correspondence: drossi@unisa.it (D.R.); cguarnaccia@unisa.it (C.G.)

Abstract: Road traffic noise is a severe environmental hazard, to which a growing number of dwellers are exposed in urban areas. The possibility to accurately assess traffic noise levels in a given area is thus, nowadays, quite important and, on many occasions, compelled by law. Such a procedure can be performed by measurements or by applying predictive Road Traffic Noise Models (RTNMs). Although the first approach is generally preferred, on-field measurement cannot always be easily conducted. RTNMs, on the contrary, use input information (amount of passing vehicles, category, speed, among others), usually collected by sensors, to provide an estimation of noise levels in a specific area. Several RTNMs have been implemented by different national institutions, adapting them to the local traffic conditions. However, the employment of RTNMs proves challenging due to both the lack of input data and the inherent complexity of the models (often composed of a Noise Emission Model–NEM and a sound propagation model). Therefore, this work aims to propose a methodology that allows an easy application of RTNMs, despite the availability of measured data for calibration. Four different NEMs were coupled with a sound propagation model, allowing the computation of equivalent continuous sound pressure levels on a dataset (composed of traffic flows, speeds, and source–receiver distance) randomly generated. Then, a Multilinear Regressive technique was applied to obtain manageable formulas for the models' application. The goodness of the procedure was evaluated on a set of long-term traffic and noise data collected in a French site through several sensors, such as sound level meters, car counters, and speed detectors. Results show that the estimations provided by formulas coming from the Multilinear Regressions are quite close to field measurements (MAE between 1.60 and 2.64 dB(A)), confirming that the resulting models could be employed to forecast noise levels by integrating them into a network of traffic sensors.

Keywords: noise emission models; Road Traffic Noise Models; multilinear regressive approach

Citation: Rossi, D.; Pascale, A.; Mascolo, A.; Guarnaccia, C. Coupling Different Road Traffic Noise Models with a Multilinear Regressive Model: A Measurements-Independent Technique for Urban Road Traffic Noise Prediction. *Sensors* **2024**, *24*, 2275. https://doi.org/10.3390/s24072275

Academic Editor: Hector Eduardo Roman

Received: 19 February 2024
Revised: 28 March 2024
Accepted: 28 March 2024
Published: 3 April 2024

1. Introduction

When dealing with actual urban area hazards, environmental noise is surely one of the most pervasive and dangerous, with road traffic noise surely being the most prominent of all [1]. As a direct consequence of urbanization increasing, the number of vehicles per inhabitant has constantly grown during the last years, significantly impacting noise pollution in both urban and extra-urban contexts [2], and the big amount of constantly passing vehicles leaves no noise-free spaces. While studies on noise in urban areas were often neglected in the past, they have recently gained remarkable attention from national and international evaluation organizations working to implement strategies for its reduction. For instance, the European Union has outlined a goal to achieve a 30% reduction in the number of people exposed to harmful noise levels by 2030 [3].

It has been undoubtedly provided that exposition to day–evening–night noise levels exceeding 55 dB(A) leads to a series of health issues, listing from the mildest to the most severe: intelligibility during conversations, irascibility, sleep deprivation, mental issues, high blood pressure, and even sudden death [4–11]. Moreover, specific sensible areas are present in urban environments, such as schools. In these places, the control of noise is even more important, since the effects of noise exposure on children can be more severe than on adults [7]. Mitigation actions for the reduction in noise levels in urban areas is a mandatory task, as established by the directive 2002/49/EC [12]. Thus, the accurate assessment of noise levels in a specific area is a fundamental procedure, important for the implementation of targeted action plans. When trying to evaluate noise levels in a given area, two approaches are possible. The most direct–and precise–is to directly measure noise levels with dedicated instrumentation (sound level meters). Nevertheless, on-field measurements are not always the fastest or most economically viable way to proceed. In many conditions, in fact, the morphological arrangement of traffic roads does not permit the installation of fixed stations for noise level monitoring, or sometimes the measurement campaign could be expensive, long, and dangerous. To overcome these issues, the implementation of an effective sensor network in urban areas, that could provide acoustic and traffic data continuously and possibly with low-cost efforts, can be a valid alternative. Such a solution is largely explored in the literature: in [13,14], a system of sensors for the discrimination of traffic noise from anomalous noise events. In [15], a low-cost implementation of urban sensors for urban noise monitoring is described. In [16], a set of wireless acoustic sensors is described for automatic audio event classification. In [17], a general review of wireless sensor systems for smart cities is described.

When such situations are not implementable, the estimation of noise through a Road Traffic Noise Model (RTNM) is preferable. RTNMs are physical models composed of a Noise Emission Model (NEM) and a sound propagation model [18]. The former assesses the source sound power levels (L_W), while the latter transforms such information into sound pressure levels at receiver points. RTNMs take several parameters as inputs such as the number of vehicles transiting in a certain time period, their categories (light-duty vehicles—LDVs, medium vehicles, heavy-duty vehicles—HDVs), the vehicles' speeds and/or their accelerations, the distance between the road and the sensible receivers [18]. More complex RTNMs can also take into account other aspects like the presence of roundabouts or intersections (that affect the noise levels due to acceleration maneuvers), the presence or absence of acoustic barriers, and even some climatic aspects like air humidity, temperature, and wind direction [19,20]. It is very interesting to note that the implementation of RTNMs and the use of sensors for data collection are not mutually exclusive. On the contrary, they can be implemented together to obtain the best results. In this idea, sensors provide—in real-time or offline–large quantities of data that are used as input for the predictive models. Some examples of this integration are reported in [21–23], and even in [24,25], where large urban area monitoring is exploited.

Different models have been set up by different national institutions, resulting in heterogeneous results when applied in the same context. Among the most used, it is worth mentioning the CoRTN model [26], which is commonly adopted in the United Kingdom, the SonRoad model which has been implemented in Switzerland [27], the NMBP model used in France [28], the ASJ in Japan [29], and the RLS90 in Germany [30], the Harmonoise [31], and Quartieri et al. [32]. Besides all these models, the European Union (EU) has implemented the CNOSSOS one [33,34], which provides a common procedure for the assessment of transportation and industrial noise levels and the consequent development of noise maps, aiming at implementing a stand-alone model for noise assessment in all the European Countries that should receive and use it (by adapting it in some aspects if necessary). Despite the EU's efforts towards harmonization, the aforementioned national models are still used, especially in academic environments.

Generally speaking, any implemented noise model suffers from some intrinsic drawbacks, reflecting a variable amount of uncertainty in the final prediction. First of all, any model needs to be calibrated starting from a set of collected data. Such an unavoidable procedure implies that a given RTNM generally performs better in predicting road traffic noise in the same area where its calibration data have been collected. Consequently, when applied in a different scenario (different country, for example), its performance could be severely impaired. Moreover, an RTNM can be generally applied to road traffic conditions similar to the ones of calibration; nonetheless, if the traffic conditions are different from the ones used in the calibration (lower traffic volumes vehicles, lower or higher speeds), the model could perform poorly [18,35].

As for the outputs, RTNMs can furnish information in terms of equivalent continuous sound levels (L_{eq}), percentile levels, or day–evening–night noise levels (L_{den}). The latter is calculated from the day (L_{day}), evening ($L_{evining}$), and night (L_{night}) sound pressure levels, as their logarithmic sum which includes a penalty for evening and night hours (the same amount of noise emitted is considered to be more annoying at evening or night than during the day).

It must be stressed that, although RTNMs represent a valid alternative to long-term noise measurement campaigns, their utilization may be affected by certain factors. Indeed, the equations, constituting the framework of RTNMs, could be difficult to apply, necessitating the development of scripts for their implementation (and the relative programming know-how). In other cases, commercial software is available for the development of noise maps, using a specific RTNM as an algorithm. Therefore, there is a need for procedures that can facilitate the straightforward application of the already-existing RTNMs in the literature, permitting fast usage and reliable results.

For these reasons, the authors implemented a multiregressive technique for traffic noise assessment by calibrating it on computed data instead of real ones. As described in [36–38], such a regressive model has the advantage of not needing real data for its calibration. Moreover, the algorithms of generation of its calibration dataset make it potentially applicable to different traffic contexts. On this basis, the authors presented, in this contribution, a new application by coupling the aforementioned multiregressive model with four different existing NEMs (REMEL [39], SonRoad [27], CNOSSOS (and its amendments) [33,34], and NMPB [28]), in turn, coupled with a sound propagation model (namely a simplified formulation of the propagation provided in the CNOSSOS final report [33]). Whereas a comparison between models has already been provided in the literature, a concomitant study on the usage of a multilinear regressive model on different NEMs and sound propagation model, furnishing a modular approach in which a part can be easily substituted by another, is a novelty aspect to the best of our knowledge. The whole code for the generation of the model has been implemented in Python, using the most common packages for data analysis and visualization [36]. It has a low computational cost in terms of memory usage and time of generation (already described in detail in [37]).

Outputs of the here-presented models are provided as $L_{\text{eq,h}}$, which is one of the most commonly used noise indicators in the literature, but the proposed methodology has the potential to express the final output as a general function of time, computing the equivalent level at whichever timespan. The validation of the models is provided by applying the equations coming from the multiregression to a set of more than 3000 data elements coming from a Long-Term Monitoring Station (LTMS) by the Université Gustave Eiffel and Unité Mixte de Recherche en Acoustique Environnementale (UMRAE), Nantes [40]. This dataset contains up to seven years of both acoustic and meteorological road traffic data (from 2002 to 2007), collected from a highway located in the city of Saint-Berthevin (France). At the end of the validation process, the $L_{\text{eq,h}}$ values from the aforementioned dataset were aggregated on an hourly time basis and compared with the estimations provided by the multiregressive linear models application.

2. Materials and Methods

The generation of the model presented in this publication can be divided into four steps: (1) computing of the dataset for the calibration, (2) calibration of the multilinear regression model according to the four considered NEMs coupled with the sound propagation model, (3) validation of the models, (4) estimation of the models, as schematized in Figure 1. Below is the detailed description of each step.

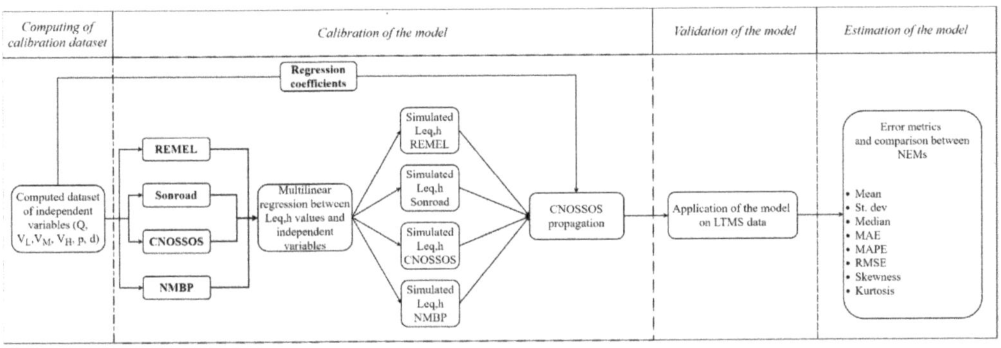

Figure 1. Flowchart of the generation of the model. The computed dataset is used to calibrate the models by computing equivalent hourly noise levels according to the four NEMs investigated, coupled with a simplified version of the sound propagation model provided in CNOSSOS. A multilinear regression technique is then applied, and with the obtained coefficients, the simulated hourly noise levels are computed and compared with the measured ones (from the LTMS dataset) to evaluate the goodness of the proposed approach.

2.1. Computing of the Dataset for the Calibration

The dataset used for the calibration of the model is entirely computed, and it is built with sequential steps. The procedure to compute the calibration dataset has been extensively described elsewhere [36,37], and here, a brief recapitulation of the process is furnished. The dataset has been built using Python 3.8, with Jupyter notebook as a compiler. The packages used to develop the code are standard packages for data analysis (*pandas, numpy*), for data plotting and visualization (*matplotlib-pyplot* and *seaborn*), and for statistical analysis (*sklearn*). The machine used is a DELL Pc (Intel® Xeon® CPU E3-1245 v5 @3.50 GHz) with 16 GB of RAM installed, 64 bit.

The first step of dataset generation is the building of a series of 200 rows having sequential values of flow, expressed as vehicle per hour (defined as variable Q), starting from 10, with incrementing of 10 vehicles at time. From now on, the following steps are intended to be repeated for each row of the dataset. The result of this first step is a column of Q spanning from 10 to 2000 vehicles per hour. The second step is the creation of a second column: speed of light vehicles (V_L) filling each row with a randomly extracted value from a minimum of 30 km/h to a maximum of 130 km/h, with a minimum range of 1 km/h. Each value has the same probability of being extracted. The third and fourth steps are the extraction of the speed of medium and heavy vehicles (V_M and V_H). The V_M value is randomly extracted from a minimum of 30 km/h to a maximum value equal to the V_L extracted in the previous step, with a minimum range of 1 km/h. Similarly, the V_H value is randomly extracted from a minimum of 30 km/h to a maximum value equal to the V_L extracted in the previous step, with a minimum range of 1 km. Both V_M and V_H values have the same probability of being extracted between the whole range. The fifth step is the random extraction of a p, which represents the percentage over the Q of the medium and heavy vehicles, which is composed of a p_{medium} and a p_{heavy} value. They are extracted as follows: p_{medium} value is randomly extracted from a minimum of 0.1% to a maximum of 20.0%%, with a minimum range of 0.1%. All values have the same probability of being extracted. Subsequently, p_{heavy} is randomly extracted from a minimum of 0.1% to a maximum value equal to 20.0% minus p_{medium}, with a minimum range of 0.1%. In such a way, the whole p value will never exceed 20.0%. The sixth step is the random extraction of d representing the source–receiver distance, which spans from a minimum of 10 m to a maximum of 100 m, with a range of 1 m. The last step is the repetition of steps from 2 to 6 for n times: in this specific application, n is equal to 20. In such a way, a dataset of 4000 rows is built.

2.2. Calibration and Validation of the Multilinear Regression Model According to the Four Considered NEMs

The independent variables, generated in the previous step, are used to calculate $L_{\mathrm{eq,h}}$ values through the four employed NEMs, coupled with a sound propagation model (retrieved from the CNOSSOS model). Particularly, the first step involved the calculation of L_W for each vehicle category, using the average speed as an input variable. It must be stressed that the REMEL and CNOSSOS models foresee a formulation for the L_W assessment of medium vehicles. For the other two NEMs, the formulation proposed to assess the L_W of HDVs was employed also for the medium vehicles. The equations adopted for the L_W calculation can be retrieved from the model-related reports. Details of such calculations can be found elsewhere [27,28,33,34,39], but for the sake of completeness, the authors report the formulations in Table 1.

Table 1. Calculations of L_W according to the four NEMs used.

REMEL [39]	$L_{\mathrm{WL}} = 31.130 \log_{10}(V_L) + 12.700$ $L_{\mathrm{WM}} = 18.765 \log_{10}(V_M) + 43.967$ $L_{\mathrm{WH}} = 12.831 \log_{10}(V_H) + 58.270$
SonRoad [27]	$L_{\mathrm{WL}} = 28.5 + 10\log_{10}\left[10^{\frac{7.3+35\times\log_{10} V_L}{10}} + 10^{60.5+\log_{10}\left(1+\left(\frac{V_L}{44}\right)^{3.5}\right)}\right]$ $L_{\mathrm{WM}} = 28.5 + 10\log_{10}\left[10^{\frac{16.3+35\times\log_{10} V_M}{10}} + 10^{74.7+\log_{10}\left(1+\left(\frac{V_M}{56}\right)^{3.5}\right)}\right]$ $L_{\mathrm{WH}} = 28.5 + 10\log_{10}\left[10^{\frac{16.3+35\log_{10} V_H}{10}} + 10^{60.5+\log_{10}\left(1+\left(\frac{V_L}{56}\right)^{3.5}\right)}\right]$
CNOSSOS [33,34]	$L_{\mathrm{W,i}} = 10\log_{10}\left(10^{\frac{L_{W,rolling,i}}{10}} + 10^{\frac{L_{W,propulsion,i}}{10}}\right)$ with $L_{W,\ rolling}$ and $L_{W,\ propulsion}$ given in [33,34] for each vehicle category and each frequency octave band (i) from 63 to 80,000 Hz
NMBP [28]	$L_W = 10\log_{10}\left(10^{\frac{L_{W,rolling}}{10}} + 10^{\frac{L_{W,propulsion}}{10}}\right) + 20\log_{10}\left(d_{ref}\right) + 10\log_{10}(2\pi)$ with $L_{W,\ rolling}$, $L_{W,\ propulsion}$, and d_{ref} given in [28] for each vehicle category

Regarding Table 1, it is worth mentioning some important differences between the NEMs used in this work. While the REMEL and SonRoad models compute L_W through a simple unique formula in which the vehicle speed is the main independent variable, the others are characterized by a more complex structure. Specifically, the CNOSSOS model assesses the propulsion and the rolling (due to the interaction between tires and road pavement) noise contributions separately in each octave band from 63 to 8000 Hz. The contributions of each octave band must be A-weighted and, therefore, logarithmically summed to obtain the overall engine and rolling sound pressure levels. These last ones can be, in turn, logarithmically summed to obtain the overall vehicle sound power level. It is worth reminding that the CNOSSOS model categorizes vehicles into five groups: light-duty vehicles, medium vehicles, heavy-duty vehicles, motorcycles, and the fifth category reserved for alternative vehicles. Since the number of hybrid and electric vehicles is growing in the EU fleet, it will be necessary, then, to update the model including this fifth category. In this regard, Licitra et al. [41] proposed coefficients for electric vehicles in the framework of the CNOSSOS model. Another approach explored in the literature is to use the CNOSSOS formulation for the LDVs by setting the propulsion coefficients to zero, as recently investigated in [42]. Finally, the NMPB model estimates the sound power level from maximum A-weighted sound pressure levels, considering both engine and rolling contributions, during single-vehicle pass-by tests at 7.5 m from the receiver. The rolling noise contribution is distinguished for three road pavement surfaces. In this study, the authors adopted the rolling noise formulation proposed for the third road pavement typology. This choice was driven by the fact that it exhibits characteristics closest to those of the site where the data for the validation process were gathered. It should be noted that correction terms related to acceleration operations, proximity to roundabouts, and intersections, among others, were neglected. The reasons behind this choice are twofold: (i) not all the employed NEMs present such correction terms; (ii) it is difficult to find a robust validation dataset in which acceleration data are available. Nonetheless, other variables as acceleration can be easily included in the proposed approach in future works. It is also noteworthy that CNOSSOS, NMPB, and SonRoad give the possibility to simulate sound power levels for different road surfaces; nevertheless, in this contribution, only the reference surface of each model has been evaluated.

The employed sound propagation method is retrieved and adapted from the CNOSSOS formulations [33]. It must be said that such a model considers the traffic flow as a linear source. At first, the hourly equivalent sound density power levels of the different vehicle categories flows (L_{WL}, L_{WM}, and L_{WH}) are calculated according to the average speeds (V_L, V_M, and V_H),

$$Lw'_{line,L} = L_{WL} + 10\log_{10}\left(\frac{Q_L}{1000 * V_L}\right) \quad (1)$$

$$Lw'_{line,M} = L_{WM} + 10\log_{10}\left(\frac{Q_M}{1000 * V_M}\right) \quad (2)$$

$$Lw'_{line,H} = L_{WH} + 10\log_{10}\left(\frac{Q_H}{1000 * V_H}\right) \quad (3)$$

and then the hourly equivalent sound pressure levels are retrieved by using the linear source propagation formulation:

$$L_{eq,\ L} = Lw'_{line,L} - 10\log_{10} d - 8 \quad (4)$$

$$L_{eq,\ M} = Lw'_{line,M} - 10\log_{10} d - 8 \quad (5)$$

$$L_{eq,\ H} = Lw'_{line,H} - 10\log_{10} d - 8 \quad (6)$$

where d is the sound–receiver distance. Therefore, the overall $L_{eq,h}$ value comes from the logarithmic sum of the partial contributions:

$$L_{eq,h} = 10\log_{10}\left[\left(10^{\frac{L_{eq,L}}{10}}\right) + \left(10^{\frac{L_{eq,M}}{10}}\right) + \left(10^{\frac{L_{eq,H}}{10}}\right)\right] \quad (7)$$

Once the $L_{eq,h}$ values are calculated according to the formulas of each NEM and to the propagation, they are used for the multilinear regression. Particularly, an Ordinary Least Squared regression is implemented between the six independent variables (Q, V_L, V_M, V_H, p, d) and the $L_{eq,h}$ by using the Python package *sklearn*. The regression formula for each NEM-sound propagation model has the same following structure:

$$L_{eq,h\ simulated} = C_1Q + C_2V_L + C_3V_M + C_4V_H + C_5p + C_6d + intercept \quad (8)$$

with C_1, C_2, etc., being the coefficients of the multilinear regression model. At this stage, the residuals of the regression are computed and analyzed (the reader can refer to Section 3.2).

The obtained regression formulas are validated by running the model on a field measurements dataset (LTMS) that will be described in the following, and comparing the estimated $L_{eq,h}$ with the measured noise levels. Please note that by applying the regression procedure, the authors faced the problem of the uncertainty of the measurement. LTMS data are, in fact, by definition, collected data, and they have an intrinsic uncertainty, which can propagate when a multilinear regression technique is applied to the data. In Section 3.3 a strategy has been implemented to consider such problems, which has also been addressed in the last part of the manuscript (Section 4.3), where the limitations of the study are presented. Moreover, the noise assessment is provided at variable distances, considering the space between the source and receiver as free, without any surrounding building that could be responsible for reflection phenomena.

2.3. Estimation of the Performances of the Model

The goodness of the regression models is established by calculating the error as the difference between the measured $L_{\text{eq,h}}$ and the computed ones, and by studying the errors distributions in terms of statistical metrics such as mean, median, standard deviation, skewness and kurtosis. In addition, the standard metric errors are calculated (Mean Absolute Error–MAE, Mean Absolute Percentage Error–MAPE, and Root-Mean-Square Error–RMSE). All the error metrics and the statistical properties have been computed by using the Python packages *numpy* and *scikitlearn*. Specifically, MAE is defined as follows:

$$MAE = \frac{1}{n}\sum_{i=1}^{n}|y_i - \hat{y}_i| \tag{9}$$

with n being the number of samples, y_i the ith measured value, and $\hat{y}_i$ the ith simulated value. MAPE has been computed by Equation (10):

$$MAPE = \frac{1}{n}\sum_{i=1}^{n}\frac{|y_i - \hat{y}_i|}{\max(\varepsilon, |y_i|)} \tag{10}$$

with ε an arbitrary small yet strictly positive number to avoid undefined results when y is zero. RMSE is computed as follows (Equation (11)):

$$RMSE = \sqrt{\frac{1}{n}\sum_{i=1}^{n}(y_i - \hat{y}_i)^2} \tag{11}$$

3. Results

3.1. Computation and Analysis of the Dataset for the Model Calibration

The first operation carried out for the generation of the multiregressive model is the computation of the original random dataset. As described in the previous section, this database is computed by joining, in rows, randomly picked values of six independent variables (Q, V_L, V_M, V_H, p, and d). This procedure has the scope of generating a robust and random database to cover a multitude of possible traffic situations. This represents a fundamental step in the model calibration, aiming to avoid potential bias due to lack of information. To augment the possibilities of obtaining a totally random database, a high number of rows is required. Based on observations described in [37], for this application, the authors chose the n factor equal to 20, obtaining a final dataset of 4000 entries. Before using it for the calibration of the model, the authors verified that the variables were independently distributed, performing a correlation analysis. The *corr* function of the *pandas* package, on details, correlates each column with all the others by using the Pearson correlation method, obtaining a final correlation value spanning from -1 (maximum inverse correlation) to 1 (maximum correlation), with 0 equal to no correlation; the method of correlation chosen is the standard correlation coefficient. In Figure 2, the correlation matrix is shown, reporting the results of the above-described procedure.

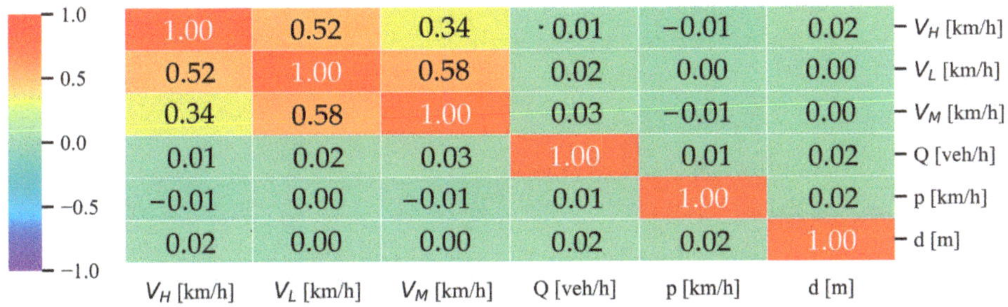

Figure 2. Correlation matrix of the randomly computed dataset for the subsequent model calibration.

The correlation matrix shows an obvious maximum correlation of the columns with themselves (central diagonal) and no correlation (green rectangles) when each variable is compared with the others. V_L, V_M, and V_H, have a moderate positive certain degree of correlation, due to the constraints used to generate the dataset. The authors, in fact, imposed that, after certain values, V_M and V_H cannot be, for every single row, higher than V_L, to avoid the unlikely situation where all the heavy vehicles, despite the limits fixed by law, run faster than common light vehicles (please refer to Section 2.1 for more details). Hence, apart from the relations between the velocities of the vehicle types, the computed database consists of uncorrelated independent variables. Another important aspect to underline is that the original database just computed corresponds to a *seed* value, which assures its reproducibility. The chosen *seed* value is the same for all the datasets used for the calibration of the model with the four different RTNMs.

3.2. Calibration of the Model and Residuals

As described in Section 2, the four NEMs are coupled with the propagation model. At this stage, the $L_{\mathrm{eq,h}}$ values are computed using input data from the randomly generated database, consisting of 4000 rows. Thus, the multiregressive model was applied using the information from the database, along with the newly computed $L_{\mathrm{eq,h}}$ values, resulting in the coefficients reported in Table 2:

Table 2. Multiregressive model coefficients.

	REMEL	SonRoad	CNOSSOS	NMBP
C_1	10.06	10.03	10.03	10.02
C_2	10.15	12.53	15.65	12.96
C_3	1.41	3.05	0.52	3.02
C_4	0.33	0.91	0.12	1.21
C_5	3.14	3.36	1.28	2.46
C_6	−12.81	−12.84	−12.85	−12.83
INT	31.87	20.51	22.43	19.65

Therefore, the residuals of this calibration process were evaluated. They are here defined as the difference between the $L_{\mathrm{eq,h}}$ values obtained by applying the models (in their basic form) to the database values and the $L_{\mathrm{eq,h}}$ computed by applying the formulas from the multiregressive technique. The statistical metrics of the residuals coming from the calibration process are shown in Table 3, while their distributions together with the autocorrelation functions are plotted in Figure 3.

Table 3. Statistical metrics of residual distributions (calibration process).

	REMEL	SonRoad	CNOSSOS	NMBP
Mean [dB(A)]	0.00	0.00	0.00	0.00
St dev [dB(A)]	0.89	0.64	0.59	0.45
Median [dB(A)]	−0.06	−0.07	−0.04	−0.05
Mode [dB(A)]	−1.43	−0.37	−0.67	−0.14
Min [dB(A)]	−2.68	−1.76	−1.27	−1.39
Max [dB(A)]	3.84	3.09	2.94	2.37
Shapiro	0.98	0.97	0.97	0.98
Skewness	0.56	0.66	0.82	0.56
Kurtosis	0.96	0.40	1.33	0.67

Residuals of calibration are well centered (Figure 3), having a mean value equal to 0.0), with low standard deviation (a minimum of 0.45 dB(A) and a maximum of 0.89 dB(A)). Median values also lie within a narrow interval (from −0.07 to −0.04 dB(A)). Shapiro–Wilk test results indicate that all the residuals are normally distributed (p-value $\geq$ 0.96). The residual distributions are characterized by a positive skewness index, due to a variable amount of data on the right side of the distribution. The kurtosis index is variable, higher for calibrations with REMEL and CNOSSOS but lower with the other RTNMs. Figure 3 reports also the autocorrelograms of the residuals for all the tested models. It is evident that no significant autocorrelation is present as a function of the lag, meaning that no information was left in the residuals and exhibiting a further endorsement of the goodness of the calibration process.

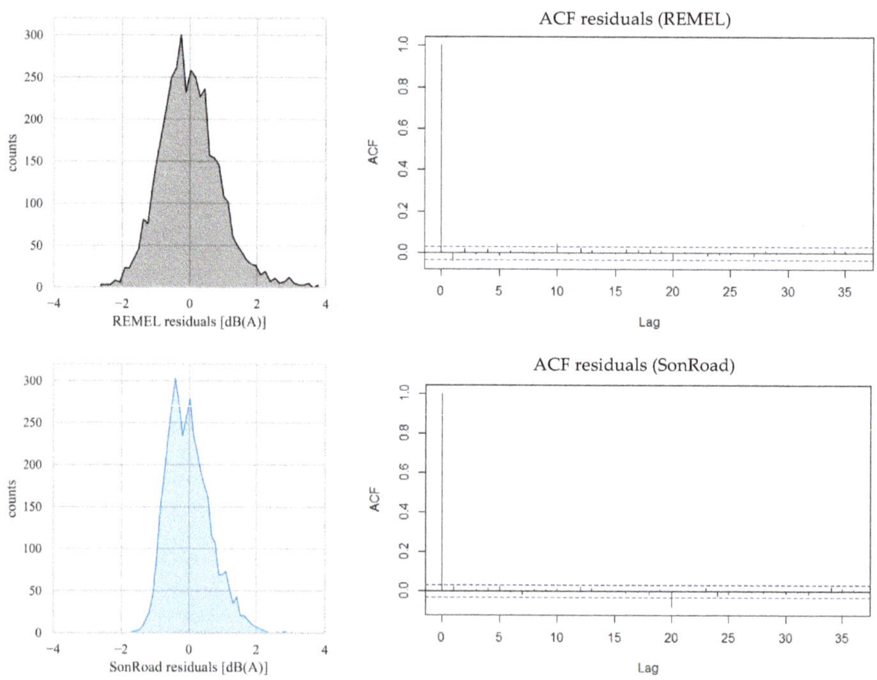

Figure 3. *Cont.*

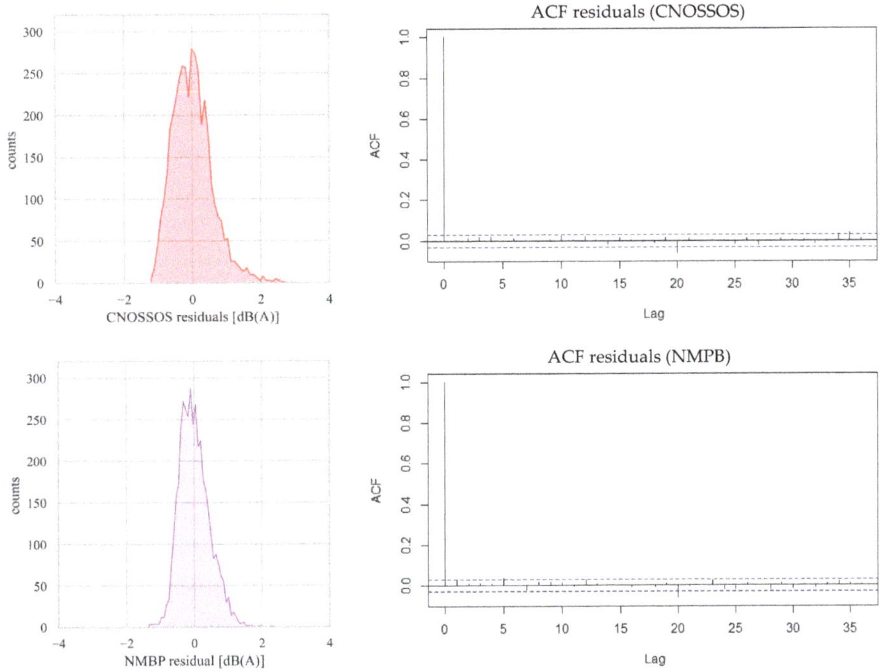

Figure 3. Distributions of the residuals (calibration process) for the different considered models and their autocorrelation plots (dotted lines indicating the level of statistical significance).

3.3. Validation of the Model

The calibration phase is followed by the validation of the models, which involves assessing error metrics using field-measured data.

The dataset used in this paper comes from a Long-Term Monitoring Station (LTMS) installed by the Université Gustave Eiffel (former IFSTTAR) and Unité Mixte de Recherche en Acoustique Environnementale (UMRAE), Nantes [40]. This project was based on the installation of both acoustic and meteorological masts that collected data continuously from 2002 to 2007, in the proximity of a highway in the city of Saint-Berthevin (France). A detailed description of the experimental site is reported in [40], and the data are available upon request. This dataset is originally created from more than 30,000 entries, reporting 15 min $L_{eq,h}$ values. For the purposes of this work, hourly $L_{eq,h}$ values are needed; therefore, the authors aggregated the data by logarithmically summing all the 15-min entries belonging to the same hour, and excluding the rows with missing values (no missing data imputation method was performed), resulting in a final dataset of 3404 rows complete of all the inputs needed to run the model. Please note that, as described in [36,37], the original LTMS dataset has to be adapted to the model, specifically for the medium and heavy vehicle flows and speeds.

Figure 4 reports the measured $L_{eq,h}$ values and the simulated ones for each model when the multiregressive linear approach is applied.

Red lines on the plot show the bisector (continue line) and an interval of ±2 dB(A) (dashed lines). It is visible how the clouds of points all have a similar shape, but their positions vary between the chosen RTNM. Specifically, 71%, 49%, 50%, and 42% of the points are in the region detected by the bisector shifted up and down by 2 dB(A) for REMEL, SonRoad, CNOSSOS, and NMPB, respectively. Such percentages become 84%, 71%, 72%, and 67%, respectively, when the bisector is shifted by 3 dB(A), corresponding to the doubling (halving) of the acoustic pressure.

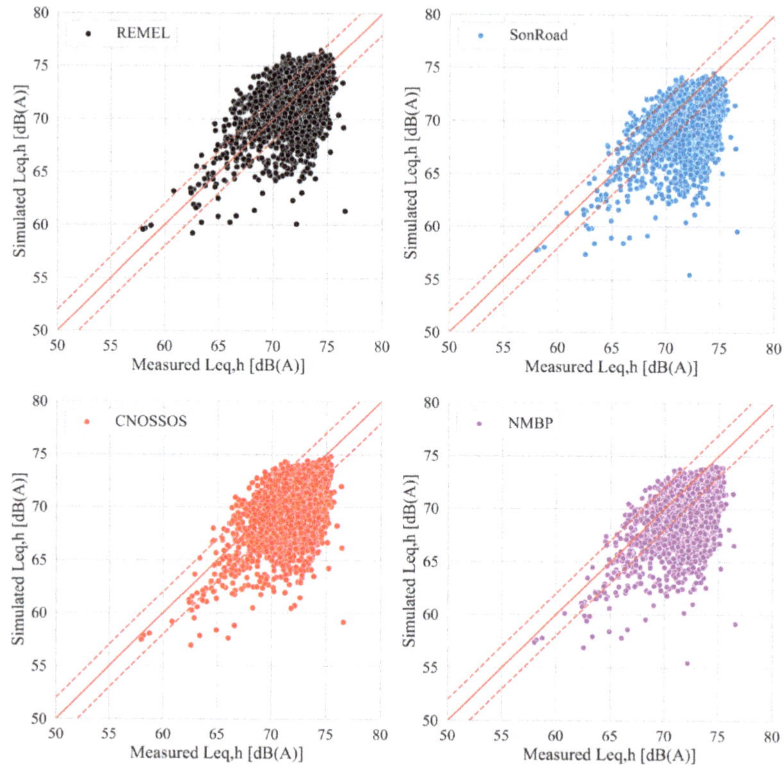

Figure 4. Scatterplots of measured vs. simulated $L_{eq,h}$ values of the LTMS dataset for all the four RTNMs after implementation of the multilinear regressive model. The dashed lines represent a ±2 dB(A) interval with respect to the bisector.

Compared to the REMEL model, the other models tend to underestimate the noise levels. As the sound propagation model is common to all the four employed NEMs, the explanation for such behavior could be attributed to the noise emission curves (expressing the relationship between the vehicle speed and the sound power level) of SonRoad, CNOSSOS, and NMPB, which are lower compared to the ones furnished in REMEL, as it is possible to ascertain from Figure 5.

The metrics related to the distributions of the errors (i.e., the difference between the measured and simulated L_{Aeq}) are reported in Table 4. REMEL is the model characterized by the lowest mean error, while CNOSSOS, NMPB, and SonRoad present similar performances. The distribution of the errors turns out to be almost symmetric (around the mean), as confirmed by the skewness values close to zero. Moreover, there is a high concentration of errors around the mean, as it is possible to note by the kurtosis values above 1.

Table 4. Metrics related to the distributions of the errors.

	REMEL	SonRoad	CNOSSOS	NMBP
Mean [dB(A)]	0.15	2.15	2.01	2.40
St dev [dB(A)]	2.15	2.19	2.24	2.18
Median [dB(A)]	−0.02	1.95	1.87	2.24

Table 4. *Cont.*

	REMEL	SonRoad	CNOSSOS	NMBP
Min [dB(A)]	−6.15	−4.12	−4.29	−3.85
Max [dB(A)]	15.20	17.00	17.42	17.42
Shapiro	0.97	0.97	0.98	0.97
Skewness	0.67	0.71	0.53	0.67
Kurtosis	1.79	1.96	1.24	1.85

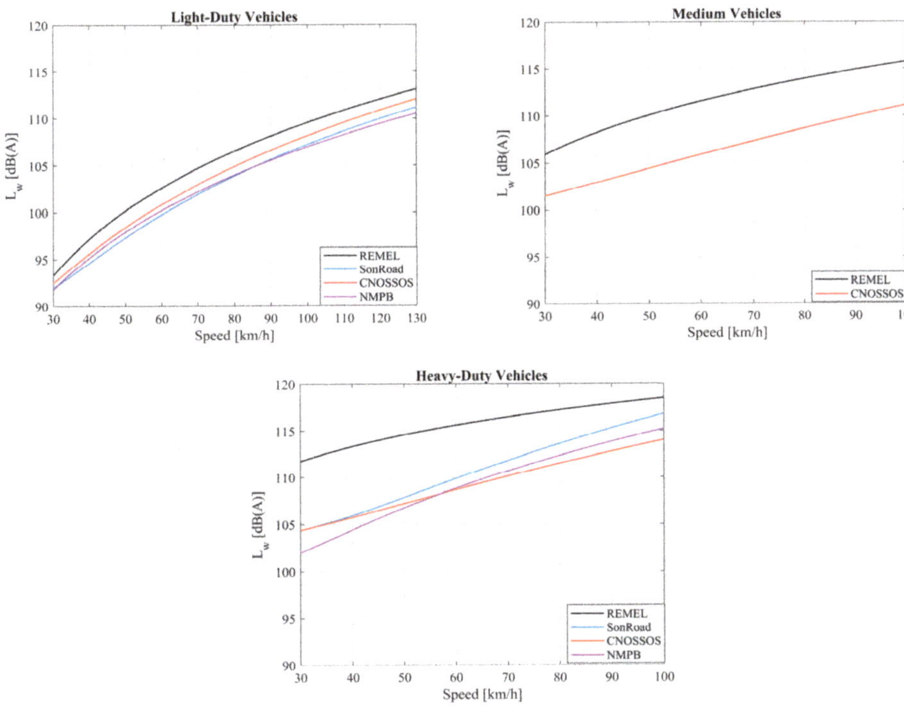

Figure 5. Noise emission curves for LDVs, medium vehicles, and HDVs.

3.4. Comparison with RTNMs Application without Regression

After obtaining simulations of $L_{eq,h}$ with multilinear regression techniques, a comparison with a straightforward application of RTNMs has been implemented and investigated. As previously stated, one of the issues of the application of the RTNMs is their difficulty of application and the requirement for programming scripts or commercial software for implementations. To overcome these problems, then, a single-time calibration of a multilinear regression technique is helpful in permitting future fast estimations of $L_{eq,h}$ values from road traffic data. However, the multiregressive technique must be reliable and present a validation efficiency comparable to that of the RTNMs themselves, so as to make the calibration effort worthwhile. Thus, to estimate the effective validity of the multiregressive approach compared to the sole applications of RTNMs, the authors performed a comparison between the two approaches. The comparison involved statistics of the distributions of simulated $L_{eq,h}$ values, error metrics, and computational time investment. This comparison has been carried out on the LTMS dataset.

3.4.1. Statistical Distributions of RTNMs Results

At first, the authors computed the simulated $L_{eq,h}$ distributions and the related statistical parameters when the RTNMs were employed with the formulation coming from the multilinear regressive technique and in their original form. Figure 6 overlaps the distributions of the simulated $L_{eq,h}$ for the four chosen RTNMs in the two aforesaid approaches, while Table 4 reports the exact values of statistical parameters of the related distributions.

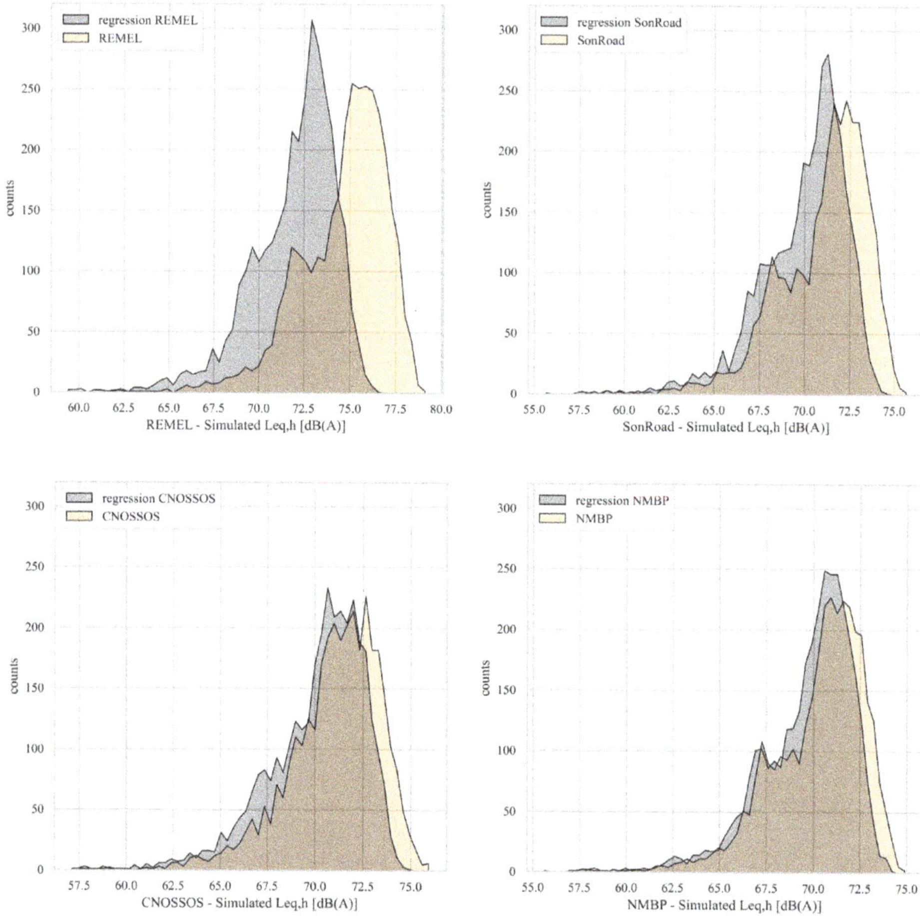

Figure 6. Distribution of simulated $L_{Aeq,h}$ from the application of RTNMs with and without the multiregressive approach.

As is evident from the graphs displayed in Figure 6, the simulated $L_{eq,h}$ values using the multilinear regressive approach tend to assume slightly lower values compared to the case where RTNMs are applied in their basic form. Consequently, the multiregressive approach may introduce underestimations of the noise levels due to the loss of information introduced by the application of the technique itself. This pattern is further highlighted by the mean values of the simulated L_{Aeq}, consistently lower when employing the multiregressive linear technique compared to simulations without this approach (the reader can refer to Table 5). In the case of REMEL, the difference between the mean values of simulated $L_{eq,h}$ is notably higher than 2 dB(A), highlighting a more pronounced effect. Regarding the shape of the distributions, similarities are observed in both cases, as confirmed by the standard deviation-, skewness-, and kurtosis-related values.

Table 5. Statistical properties of measured and simulated $L_{eq,h}$ distributions for the RTNMs with and without the multiregressive approach.

	Measured	REMEL		SonRoad		CNOSSOS		NMBP	
		Mult. Regr.	w/o Mult. Regr.	Mult. Regr.	w/o Mult. Regr.	Mult. Regr.	w/o Mult. Regr.	Mult. Regr.	w/o Mult. Regr.
Mean [dB(A)]	72.09	71.93	74.59	69.94	71.03	70.08	70.97	69.68	70.23
Std [dB(A)]	2.00	2.35	2.42	2.38	2.46	2.53	2.48	2.42	2.47
Median [dB(A)]	72.47	72.46	75.10	70.48	71.58	70.62	71.33	70.23	70.74
Shapiro	0.89	0.92	0.93	0.92	0.93	0.93	0.94	0.92	0.93
Skewness	−1.69	−1.25	−1.14	−1.25	−1.11	−1.15	−1.14	−1.23	−1.10
Kurtosis	4.87	2.49	2.05	2.50	1.78	1.83	2.22	2.31	1.77

3.4.2. Error Metrics

A comparison between the two approaches was performed also through important error metrics such as MAE, MAPE, MSE, and RMSE (Table 6), computed based on the errors committed for the simulation of the $L_{eq,h}$ values on the LTMS dataset.

Table 6. Error metrics of simulated $L_{eq,h}$ for the RTNMs with and without the multiregressive approach.

	REMEL		SonRoad		CNOSSOS		NMBP	
	Mult. Regr.	w/o Mult. Regr.	Mult. Regr.	w/o Mult. Regr.	Mult. Regr.	w/o Mult. Regr.	Mult. Regr.	w/o Mult. Regr.
MAE [dB(A)]	1.60	2.89	2.44	1.85	2.39	1.88	2.64	2.29
MAPE [%]	0.02	0.04	0.03	0.03	0.03	0.03	0.04	0.03
RMSE [dB(A)]	2.16	3.33	3.07	2.47	3.00	2.41	3.24	2.89

As it is possible to note, the MAE values associated with SonRoad, CNOSSOS, and NMPB are slightly lower when the models are applied in their basic form (less than 0.6 dB(A) than in the case in which the multiregression is applied). This is attributed, as mentioned in the previous subsection, to the slight underestimation that the multilinear regressive approach may introduce due to the loss of information during its application. The only exception is REMEL, which appears to experience fewer underestimation issues, at least for the selected case study. Similar trends are observed for RMSE values. In contrast, MAPE values remain consistent across the four considered models.

In general, the performance of the models when the multiregressive approach is applied remains in line with the cases where RTNMs are applied in their basic form, confirming the goodness of the presented methodology.

3.4.3. Computational Efforts Required–CPU Time and Wall Time

The advantage in the implementation of a multiregressive approach can also be found in the computational efforts required to perform the simulation of given $L_{eq,h}$ values coming from a set of traffic data. In this subsection, the authors present an evaluation of the time required to compute a fixed number of $L_{eq,h}$ with and without multilinear regression implementation for all the four RTNMs investigated. The computer on which the following tests have been performed is the same one described in Section 2, and the tests have been run without any other non-necessary running programs in the background. Two types of time have been evaluated: CPU time (also known as "Execution time"), which is defined as the time needed for the effective execution of the code lines, and wall time, which is the time elapsed from the beginning of the operation to the visualization of the result. These times were evaluated five times for each model (for the estimations of $L_{eq,h}$ on the same set of input data), and then the average was computed. It is very important to remember that the implementation of the multiregressive approach is divided into two steps: a calibration step and a validation step. The calibration step, which demands a higher computational effort (increasing with the dimensions of the calibration dataset), only needs to be implemented once. This is because the multiregression coefficients generated can be saved and subsequently used for the validation step. Thus, the authors only compared the validation time of the multiregressive approach with the time needed for the simulation of data by application of each single RTNM. To be complete, indications regarding the calibration time are provided anyway. The time for calibration of the multiregressive model is variable, as shown in Figure 7.

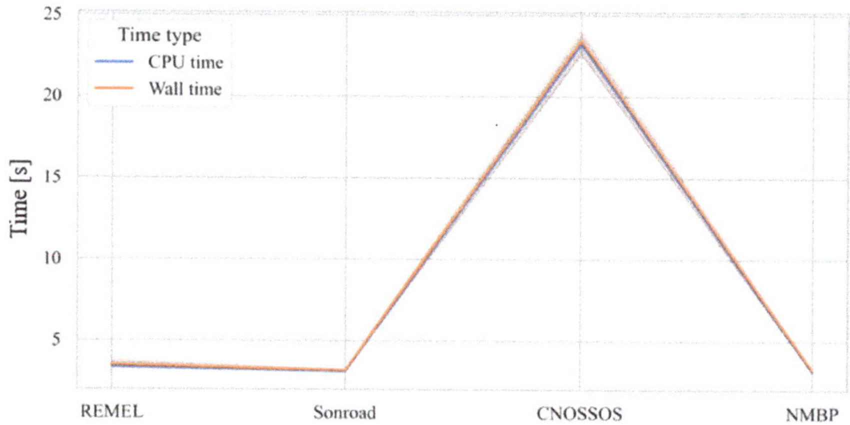

Figure 7. CPU and wall time for implementation of the calibration of the multiregressive approach with the four different RTNMs considered.

The implementation of the regressive model, in fact, requires a comparable time with three RTNMs (REMEL, SonRoad, and NMBP), with an average time of 3.27 ± 0.2 s. Calibration time rises with CNOSSOS, which requires 23.39 ± 0.75 s. This can be explained by the fact that the latter is characterized by more complex equations for evaluating the sound power level, resulting in an increased computational burden compared to the other models. The calibration process, then, requires a variable time in the order of seconds. The other part of the process involves using the obtained coefficients to simulate the $L_{eq,h}$ values, which are, of course, independent from the RTNMs used for the calibration.

The simulation of the $L_{Aeq,h}$ values starting from the coefficient obtained from the multiregression requires less CPU time than the RTNMs alone which, as remembered in Section 1, can be difficult to implement or require dedicated software. The difference is in the order of milliseconds, which may seem to be irrelevant, but it can become significant when the number of $L_{eq,h}$ values to be simulated increases. It also has to be noted that the variation in the time needed for the calculation of $L_{Aeq,h}$ is more stable when implementing regression than when applying only RTNMs. This may be due to the higher number of lines to be read from the compiler than the ones of the regression technique.

4. Discussion

4.1. Dataset for Calibration

The simulation of noise levels coming from road traffic data has dramatic importance when real on-field measurements cannot be implemented. Many models have been implemented over time to best transform the input data to noise levels close to real ones. In this contribution, an approach based on a multiregressive technique has been implemented, to retrieve a model that can result in a reliable output. The calibration of the model has been based on a computed dataset of six independent variables; this approach has a double meaning: (i) it can help in conditions where no real field data can be collected, and (ii) it helps in virtually simulating any type of traffic situation that could occur in a given scenario. Such dataset length can be varied according to the necessity, and it has been established in 4000 entries to assure reliability and repeatability. A smaller dataset, in fact, could help in a reduction in the final total computation time but results in more unstable results since the output coefficients would fluctuate over the repetitions. In the present manuscript, we also demonstrated that the six variables used for the simulation of $L_{eq,h}$ are independent between them, which is a mandatory condition for a correct multilinear regression.

4.2. Model Performances

Error metrics used to evaluate the model have shown interesting aspects of the multilinear regression technique when compared to the RTNMs in their basic formulation. At first, the final results are impaired by an error that is similar to or slightly higher than the one of the RTNMs applications. This loss of accuracy in the regression model can be explained by the multilinear regression technique itself, which adjusts all the coefficients to best minimize the error of each linear regression. Due to the number of variables, such fitting inevitably requires relative adjustments that may lead to the loss of information. Apparently, this procedure could also require a very high amount of time to be implemented, and thus ultimately not be convenient over the application of the RTNMs as they are. However, the simulation of the noise levels (in our application hourly levels) is faster once the regression coefficients have been established. This may be a high advantage when simulating a very high number of road traffic data, which is more and more common with the emerging recording techniques. Another aspect to take into consideration is the simplicity of the simulation of noise levels by using the multiregression coefficients compared to the application of RTNMs, which requires a lower computation time.

4.3. Connections with Sensors Networks

The road traffic noise model proposed can be calibrated on a computed dataset to cover multiple traffic conditions, as presented in this paper, or on any field measurements dataset. The latter option, of course, may be affected by the measurement location features. Anyway, in both cases, the model needs to be validated on a large dataset collected by sensors networks, as was done in the paper using the LTMS sensors data. Thus, the outputs of monitoring networks and digital infrastructures are essential for a proper development of the proposed methodology.

Moreover, the idea of building an IoT framework for assessing noise impact on a given area with this approach can surely be developed. A network of sensors continuously collecting road traffic data related to variables used in the regression could be interfaced with the proposed methodology, to output equivalent noise levels in near real-time, thanks to the very low computational cost. The outputs can then be pivoted to any software able to spatialize the data, such as any Geographic Information System (GIS) framework, to produce noise maps.

4.4. Final Evaluation of the Model and Its Limitations

In a comprehensive evaluation of all the aspects of this research, the implementation and usage of the multilinear regression technique is finally advantageous since it is reliable, simple, and based on a solid calibration dataset that does not require real measurement to be built. The calibration dataset used is a key point of the whole procedure since it gives the possibility to build a solid model for road traffic noise simulation without any on-field measurements. This is important in situations where measurements are difficult to be carried out but also when the evaluation of a future noise impact is at stake. Properly simulating the independent road traffic variables could, in fact, aid in forecasting the impact of traffic, facilitating the accurate evaluation of the infrastructure arrangement. The presented model also presents drawbacks, and they are mainly addressed in the loss of information during the generation of the coefficients of the regression. During this operation, in fact, an amount of information is inevitably sacrificed for the sake of simplicity. By observing the error metrics, moreover, it can be seen that the final accuracy of the model is strongly dependent on the NEMs used for the calibration. Up to now, this forces the user to conduct multiple calibrations to find the best NEMs for the fitting (just like the application of more than one RTNM is often required for the best results). A second intrinsic limitation of the study is that the application of a regressive model on collected data has to inevitably face the problem of uncertainty of measure, already mentioned in Section 2.2. Future steps of this work will deepen the statistical analysis and the interval of confidence approach to assure a more coherent comparison with the real data used for model validation. Another aspect to bear in mind is that the validation process, at this stage, has only been pursued on a single database. One of the first next steps of this research, then, will be testing the validity of the model on different traffic conditions, following the incorporation of additional variables such as different road surfaces into the noise emission models, as well as ground and obstacle reflections, atmospheric absorption, among others, into the sound propagation model. A last limitation aspect to take into consideration is that the employed NEMs have all been built in the framework of a combustion engine fleet of vehicles, but recently, the composition of fleet is changing due to a growing number of hybrid and electric cars. Anyway, the modular structure of the proposed approach allows to easily integrate new versions of noise emission models that will consider the different emission curves for electric vehicles as soon as they become available.

5. Conclusions

The multilinear regressive approach presented in this study yields robust simulations of $L_{eq,h}$ values. A computed dataset was employed to calibrate the models, while the validation process was performed by using robust and reliable traffic and noise data from a large database, available in the literature. A detailed comparison has been presented by using four different RTNMs for the calibration (resulting from the combination of four NEMs and a simplified sound propagation model). A validation on a field measurement dataset, built with the adoption of several sensors, has been performed. The results demonstrated that the proposed approach is suitable for the estimation of noise levels (MAE ranging between 1.60 and 2.64 dB(A)), particularly when compared with the application of the models in their basic form (MAE values between 1.85 and 2.89 dB(A)). While the multilinear regression approach may result in a loss of information, causing a slight underestimation of the noise levels on one side, on the other side, it leads to obtaining easy formulas to be applied after an initial calibration process. This also has repercussions on the computational burden associated with the applications of the models.

Finally, it must be stressed that the proposed methodology could serve as support for a network of traffic sensors (collecting data in terms of traffic volumes and speed), allowing a fast and online estimation of noise levels, without the aid of sound level meters.

Author Contributions: Conceptualization, D.R. and C.G.; methodology, D.R. and A.P.; software, D.R.; validation, D.R. and A.M.; formal analysis, D.R., A.P., A.M. and C.G.; investigation, D.R. and C.G.; resources, C.G.; data curation, D.R. and A.M.; writing—original draft preparation, D.R.; writing—review and editing, D.R., A.P., A.M. and C.G.; visualization, D.R. and C.G.; supervision, C.G.; funding acquisition, C.G. All authors have read and agreed to the published version of the manuscript.

Funding: This study was carried out within the MOST–Sustainable Mobility National Research Center and received funding from the European Union Next-GenerationEU (PIANO NAZIONALE DI RIPRESA E RESILIENZA (PNRR)–MISSIONE 4 COMPONENTE 2, INVESTIMENTO 1.4–D.D. 1033 17/06/2022, CN00000023). This manuscript reflects only the authors' views and opinions; neither the European Union nor the European Commission can be considered responsible for them. A. Pascale acknowledges the support of the following projects: UIDB/00481/2020 (https://doi.org/10.54499/UIDB/00481/2020), UIDP/00481/2020–FCT (https://doi.org/10.54499/UIDP/00481/2020), and CENTRO-01-0145-FEDER-022083. A. Pascale acknowledges the support of FCT for the Scholarship 2020.05106.BD.

Institutional Review Board Statement: Not applicable.

Informed Consent Statement: Not applicable.

Data Availability Statement: The data used for the calibration are available upon request from the corresponding authors. The data used for the validation are available upon request at http://ltms2002-2007.ifsttar.fr/ (accessed on 17 September 2023) [31].

Conflicts of Interest: The authors declare no conflicts of interest.

References

1. European Environment Agency. *The European Environment—State and Outlook 2020*; European Environment Agency: Copenhagen, Denmark, 2019; ISBN 9789292131142.
2. European Commission. *Statistical Pocketbook 2023—EU Transport in Figures*; European Commission: Brussels, Belgium, 2023. [CrossRef]
3. European Environmental Bureau. Noise Pollution. Available online: https://eeb.org/work-areas/air-and-noise-pollution/noise-pollution/ (accessed on 20 October 2023).
4. Muzet, A. Environmental Noise, Sleep and Health. *Sleep Med. Rev.* **2007**, *11*, 135–142. [CrossRef]
5. Singh, D.; Kumari, N.; Sharma, P. A Review of Adverse Effects of Road Traffic Noise on Human Health. *Fluct. Noise Lett.* **2018**, *17*, 1830001. [CrossRef]
6. Erickson, L.C.; Newman, R.S. Influences of Background Noise on Infants and Children. *Curr. Dir. Psychol. Sci.* **2017**, *26*, 451–457. [CrossRef]
7. Minichilli, F.; Gorini, F.; Ascari, E.; Bianchi, F.; Coi, A.; Fredianelli, L.; Licitra, G.; Manzoli, F.; Mezzasalma, L.; Cori, L. Annoyance Judgment and Measurements of Environmental Noise: A Focus on Italian Secondary Schools. *Int. J. Environ. Res. Public Health* **2018**, *15*, 208. [CrossRef]
8. Petri, D.; Licitra, G.; Vigotti, M.A.; Fredianelli, L. Effects of Exposure to Road, Railway, Airport and Recreational Noise on Blood Pressure and Hypertension. *Int. J. Environ. Res. Public Health* **2021**, *18*, 9145. [CrossRef]
9. Banerjee, D. Road Traffic Noise Exposure and Annoyance: A Cross-Sectional Study among Adult Indian Population. *Noise Health* **2013**, *15*, 342–346. [CrossRef] [PubMed]
10. Halonen, J.I.; Dehbi, H.M.; Hansell, A.L.; Gulliver, J.; Fecht, D.; Blangiardo, M.; Kelly, F.J.; Chaturvedi, N.; Kivimäki, M.; Tonne, C. Associations of Night-Time Road Traffic Noise with Carotid Intima-Media Thickness and Blood Pressure: The Whitehall II and SABRE Study Cohorts. *Environ. Int.* **2017**, *98*, 54–61. [CrossRef] [PubMed]
11. Sørensen, M.; Hvidtfeldt, U.A.; Poulsen, A.H.; Thygesen, L.C.; Frohn, L.M.; Khan, J.; Raaschou-Nielsen, O. Long-Term Exposure to Transportation Noise and Risk of Type 2 Diabetes: A Cohort Study. *Environ. Res.* **2023**, *217*, 114795. [CrossRef] [PubMed]
12. European Commission. *Directive 2002/49/EC Relating to the Assessment and Management of Environmental Noise*; European Commission: Brussels, Belgium, 2002.
13. Socoró, J.C.; Alías, F.; Alsina-Pagès, R.M. An Anomalous Noise Events Detector for Dynamic Road Traffic Noise Mapping in Real-Life Urban and Suburban Environments. *Sensors* **2017**, *17*, 2323. [CrossRef]
14. Sevillano, X.; Socoró, J.C.; Alías, F.; Bellucci, P.; Peruzzi, L.; Radaelli, S.; Coppi, P.; Nencini, L.; Cerniglia, A.; Bisceglie, A.; et al. DYNAMAP—Development of Low Cost Sensors Networks for Real Time Noise Mapping. *Noise Mapp.* **2016**, *3*, 172–189. [CrossRef]
15. Picaut, J.; Can, A.; Fortin, N.; Ardouin, J.; Lagrange, M. Low-Cost Sensors for Urban Noise Monitoring Networks—A Literature Review. *Sensors* **2020**, *20*, 2256. [CrossRef]
16. Navarro, J.; Vidaña-Vila, E.; Alsina-Pagès, R.M.; Hervás, M. Real-Time Distributed Architecture for Remote Acoustic Elderly Monitoring in Residential-Scale Ambient Assisted Living Scenarios. *Sensors* **2018**, *18*, 2492. [CrossRef]
17. Alías, F.; Alsina-Pagès, R.M. Review of Wireless Acoustic Sensor Networks for Environmental Noise Monitoring in Smart Cities. *J. Sens.* **2019**, *2019*, 7634860. [CrossRef]
18. Pascale, A.; Fernandes, P.; Guarnaccia, C.; Coelho, M.C. A Study on Vehicle Noise Emission Modelling: Correlation with Air Pollutant Emissions, Impact of Kinematic Variables and Critical Hotspots. *Sci. Total Environ.* **2021**, *787*, 147647. [CrossRef]
19. Pascale, A.; Macedo, E.; Guarnaccia, C.; Coelho, M.C. Smart Mobility Procedure for Road Traffic Noise Dynamic Estimation by Video Analysis. *Appl. Acoust.* **2023**, *208*, 109381. [CrossRef]

20. Pascale, A.; Guarnaccia, C.; Macedo, E.; Fernandes, P.; Miranda, A.I.; Sargento, S.; Coelho, M.C. Road Traffic Noise Monitoring in a Smart City: Sensor and Model-Based Approach. *Transp. Res. Part D Transp. Environ.* **2023**, *125*, 103979. [CrossRef]
21. Baclet, S.; Khoshkhah, K.; Pourmoradnasseri, M.; Rumpler, R.; Hadachi, A. Near-Real-Time Dynamic Noise Mapping and Exposure Assessment Using Calibrated Microscopic Traffic Simulations. *Transp. Res. Part D Transp. Environ.* **2023**, *124*, 103922. [CrossRef]
22. Benocci, R.; Molteni, A.; Cambiaghi, M.; Angelini, F.; Roman, H.E.; Zambon, G. Reliability of Dynamap Traffic Noise Prediction. *Appl. Acoust.* **2019**, *156*, 142–150. [CrossRef]
23. Smiraglia, M.; Benocci, R.; Zambon, G.; Roman, H.E. Predicting Hourly Trafic Noise from Trafic Flow Rate Model: Underlying Concepts for the DYNAMAP Project. *Noise Mapp.* **2016**, *3*, 130–139. [CrossRef]
24. Zambon, G.; Roman, H.E.; Smiraglia, M.; Benocci, R. Monitoring and Prediction of Traffic Noise in Large Urban Areas. *Appl. Sci.* **2018**, *8*, 251. [CrossRef]
25. Benocci, R.; Confalonieri, C.; Roman, H.E.; Angelini, F.; Zambon, G. Accuracy of the Dynamic Acoustic Map in a Large City Generated by Fixed Monitoring Units. *Sensors* **2020**, *20*, 412. [CrossRef]
26. Hood, R.A. Accuracy of Calculation of Road Traffic Noise. *Appl. Acoust.* **1987**, *21*, 139–146. [CrossRef]
27. Heutschi, K. SonRoad: New Swiss Road Traffic Model. *Acta Acust. United Acust.* **2004**, *90*, 548–554.
28. Dutilleux, G.; Defrance, J.; Ecotière, D.; Gauvreau, B.; Bérengier, M.; Besnard, F.; Le Duc, E. NMPB-Routes-2008: The Revision of the French Method for Road Traffic Noise Prediction. *Acta Acust. United Acust.* **2010**, *96*, 452–462. [CrossRef]
29. Sakamoto, S. Road Traffic Noise Prediction Model "ASJ RTN-Model 2018": Report of the Research Committee on Road Traffic Noise. *Acoust. Sci. Technol.* **2020**, *41*, 529–589. [CrossRef]
30. *RLS Richtlinien für den Lärmschutzan Strassen*; BM für Verkehr: Bonn, Germany, 1990.
31. Watts, G. *Harmonoise Prediction Model for Road Traffic Noise*; Published Project Report PPR034; TRL: Wokingham, UK, 2005.
32. Quartieri, J.; Iannone, G.; Guarnaccia, C. On the Improvement of Statistical Traffic Noise Prediction Tools. In Proceedings of the 11th WSEAS International Conference on "Acoustics & Music: Theory & Applications" (AMTA'10), Iasi, Romania, 13–15 June 2010; pp. 201–207.
33. Kephalopoulos, S.; Paviotti, M.; Anfosso-Lédée, F. *Common Noise Assessment Methods in Europe (CNOSSOS-EU)*; Publications Office of the European Union: Luxembourg, 2012; ISBN 9789279252815.
34. Kok, A.; van Beek, A. *Amendments for CNOSSOS-EU*; RIVM: Bilthoven, The Netherlands, 2019. [CrossRef]
35. Guarnaccia, C.; Bandeira, J.; Coelho, M.C.; Fernandes, P.; Teixeira, J.; Ioannidis, G.; Quartieri, J. Statistical and Semi-Dynamical Road Traffic Noise Models Comparison with Field Measurements. *AIP Conf. Proc.* **2018**, *1982*, 020039. [CrossRef]
36. Rossi, D.; Mascolo, A.; Guarnaccia, C. Calibration and Validation of a Measurements-Independent Model for Road Traffic Noise Assessment. *Appl. Sci.* **2023**, *13*, 6168. [CrossRef]
37. Rossi, D.; Mascolo, A.; Guarnaccia, C. Optimization of Dataset Generation for a Multilinear Regressive Road Traffic Noise Model. *Wseas Trans. Environ. Dev.* **2023**, *19*, 1145–1159. [CrossRef]
38. Rossi, D.; Mascolo, A.; Guarnaccia, C. Road Traffic Noise Predictions by Means of L10 Modelling with a Multilinear Regression Calibrated on Simulated Data. *Int. J. Mech.* **2023**, *17*, 51–56. [CrossRef]
39. Wayson, R.L.; Ogle, T.W.A.; Lindeman, W. Development of Reference Energy Mean Emission Levels for Highway Traffic Noise in Florida. In *Transportation Research Record*; Transportation Research Board: Washington, DC, USA, 1993; pp. 82–91.
40. Gauvreau, B. Long-Term Experimental Database for Environmental Acoustics. *Appl. Acoust.* **2013**, *74*, 958–967. [CrossRef]
41. Licitra, G.; Marco, B.; Ricardo, M.; Francesco, B.; Fredianelli, L. CNOSSOS-EU Coefficients for Electric Vehicle Noise Emission. *Appl. Acoust.* **2023**, *211*, 109511. [CrossRef]
42. Nygren, J.; Boij, S.; Rumpler, R.; O'Reilly, C.J. Vehicle-Specific Noise Exposure Cost: Noise Impact Allocation Methodology for Microscopic Traffic Simulations. *Transp. Res. Part D Transp. Environ.* **2023**, *118*, 103712. [CrossRef]

sensors

Article

Numerical Analysis of the Mitigation Performance of a Buried PT-WIB on Environmental Vibration

Lei Gao [1,2], Chenzhi Cai [1], Chao Li [1,*] and Cheuk Ming Mak [2,*]

1 School of Civil Engineering, Central South University, Changsha 410000, China; l.gao@connect.polyu.hk (L.G.); chenzhi.cai@csu.edu.cn (C.C.)
2 Department of Building Environment and Energy Engineering, The Hong Kong Polytechnic University, Hong Kong
* Correspondence: lichaocsu@csu.edu.cn (C.L.); cheuk-ming.mak@polyu.edu.hk (C.M.M.)

Abstract: Environmental vibration pollution has serious negative impacts on human health. Among the various contributors to environmental vibration pollution in urban areas, rail transit vibration stands out as a significant source. Consequently, addressing this issue and finding effective measures to attenuate rail transit vibration has become a significant area of concern. An infilled trench can be arranged periodically along the propagation paths of the waves in the soil to attenuate vibration waves in a specific frequency range. However, the periodic infilled trench seems to be unsatisfactory for providing wide band gaps at low and medium frequencies. To improve the isolation performance of wave barriers at low to medium frequencies, a buried PT-WIB consisting of a periodic infilled trench and a wave impedance block barrier has been proposed in this paper. A three-dimensional finite element model has been developed to evaluate the isolation performance of three wave barriers. The influence of the PT-WIB's parameters on isolation performance has been analyzed. The results indicate that the combined properties of the periodic structure and the wave impedance block barrier can effectively achieve a wide attenuation zone at low and medium frequencies, enhancing the isolation performance for mitigating environmental vibration pollution.

Keywords: environmental vibration; attenuation zone; periodic structure; wave impedance block; finite element

Citation: Gao, L.; Cai, C.; Li, C.; Mak, C.M. Numerical Analysis of the Mitigation Performance of a Buried PT-WIB on Environmental Vibration. *Sensors* **2023**, *23*, 7666. https://doi.org/10.3390/s23187666

Academic Editor: Hector Eduardo Roman

Received: 15 August 2023
Revised: 4 September 2023
Accepted: 4 September 2023
Published: 5 September 2023

1. Introduction

Environmental vibration pollution arising from traffic, machines, and construction blasting has been the subject of increasing concern in recent years, especially train-induced environmental vibration pollution. With the development of rail transportation systems in urban settings, the accompanied vibration from rail systems brings about a negative impact on the surrounding area. The environmental vibrations may become an annoyance issue to surrounding residents in both physiological and psychological aspects [1–5]. Therefore, a lot of attention has been paid to environmental vibration isolation measures. Passive isolation solutions such as open trenches, infilled trenches, and pile barriers [6–10] have been investigated numerically and experimentally. Since these wave barriers may intercept, scatter, or diffract incoming waves, they are generally installed along the propagation paths of the waves in the soil for environmental vibration isolation. However, a broadband attenuation band for environmental vibration has not been achieved through these previous investigations. A relatively broad isolation frequency range is generally required in practical engineering. Thus, it is significant to obtain a specific broadband isolation frequency range.

In recent years, the concept of acoustic metamaterials (AMs) has attracted increasing research attention worldwide due to their peculiar wave dispersion characteristics. The AMs can be regarded as a kind of functional material consisting of identical unit cells with periodic distributions in another medium [11]. The energy of a wave in a particular frequency range can be attenuated through its propagation in the AMs, which is considered to be a band gap or attenuation zone (AZ). The dispersion properties of AMs in physics open up a new horizon for environmental vibration isolation. Subsequently, a large number of investigations, including theoretical analysis, numerical simulations, and experiments, have been conducted to reveal the mechanism of periodic structures in environmental vibration reduction [12–21]. Huang et al. [15] proposed a layered periodic structure and investigated its frequency zone of vibration reduction using FEM. The simulated AZs were consistent with the results from Bloch theory. Pu and Shi [17] arranged piles in a periodic way and investigated their isolation effects for Rayleigh waves from the perspective of the periodic theory. Huang et al. [19] analyzed the vibration isolation performance of periodic barriers under different excitations through some field experiments. The results indicated that the frequency range of band gaps in the periodic structures can be identified.

Most of the aforementioned studies focused on vibration isolation at medium frequencies. However, the concerning train-induced environmental vibration concentrates on both low and medium frequencies, with a relatively broad range. The dominant frequency of train-induced environmental vibration usually exists in the range of 30–60 Hz. However, the accompanied low-frequency vibration below 10 Hz can travel over a long distance with less attenuation, which may cause more serious impacts on the surrounding area. The attenuation mechanisms of periodic structures are classified as Bragg scattering mechanisms and local resonance mechanisms. For Bragg scattering, the formation of a band gap is based on a complex result of elastic wave reflection and refraction at the interfaces of different materials. It is suitable for medium- and high-frequency vibration reduction, but it has difficulty formulating the low-frequency band gaps [13,14]. The wavelength attenuated by the locally resonant band gap can reach two orders higher than the lattice constant [22], which allows relative material structures to isolate lower-frequency vibration. The local resonance mechanism of periodic structures overcomes the limitations of Bragg scattering theory, and numerous researchers have utilized this mechanism to obtain low-frequency vibration isolation [23–26]. However, the local resonance mechanism is contradictory to the broadband frequency gaps [26].

This paper aims to identify a broadband attenuation zone for environmental vibration at both low and medium frequencies. The wave-impeding block (WIB) is an efficient and cost-effective method for low-frequency vibration reduction. It is usually embedded at a certain depth in the ground under the vibration source and has been widely studied for environmental vibration control [27–31]. The principle of the WIB is to introduce artificially stiffened horizontal layers for the sake of changing the wave propagation mechanism in the ground. Thus, the wave propagation relies on the relationship between the source excitation frequency and the cutoff frequency of the overlaying soil above the WIB. Therefore, a buried PT-WIB (periodic infilled trench–wave impedance block barrier) consisting of a periodic infilled trench and wave impedance block barrier has been proposed in this paper to realize a wide attenuation zone at both low and medium frequencies. A three-dimensional finite element model was developed to analyze the vibration isolation performance of the WIB, periodic infilled trenches, and PT-WIB in both the frequency and time domains. The influence of the different parameters of PT-WIB on the vibration isolation performance is revealed. It is hoped that the present study can be applied to reduce environmental vibration in practical engineering.

2. Model and Methods

The purpose of this paper is to investigate the environmental vibration isolation performance of the proposed PT-WIB. As illustrated in Figure 1, the WIB is installed below the vibration source (under the railway), and the periodic infilled trenches are arranged between the vibration source and the protected objects. A slice of the model includes both the load actions and wave barriers. Thus, the overall effect of the infinite propagation domain can be represented by a slice along the longitudinal direction of the load [16].

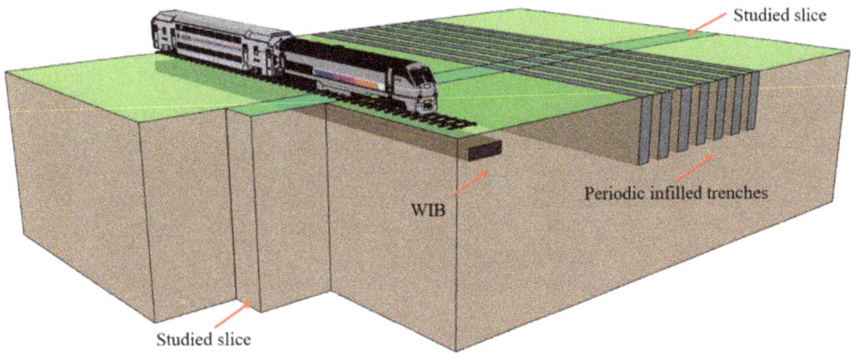

Figure 1. A schematic diagram of the PT-WIB.

The finite element method using the software COMSOL Multiphysics has been adopted to analyze the isolation performance of wave barriers (WIB, periodic infilled trenches, and PT-WIB) in an elastic half-space. Figure 2 shows the schematic diagram of a 3D model used for the vibration isolation analysis. A pair of periodic boundary conditions (PBCs) were applied to the model in the y direction to reduce calculation time. The effectiveness of this simplified model can also be found in related studies [16,17,32,33]. The dimensions of the considered model, $l \times m \times n$, are 40 m, 20 m, and 1 m, respectively. The widths of the WIB and infilled trench are w_1 and w_2; the depths of the WIB and infilled trench are d_1 and d_2, respectively; l_1 is the distance from the vertical harmonic line source to the first row of infilled trenches; l_2 is the distance from the vertical harmonic line source to the observation area of the vibration response; l_3 is the length of the observation area for the vibration response; and h is the embedded depth of the WIB. The spacing of the periodic infilled trenches is b. The materials used for the periodic infilled trenches and WIB are geofoam and concrete, respectively. Low-density geofoam exhibits significant energy dissipation capacity and excels as a vibration isolation infill material, offering numerous advantages over alternative infill materials. This superiority stems from its lightweight nature, economic viability, and exceptional isolation efficiency. Concrete, renowned for its high-strength properties, is frequently employed as a widely used material for WIB. The properties of the soil and wave barrier materials were considered based on some previous studies [18,34,35], as shown in Table 1. It was assumed that the materials of the soil and wave barriers are elastic, isotropic, and homogeneous. The damping factor of the soil was 0.05. The perfectly matched layer (PML), as an effective absorption boundary condition, was added on the left, right, and bottom of the considered model to simulate semi-infinite media in the frequency domain analysis [36]. The velocity of the Rayleigh waves V_{R} was calculated in view of the propagation of surface waves in the soil, as follows:

$$V_R = \frac{\sqrt{E/p}}{\sqrt{2(1+v)}} \cdot \frac{0.87 + 1.12v}{1+v}, \quad (1)$$

where E, p, and v represent the Young modulus, density, and Poisson ratio of the soil, respectively. The Rayleigh wavelength λ_R was calculated to be 2 m at a frequency of 45 Hz.

Table 1. The material parameters of the soil and the barriers.

Material	Young Modulus E (MPa)	Poisson Ration v	Density p (kg/m^3)
Soil	46	0.25	1800
Geofoam	37	0.32	60
Concrete	25,500	0.20	2500

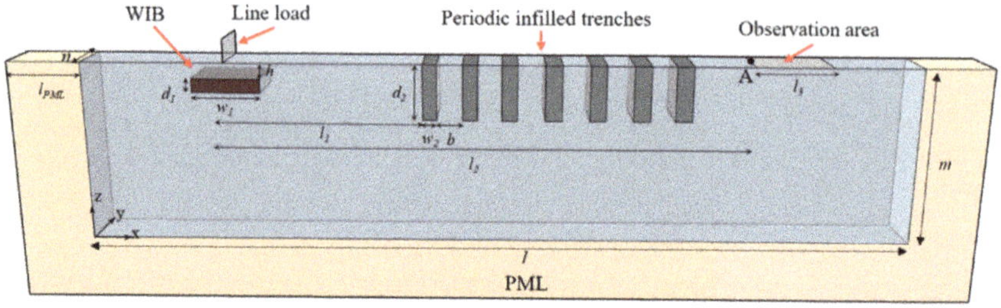

Figure 2. A schematic diagram of the 3D model used for the numerical analysis.

The amplitude reduction ratio A_R, which is represented as the ratio between surface displacement amplitude with and without a wave barrier, was used to assess the vibration isolation performance of wave barriers, and it can be expressed as [6]:

$$A_R = \frac{u_{y1}}{u_{y0}}, \quad (2)$$

where u_{y1} and u_{y0} contribute the surface displacement amplitude with and without a wave barrier, respectively. The area-averaged frequency response function can be described as:

$$\overline{A_R} = \frac{1}{nl_3} \int_0^n \int_0^{l_3} A_R dxdy. \quad (3)$$

The wave attenuation through the trenches would appear when $\overline{A_R}$ is less than 1. Thus, $\overline{A_R}$ was adopted to assess the vibration isolation performance in relation to the barriers.

A mesh size convergence study was carried out to verify the accuracy of the numerical model. The present numerical simulation results in respect to four different mesh sizes are compared with the solutions of [18], as shown in Figure 3a. It is indicated that the simulation results of vibration isolation by periodic geofoam-filled trenches converged to the exact solution as the mesh size decreased. A tetrahedral mesh size up to $\lambda_R/8$ was sufficient to satisfy the calculation requirements, and further reduction of the mesh size would not affect the accuracy of the results [36]. Hence, the model can be discretized into tetrahedral elements with a size of $\lambda_R/8$ for subsequent studies. For the sake of substantiating the numerical model further, the vibration isolation performance of an open trench with 1 λ_R depth and 0.1 λ_R width was utilized as a reference, which was located 5 λ_R away from the excitation source. It can be seen from Figure 3b that the present results are identical to the previous research results of Yang and Huang [34] and Bordon et al. [37]. As a consequence of the comparison research, the present grid precision is appropriate for numerical simulations.

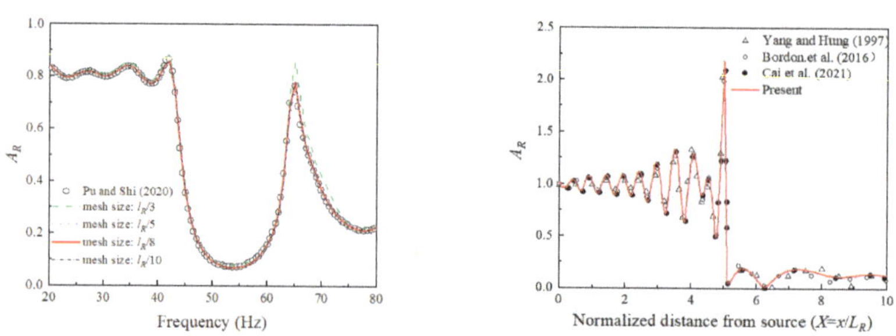

(a) (b)

Figure 3. Comparative study for vibration isolation: (**a**) periodic geofoam-filled trenches; (**b**) open trench. Retrieved from refs. [20,34,37].

3. Results and Discussions

In order to validate the isolation effectiveness of the PT-WIB, extensive analyses were conducted to investigate the vibration isolation performance of three kinds of wave barriers (WIB, periodic infilled trenches, and PT-WIB) under different excitations in both the frequency domain and time domain. Whereafter, a parametric analysis is proposed to investigate the behavior of the PT-WIB.

3.1. Isolation Characteristics of Three Kinds of Wave Barriers in the Frequency Domain

For the sake of illustrating the vibration isolation performance of these wave barriers in a more direct way, the parameters of periodic infilled trenches are referred to in previous studies [18] as: $d_2 = 2$ m, $w_2 = 0.3$ m, $b = 0.7$ m. The number of rows for infilled trenches is seven. The parameters of the WIB for the analysis are $d_1 = 0.3$ m, $w_1 = 3$ m, $h = 0.45$ m. The PT-WIB is a combination of the WIB and periodic in-filled trenches; thus, the parameters of the PT-WIB are the same as those of the above wave barriers. In addition, $l_1 = 10$ m, $l_2 = 20$ m, and $l_3 = 6$ m are chosen.

Figure 4a,b show the vibration isolation performance with respect to three kinds of wave barriers at low excitation frequencies (8 Hz and 15 Hz). The isolation effectiveness of the PT-WIB for low-frequency vibrations was better than that of the WIB and periodic infilled trenches, especially in the region behind the trenches ($x > 10$ m). The vibration isolation effectiveness of periodic infilled trenches is dissatisfactory at low excitation frequencies. This is because wavelength is longer at a lower frequency, and shallow trenches have difficulty achieving acceptable isolation performance. The results also show that the WIB had the advantage of reducing the low-frequency vibration, which is consistent with other investigation results [29–31]. The nephogram of the vertical displacement amplitude field with three different kinds of wave barriers at the excitation frequency of 15 Hz is shown in Figure 5.

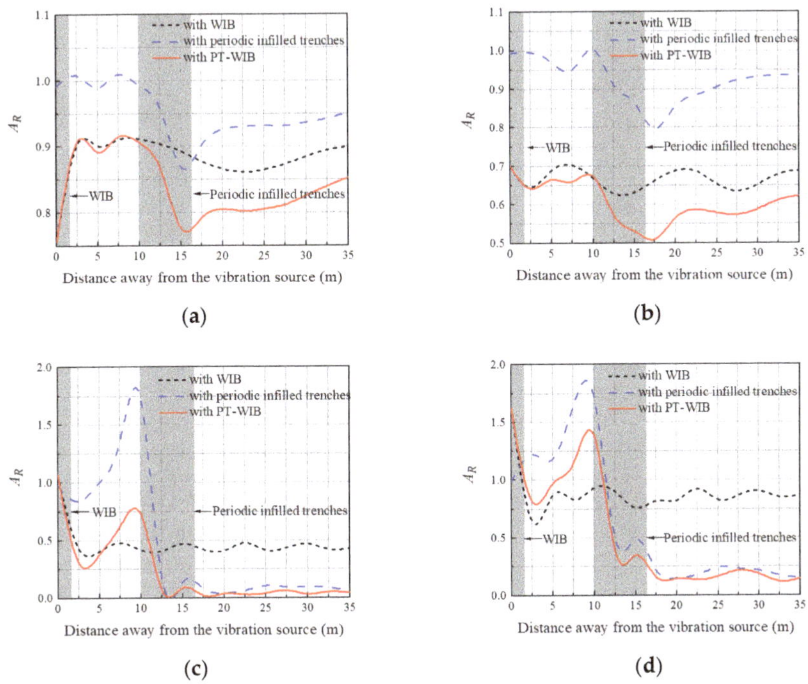

Figure 4. Vibration isolation performances with respect to three kinds of wave barriers at different fixed frequency excitation: (**a**) 8 Hz; (**b**) 15 Hz; (**c**) 55 Hz; (**d**) 80 Hz.

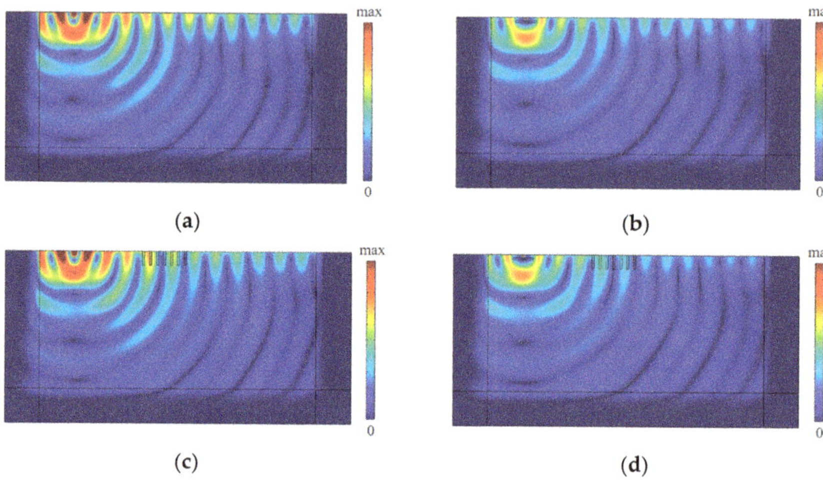

Figure 5. The nephogram of vertical displacement amplitude field at f = 15 Hz: (**a**) without wave barriers; (**b**) with WIB; (**c**) with periodic infilled trenches; (**d**) with PT-WIB.

Figure 4c,d compare the vibration isolation performances with respect to three kinds of wave barriers at medium- and high-excitation frequencies (55 Hz and 80 Hz). It can be found that the vibration isolation effectiveness of the periodic infilled trenches was rather high in the region behind the periodic infilled trenches at medium- and high-excitation frequencies. This is because the designed periodic infilled trenches could isolate the surface waves of the attenuation zone (45–62 Hz) effectively, and there was a region of evanescent surface waves above the attenuation zone, namely, leaky surface modes [18]. Thus, the surface waves of 55 Hz and 80 Hz could also be isolated effectively, whereas some vibration amplification was observed in front of the periodic infilled trenches on account of the vibration reflection at the interface of the geofoam and the ground. Whereas WIB performed better in the area in front of the trenches ($x < 10$ m), its isolation effectiveness was inferior to that of the periodic infilled trenches in the region behind the trenches ($x > 10$ m). It should be noted that the PT-WIB combined the vibration isolation advantages of both the WIB and the periodic infilled trenches. It could isolate specific surface waves of the attenuation zone due to the presence of the periodic infilled trenches and take advantage of the WIB to weaken the vibration amplification in the front region of the periodic infilled trenches. The difference in vibration isolation for these kinds of wave barriers at an excitation frequency of 55 Hz can also be observed in Figure 6.

Figure 7 illustrates vibration isolation performance with respect to three kinds of wave barriers in the frequency domain. The observation area for vibration response is shown in Figure 2. It can be noticed from Figure 7 that the vibration isolation performance of the periodic infilled trenches was effective at medium- and high-excitation frequencies, especially in the attenuation zone (45–62 Hz). The WIB was more efficient in reducing low- and medium-frequency vibration, but it was limited in isolating high-frequency vibration. The PT-WIB can reduce low-frequency vibration as well as the isolation of surface waves in some specific ranges, which are expected to be the frequency band gaps for the periodic structure. Thus, a broadband attenuation zone for the vibration could be achieved, which is difficult to achieve by other single measures. Moreover, the countermeasure was easy to apply in practical engineering.

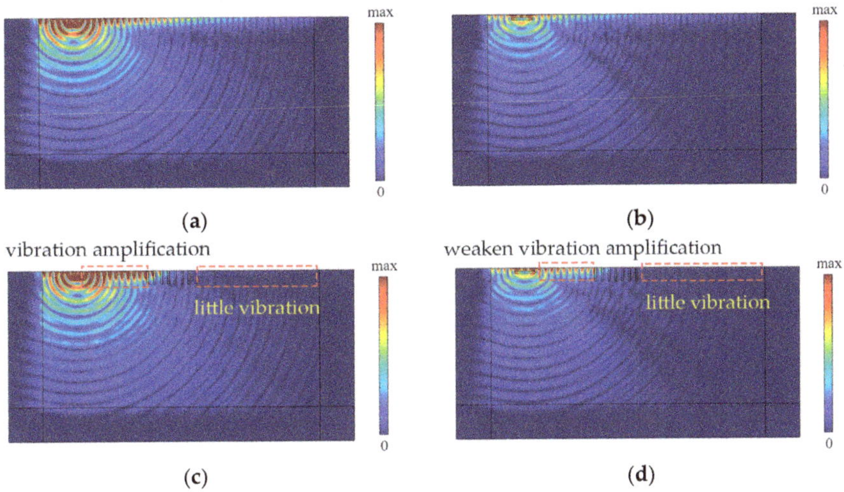

Figure 6. The nephogram of vertical displacement amplitude field at f = 55 Hz: (**a**) without wave barriers; (**b**) with WIB; (**c**) with periodic infilled trenches; (**d**) with PT-WIB.

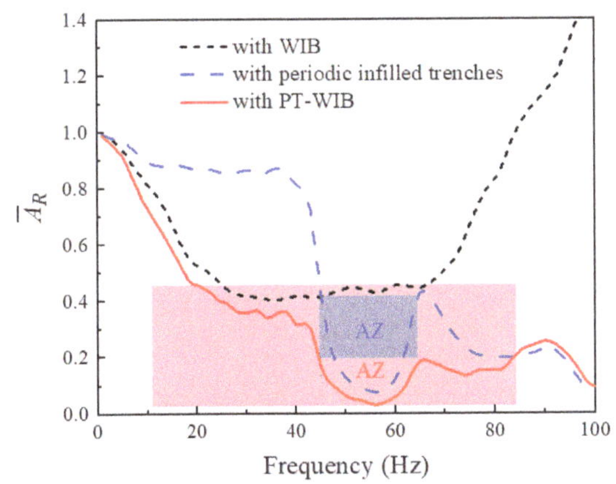

Figure 7. Vibration isolation performance with frequency with respect to three kinds of wave barriers.

3.2. Isolation Characteristics of Three Kinds of Wave Barriers in the Time Domain

In this section, the isolation effectiveness with respect to these three kinds of wave barriers is investigated in the time domain. A field test to measure traffic-induced environmental vibrations was conducted in Dongguan (Guangdong Province, China), as shown in Figure 8a. The type of daily operating elevated intercity train was CRH6A. The 941B-type ultra-low accelerometers were used to collect the accelerometer signal, and the INV3062-type vibration signal acquisition instrument was used to process and analyze the signals. The location of the test was 10 m away from the pier in a field. Figure 8b,c shows the measured acceleration record and the corresponding Fourier spectrum when a train ran by at approximately 100 km/h, respectively. The measured acceleration data were integrated twice to acquire the displacement excitation, as described in Figure 8d, which was applied at the load location shown in Figure 2. The duration of the railway excitation was 20 s; the dynamic sub-step was 20,000; and the time step of integration was approximately 0.001 s.

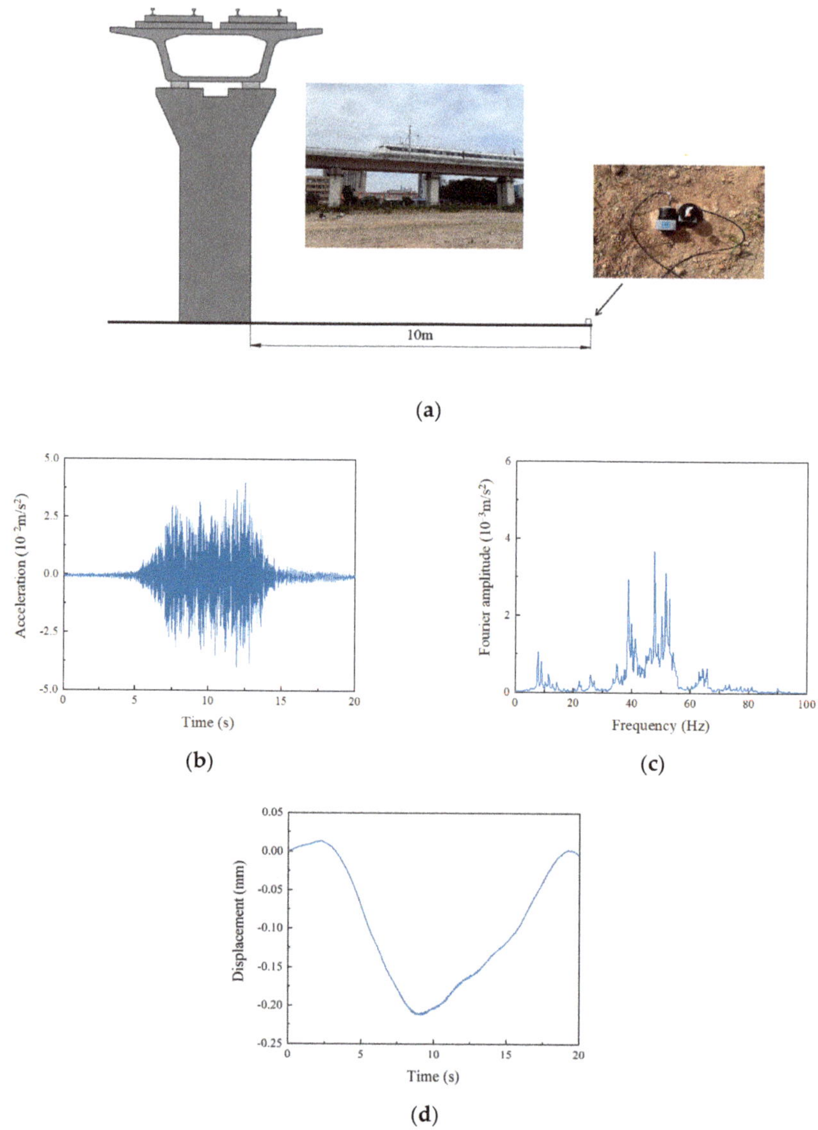

Figure 8. Vibration response in the field caused by the intercity railway: (**a**) field measure; (**b**) vertical acceleration record; (**c**) Fourier spectrum; (**d**) the corresponding displacement excitation.

The geometries and arrangements of three types of wave barriers were the same as described in Section 3.1. PMLs are typically not employed in time-domain analysis. Instead, in the model, a low-reflective boundary condition was implemented to effectively mitigate the reflection of vibration waves during time-domain analysis. Figure 9 shows vertical acceleration responses and the corresponding Fourier spectrum at the detection point A (x = 20 m) with three kinds of wave barriers and without barriers subjected to railway excitation. It can be seen that the PT-WIB provided better isolation performance than other wave barriers; the acceleration amplitude with the PT-WIB was reduced by 74.8% when compared to that without a wave barrier. And the acceleration amplitudes with the WIB and periodic infilled trenches were reduced by 54.6% and 52.4%, respectively. Meanwhile, it can be observed from the Fourier spectra of the PT-WIB that vibrations were attenuated over the whole observed frequency range. The analysis results in the time domain reveal the feasibility of isolating environmental vibration with PT-WIB in practice.

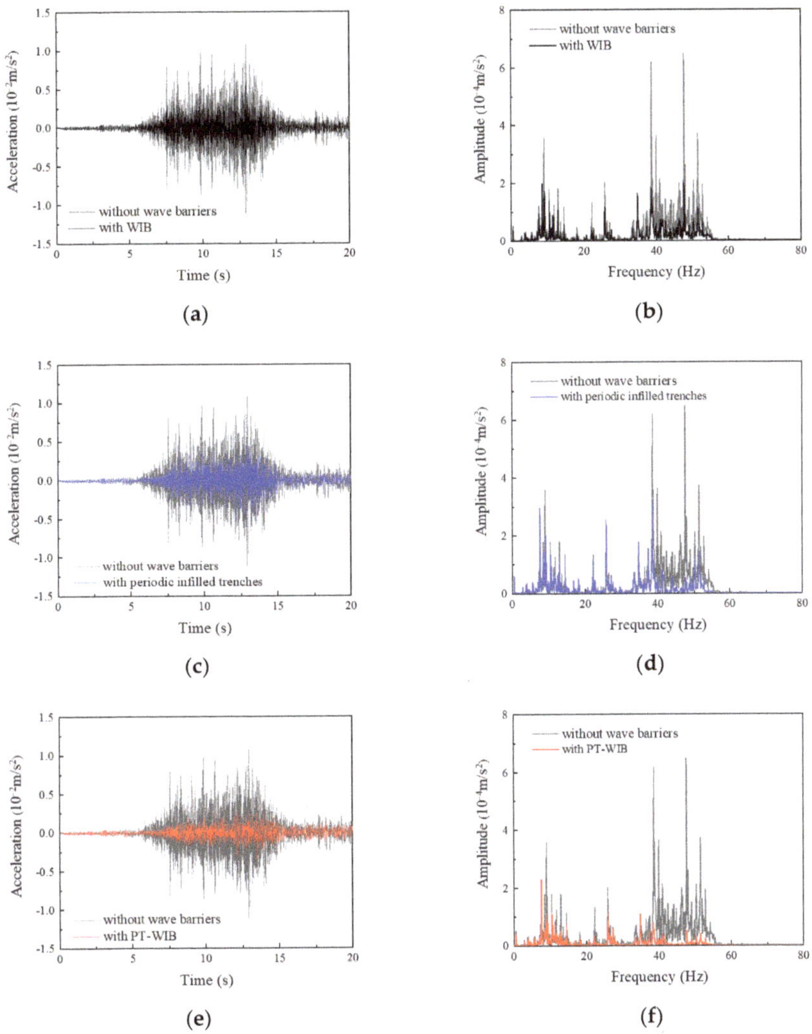

Figure 9. Vertical acceleration responses at the detection point and corresponding Fourier spectrum with and without barriers: (**a**,**b**) with and without WIB; (**c**,**d**) with and without periodic infilled trenches; (**e**,**f**) with and without PT-WIB.

3.3. Parametric Study of the PT-WIB

The vibration isolation performances of three different wave barriers have been analyzed in both the frequency domain and the time domain. The proposed PT-WIB shows the advantage of achieving a wide attenuation zone in environmental vibration isolation. A parametric study was carried out to analyze the influence of these effects on the vibration isolation performance of the PT-WIB. A large number of investigations, including theoretical analysis, numerical simulations, and experiments, have been conducted to reveal the mechanism of periodic structures in environmental vibration reduction, and a reasonable design process for isolating vibration in a specific frequency range has been proposed. The purpose of this study was to extend the attenuation zone and improve the vibration isolation performance by introducing the WIB into the periodic infilled trenches. Therefore, only the influences of parameters associated with the WIB (width, thickness, embedded depth, and distance from the source) on the isolation performance were analyzed. The parameters of the periodic infilled trenches remain unchanged, and the parameters (d_2 = 2 m, w_2 = 0.3 m, b = 0.7 m) are referring to [18]. The observation area for the vibration response is shown in Figure 2.

Figure 10a,b show that the average amplitude reduction ratio $\overline{A_R}$ of the observation area varied with the WIB's width and thickness, respectively. It can be observed that the vibration isolation effectiveness of the PT-WIB increased with the width and thickness of the WIB, especially in the frequency region outside the attenuation zone (45–62 Hz) of the periodic infilled trenches. The vibration isolation effectiveness increased with a decrease in the embedded depth of the WIB, as shown in Figure 10c. The embedded depth of the WIB varied from 0.2 m to 0.6 m, and there was an obvious increase in vibration isolation effectiveness, particularly in the high-frequency region. The effect of the distance from the excitation source of the WIB on $\overline{A_R}$ is shown in Figure 10d. It is evident that the WIB installed below the vibration source (x = 0) was more efficient for isolating low- and medium-frequency waves. A better performance for high-frequency vibration isolation can be realized by placing the WIB in the appropriate position between the vibration source and periodic infilled trenches.

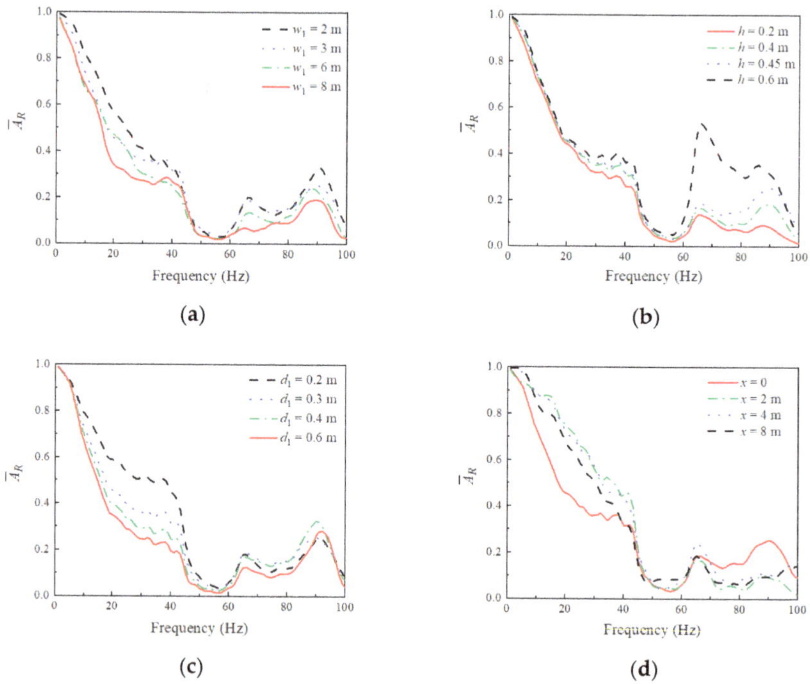

Figure 10. Influence of WIB's parameters on ($\overline{A_R}$): (**a**) width (d_1 = 0.3 m, h = 0.45 m, x = 0); (**b**) thickness (w_1 = 3 m, h = 0.45 m, x = 0); (**c**) embedded depth (w_1 = 3 m, d_1 = 0.3 m, x = 0); (**d**) distance from excitation source of WIB (d_1 = 0.3 m, w_1 = 3 m, h = 0.45 m).

Furthermore, the nephogram of the vertical displacement amplitude field for the PT-WIB with respect to different parameters (WIB's width, thickness, and embedded depth) is illustrated in Figure 11 in order to demonstrate these parameters' effects on the vibration isolation throughout the region in a more direct way. Since the effects of different parameters on the observation area after the periodic infilled trenches have been discussed above, attention has been paid to the region before the periodic infilled trenches ($0 < x < 10$)

here. The w_1 varied from 2 m to 6 m, the d_1 varied from 0.2 m to 0.6 m, and the h varied from 0.2 m to 0.6 m. It can be seen from Figure 11 that the embedded depth of the WIB played a significant part in the vibration isolation of the region before the periodic infilled trenches. The vibration isolation performance could be distinctly improved with a decrease in the embedded depth of the WIB. Furthermore, the first row and column of the nephogram in Figure 11 reveal that the isolation performance improvement of the PT-WIB was not so obvious when increasing the width and thickness of the WIB. This indicates that decreasing the embedded depth of WIB is the most effective way to improve the vibration isolation performance of the PT-WIB.

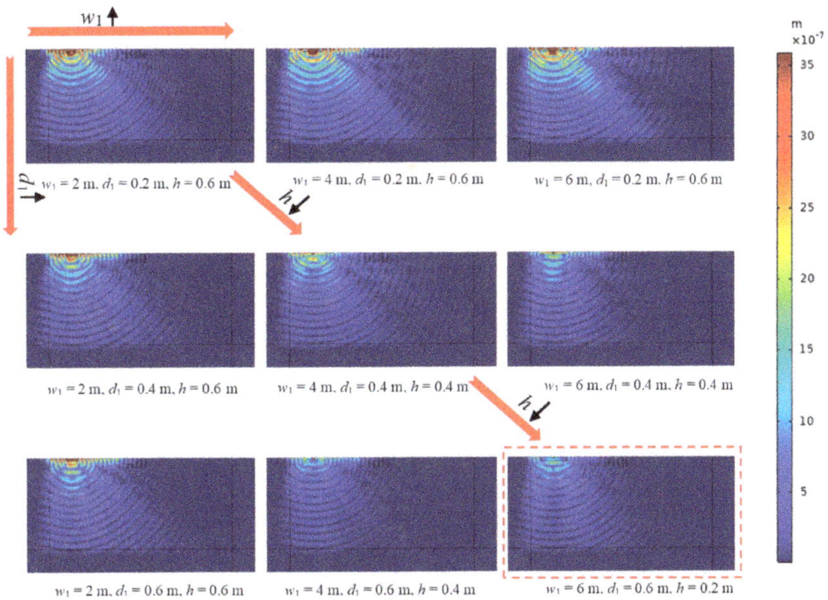

Figure 11. The nephogram of vertical displacement amplitude field for the PT-WIB with different parameters (width, thickness, and embedded depth of WIB) at f = 55 Hz. (The direction indicated by the red arrow represents the direction of parameter change. The nephogram within the red dashed line box visually demonstrates that the most significant vibration reduction effect is achieved when the parameters are combined in a specific manner.)

3.4. The Application of PT-WIB in Layered Ground

In practice, the soil is generally layered rather than single-layered. The soil medium with layered soil [27] was adopted to replace the single-layer soil. The properties of the layered soil are referred to in [27]. The vibration isolation performances with respect to these three wave barriers in the layered soil were compared. The geometric dimensions, number of rows, and material properties of the periodic infilled trenches and WIB are the same as those described in Section 3.1. It can be observed from Figure 12 that the vibration isolation performance of PT-WIB is excellent in the layered-soil medium, which is consistent with the results in Section 3.1. Simultaneously, a parametric study, which is similar to Section 3.3, was carried out, as shown in Figure 13. It indicates that the influence of different WIB parameters on vibration isolation is similar to that of Section 3.3.

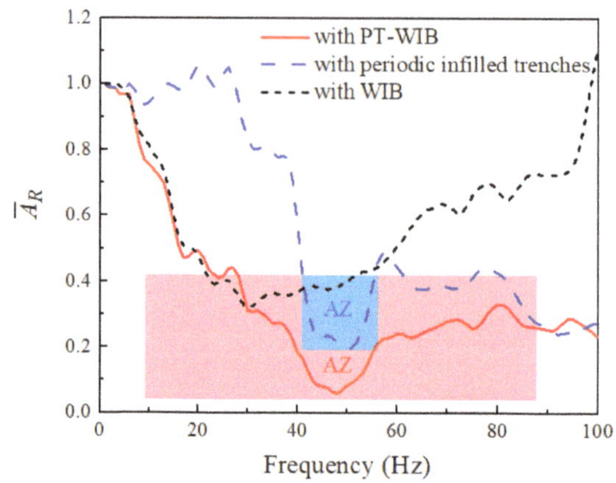

Figure 12. Vibration isolation performance with respect to three kinds of wave barriers in layered-soil medium.

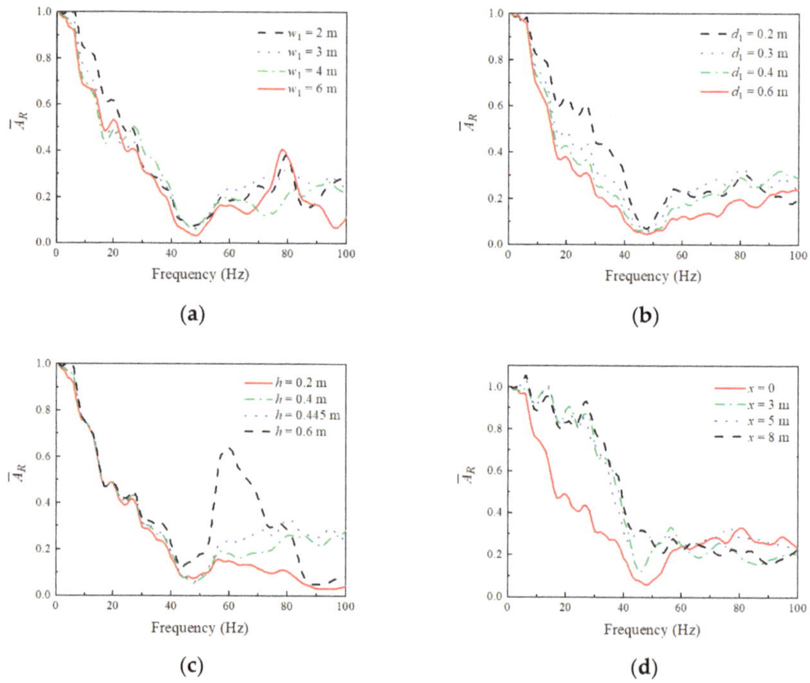

Figure 13. Influence of WIB's parameters $\overline{A_R}$ in layered-soil medium: (**a**) width (d_1 = 0.3 m, h = 0.45 m, x = 0); (**b**) thickness (w_1 = 3 m, h = 0.45 m, x = 0); (**c**) embedded depth (w_1 = 3 m, d_1 = 0.3 m, x = 0); (**d**) distance from excitation source of WIB (d_1 = 0.3 m, w_1 = 3 m, h = 0.45 m).

4. Conclusions

In rail transit, vibration is the main source of environmental pollution. This paper proposes a buried PT-WIB consisting of periodic infilled trenches and a wave-impedance block barrier to achieve a broadband attenuation zone for rail transit vibration isolation. The main conclusions are summarized as follows:

- The WIB is positioned beneath the railway tracks, while periodic infilled trenches are strategically placed between the railway and the protected buildings. In situations where the vibration isolation requirements cannot be met solely by the periodic trenches, typically due to the limited effectiveness of a narrow band gap at low and medium frequencies, the newly proposed PT-WIB offers a practical and viable solution. This innovative approach demonstrates the convenience and feasibility of creating a broadband attenuation zone, effectively addressing the limitations encountered in traditional setups.

- The occurrence of vibration amplification phenomena is observed in the vicinity of periodic infilled trenches and is primarily attributed to wave reflections at the interface between the geofoam and the ground. However, the implementation of the WIB effectively mitigates these vibration amplifications. Consequently, the newly proposed PT-WIB offers a notable advantage by providing a relatively consistent and stable environmental vibration isolation performance throughout varying distances from the vibration source.
- Although an increase in the width and thickness of the WIB can improve the vibration isolation performance of the PT-WIB, a decrease in the embedded depth of the WIB is a more effective way to improve the vibration isolation performance of the PT-WIB. Moreover, the PT-WIB can also be applied to a layered ground for the improvement of vibration isolation performance.

Author Contributions: Conceptualization, L.G. and C.C.; methodology, L.G. and C.L.; software, L.G.; validation, C.C., C.L. and C.M.M.; investigation, L.G.; resources, C.C.; data curation, L.G.; writing—original draft preparation, L.G.; writing—review and editing, C.C.; visualization, L.G.; supervision, C.M.M.; funding acquisition, C.C. All authors have read and agreed to the published version of the manuscript.

Funding: The work described in this paper was fully supported by grants from the Hunan Provincial Natural Science Foundation of China (Project No. 2023JJ30665).

Institutional Review Board Statement: Not applicable.

Informed Consent Statement: Not applicable.

Data Availability Statement: Data are available upon request.

Conflicts of Interest: The authors declare no conflict of interest.

References

1. Connolly, D.P.; Marecki, G.P.; Kouroussis, G.; Thalassinakis, I.; Woodward, P.K. The growth of railway ground vibration problems—A review. *Sci. Total Environ.* **2016**, *568*, 1276–1282. [CrossRef] [PubMed]
2. Smith, M.G.; Ogren, M.; Morsing, J.A.; Waye, K.P. Effects of ground-borne noise from railway tunnels on sleep: A polysomnographic study. *Build. Environ.* **2019**, *149*, 288–296. [CrossRef]
3. Hao, H.; Ang, T.C.; Shen, J. Building vibration to traffic-induced ground motion. *Build. Environ.* **2001**, *36*, 321–336. [CrossRef]
4. Zou, C.; Moore, J.A.; Sanayei, M.; Tao, Z.; Wang, Y. Impedance model of train-induced vibration transmission across a transfer structure into an overtrack building in a metro depot. *J. Struct. Eng.* **2022**, *148*, 04022187. [CrossRef]
5. Xu, L.; Ma, M. Analytical solution of ground-borne vibration due to a spatially periodic harmonic moving load in a tunnel embedded in layered soil. *J. Zhejiang Univ.-Sci. A* **2023**, *24*, 637–652. [CrossRef]
6. Woods, R.D. Screening of surface waves in soils. *J. Soil Mech. Found. Eng. Div.* **1968**, *94*, 951–979. [CrossRef]
7. Celebi, E.; Firat, S.; Beyhan, G.; Cankaya, I.; Vural, I.; Kirtel, O. Field experiments on wave propagation and vibration isolation by using wave barriers. *Soil Dyn. Earthq. Eng.* **2009**, *29*, 824–833. [CrossRef]
8. Alzawi, A.; El Naggar, M.H. Full scale experimental study on vibration scattering using open and in-filled (Geofoam) wave barriers. *Soil Dyn. Earthq. Eng.* **2011**, *31*, 306–317. [CrossRef]
9. Toygar, O.; Ulgen, D. A full-scale field study on mitigation of environmental ground vibrations by using open trenches. *Build. Environ.* **2021**, *203*, 108070. [CrossRef]
10. Ibrahim, Y.E.; Nabil, M. Finite element analysis of multistory structures subjected to train-induced vibrations considering soil-structure interaction. *Case Stud. Constr. Mater.* **2021**, *15*, e00592. [CrossRef]
11. Lu, M.; Feng, L.; Chen, Y. Phononic crystals and acoustic metamaterials. *Mater. Today* **2009**, *12*, 34–42. [CrossRef]
12. Shi, Z.; Cheng, Z.; Xiang, H. *Periodic Structures: Theory and Applications to Seismic Isolation and Vibration Reduction*; Science Press Ltd.: Beijing, China, 2017. (In Chinese)

13. Huang, J.; Shi, Z. Application of Periodic Theory to rows of piles for horizontal vibration attenuation. *Int. J. Geomech.* **2013**, *13*, 132–142. [CrossRef]
14. Huang, J.; Shi, Z. Attenuation zones of periodic pile barriers and its application in vibration reduction for plane waves. *J. Sound. Vib.* **2013**, *332*, 4423–4439. [CrossRef]
15. Huang, J.; Liu, W.; Shi, Z. Surface-wave attenuation zone of layered periodic structures and feasible application in ground vibration reduction. *Constr. Build. Mater.* **2017**, *141*, 1–11. [CrossRef]
16. Albino, C.; Godinho, L.; Amado-Mendes, P.; Alves-Costa, P.; Dias-da-Costa, D.; Soares, D.S., Jr. 3D FEM analysis of the effect of buried phononic crystal barriers on vibration mitigation. *Eng. Struct.* **2019**, *196*, 109340. [CrossRef]
17. Pu, X.; Shi, Z. Periodic pile barriers for Rayleigh wave isolation in a poroelastic half-space. *Soil Dyn. Earthq. Eng.* **2019**, *121*, 75–86. [CrossRef]
18. Pu, X.; Shi, Z. Broadband surface wave attenuation in periodic trench barriers. *J. Sound. Vib.* **2020**, *468*, 115130. [CrossRef]
19. Huang, H.W.; Zhang, B.; Wang, J.; Menq, F.; Nakshatrala, K.B.; Mo, Y.L.; Stokoe, K.H. Experimental study on wave isolation performance of periodic barriers. *Soil Dyn. Earthq. Eng.* **2021**, *144*, 106602. [CrossRef]
20. Cai, C.; Gao, L.; He, X.; Zou, Y.; Yu, K.; Wu, D. The surface wave attenuation zone of periodic composite in-filled trenches and its isolation performance in train-induced ground vibration isolation. *Comput. Geotech.* **2021**, *139*, 104421. [CrossRef]
21. Gao, L.; Cai, C.; Mak, C.M.; He, X.; Zou, Y.; Wu, D. Surface wave attenuation by periodic hollow steel trenches with Bragg band gap and local resonance band gap. *Constr. Build. Mater.* **2022**, *356*, 129289. [CrossRef]
22. Liu, Z.; Zhang, X.; Mao, Y.; Zhu, Y.Y.; Yang, Z.; Chan, C.T.; Sheng, P. Locally resonant sonic materials. *Science* **2000**, *289*, 1734. [CrossRef]
23. Huang, J.; Shi, Z. Vibration Reduction of Plane Waves Using Periodic In-Filled Pile Barriers. *J. Geotech. Geoenviron Eng.* **2015**, *141*, 04015018. [CrossRef]
24. Meng, L.; Cheng, Z.; Shi, Z. Vibration mitigation in saturated soil by periodic in-filled pipe pile barriers. *Comput. Geotech.* **2020**, *124*, 103633. [CrossRef]
25. Jiang, Y.; Meng, F.; Chen, Y.; Zheng, Y.; Chen, X.; Zhang, J.; Huang, X. Vibration attenuation analysis of periodic underground barriers using complex band diagrams. *Comput. Geotech.* **2020**, *128*, 103821. [CrossRef]
26. Miao, L.; Li, C.; Lei, L.; Fang, H.; Liang, X. A new periodic structure composite material quasi-phononic crystals. *Phys. Lett. A* **2020**, *384*, 7. [CrossRef]
27. Chouw, N.; Le, R.; Schmid, G. Propagation of vibration in a soil layer over bedrock. *Eng. Anal. Bound. Elem.* **1991**, *8*, 125–131. [CrossRef]
28. Takemiya, H.; Fujiwara, A. Wave propagation/impediment in a stratum and wave impeding block (WIB) measured for SSI response reduction. *Soil Dyn. Earthq. Eng.* **1994**, *13*, 49–61. [CrossRef]
29. Gao, G.; Li, N.; Gu, X. Field experiment and numerical study on active vibration isolation by horizontal blocks in layered ground under vertical loading. *Soil Dyn. Earthq. Eng.* **2015**, *69*, 251–261. [CrossRef]
30. Gao, G.; Chen, J.; Gu, X.; Song, J.; Li, S.; Li, N. Numerical study on the active vibration isolation by wave impeding block in saturated soils under vertical loading. *Soil Dyn. Earthq. Eng.* **2017**, *93*, 99–112. [CrossRef]
31. Gao, G.; Zhang, Q.; Chen, J.; Chen, Q. Field experiments and numerical analysis on the ground vibration isolation of wave impeding block under horizontal and rocking coupled excitations. *Soil Dyn. Earthq. Eng.* **2018**, *115*, 507–512. [CrossRef]
32. Khelif, A.; Achaoui, Y.; Benchabane, S.; Laude, V.; Aoubiza, B. Locally resonant surface acoustic wave band gaps in a two-dimensional phononic crystal of pillars on a surface. *Phys. Rev. B* **2010**, *81*, 214303. [CrossRef]
33. Meng, Q.; Shi, Z. Vibration Isolation of Plane Waves by Periodic Pipe Pile Barriers in Saturated Soil. *J. Aerosp. Eng.* **2019**, *32*, 04018114. [CrossRef]
34. Yang, Y.; Hung, H.H. A parametric study of wave barriers for reduction of train-induced vibrations. *Int. J. Numer. Methods Eng.* **1997**, *40*, 3729–3747. [CrossRef]
35. Saikia, A. Numerical study on screening of surface waves using a pair of softer backfilled trenches. *Soil Dyn. Earthq. Eng.* **2014**, *65*, 206–213. [CrossRef]
36. Jones, S. Harmonic response of layered half space using reduced finite element model with perfectly-matched layer boundaries. *Soil Dyn. Earthq. Eng.* **2017**, *92*, 1–8. [CrossRef]
37. Bordon, J.D.R.; Aznarez, J.J.; Maeso, O. Two-dimensional numerical approach for the vibration isolation analysis of thin walled wave barriers in poroelastic soils. *Comput. Geotech.* **2016**, *71*, 168–179. [CrossRef]

sensors

MDPI

Article

Graph-Based Audio Classification Using Pre-Trained Models and Graph Neural Networks

Andrés Eduardo Castro-Ospina [1,*], Miguel Angel Solarte-Sanchez [1], Laura Stella Vega-Escobar [1], Claudia Isaza [2] and Juan David Martínez-Vargas [3]

1 Grupo de Investigación Máquinas Inteligentes y Reconocimiento de Patrones, Instituto Tecnológico Metropolitano, Medellín 050013, Colombia; miguelsolarte244621@correo.itm.edu.co (M.A.S.-S.); lauravega@itm.edu.co (L.S.V.-E.)
2 SISTEMIC, Electronic Engineering Department, Universidad de Antioquia-UdeA, Medellín 050010, Colombia; victoria.isaza@udea.edu.co
3 GIDITIC, Universidad EAFIT, Medellín 050022, Colombia; jdmartinev@eafit.edu.co
* Correspondence: andrescastro@itm.edu.co

Abstract: Sound classification plays a crucial role in enhancing the interpretation, analysis, and use of acoustic data, leading to a wide range of practical applications, of which environmental sound analysis is one of the most important. In this paper, we explore the representation of audio data as graphs in the context of sound classification. We propose a methodology that leverages pre-trained audio models to extract deep features from audio files, which are then employed as node information to build graphs. Subsequently, we train various graph neural networks (GNNs), specifically graph convolutional networks (GCNs), GraphSAGE, and graph attention networks (GATs), to solve multi-class audio classification problems. Our findings underscore the effectiveness of employing graphs to represent audio data. Moreover, they highlight the competitive performance of GNNs in sound classification endeavors, with the GAT model emerging as the top performer, achieving a mean accuracy of 83% in classifying environmental sounds and 91% in identifying the land cover of a site based on its audio recording. In conclusion, this study provides novel insights into the potential of graph representation learning techniques for analyzing audio data.

Keywords: ecoacoustics; environmental sound classification; graph neural networks; graph representation learning; node classification; pre-trained models

Citation: Castro-Ospina, A.E.; Solarte-Sanchez, M.A.; Vega-Escobar, L.S.; Isaza, C.; Martínez-Vargas, J.D. Graph-Based Audio Classification Using Pre-Trained Models and Graph Neural Networks. *Sensors* **2024**, *24*, 2106. https://doi.org/10.3390/s24072106

Academic Editor: Hector Eduardo Roman

Received: 21 February 2024
Revised: 22 March 2024
Accepted: 23 March 2024
Published: 26 March 2024

1. Introduction

Graphs are powerful mathematical structures that have been extensively employed to model and analyze complex relationships and interactions across various domains [1]. In passive acoustic monitoring applications, which help to create conservation plans, ecoacoustics has recently gained great importance as a cost-effective tool to analyze species conservation and ecosystem alteration. In this field, it is necessary to analyze a large amount of acoustic data to assess variations in the ecosystem. Moreover, in recent years, the field of graph representation learning has grown due to the increased interest in using these graph structures for learning and inference tasks [2]. To learn from graphs, it is crucial to develop algorithms and models that can efficiently capture and make use of the detailed structural information present in graph data. These approaches have found applications in diverse fields, including bioinformatics, computer vision, recommendation systems, and social network analysis [3–6].

Graph neural networks (GNNs) have emerged as a prominent class of models for learning on graphs, offering distinct advantages over traditional artificial intelligence techniques [7]. Unlike traditional methods that operate on independent data points, GNNs use the inherent connectivity and dependencies within the graph structure to learn and propagate information across nodes. By recursively aggregating and transforming node features based on their local neighborhood, GNNs can capture both local and global patterns, enabling them to model complex relationships in graph data effectively. Notably, significant advancements in tasks such as node classification, link prediction, and graph generation have been made by leveraging their ability to capture structural dependencies [8].

Automatic audio classification tasks have attracted attention in recent years, specifically the classification of environmental sounds [9], enabling applications ranging from speech recognition [10,11] to soundscape ecology [12,13]. Traditional classification techniques such as *k*-nearest neighbors, support vector machines, and neural network classifiers have been used [14–17]. However, its performance mostly relies on hand-crafted features from representations as temporal, spectral, or spectro-temporal domains. Moreover, deep learning techniques using 1D (raw waveform) [18–21] or 2D (spectrograms) [22–25] convolutional neural networks (CNN) have shown significant improvements over hand-crafted methods. Nevertheless, these networks do not consider the relationships that may exist between different environmental sounds. Recurrent Neural Networks were initially proposed to capture feature dependencies from audio data [26–28]. More recently, Transformer models have emerged to model longer feature dependencies and leverage parallel processing [29–32]. Transformer models can handle variable input lengths and utilize attention mechanisms, making them aware of the global context and allowing their application on audio classification tasks.

Although graphs have been widely employed to represent and analyze visual and textual data, their potential to represent audio data has received relatively less attention [33–35]. Nonetheless, audio data, ranging from speech signals to music recordings, inherently exhibit temporal dependencies and complex patterns that can be effectively captured and modeled using graph-based representations. Working with graphs presents several challenges in their construction and subsequent processing. Determining how to generate feature information for each node and establishing connections between nodes in the network remain open problems. In this study, we propose utilizing pre-trained audio models to extract informative features from audio files, enabling the building of graphs that capture the inherent relationships and temporal dependencies present in the audio data.

Specifically, this study aims to address the problem of audio classification as a node classification task over graphs. To achieve this, we propose the following approach: (i) characterizing each audio with pre-trained networks to leverage transfer learning from models trained on large amounts of similar data, (ii) constructing graphs with each set of generated features, and (iii) utilizing the constructed graphs to classify nodes into predefined categories, taking advantage of their relationship. To accomplish this, we will use two datasets, a public one and one acquired in a passive acoustic monitoring study. We will evaluate the performance of three state-of-the-art GNNs: convolutional graph networks (GCN), graph attention networks (GAT), and GraphSAGE. These models leverage the rich structural information encoded in audio graphs in a transductive manner to learn discriminative representations capable of efficiently distinguishing between different audio classes. By comparing the performance of these models, we attempt to evaluate which of the graph models performs better on audio classification tasks.

In conclusion, this study contributes to the emerging field of graph representation learning by exploring the application of GNNs for audio classification. In particular, we demonstrate the effectiveness of pre-trained audio models to generate node information for graph representations and compare the performance of three GNN architectures. The results not only advance the state-of-the-art in audio classification but also emphasize the potential of graph-based approaches for modeling and analyzing complex audio data.

2. Graph Neural Networks

A graph is a widely used data structure, denoted as $\mathcal{G} = (\mathcal{V}, \mathcal{E})$, consisting of nodes $\left(\mathcal{V} = \left\{v_1, v_2, \ldots, v_{|\mathcal{V}|}\right\}\right)$ and edges $(\mathcal{E} = \{e_{ij}\})$ representing a link between node $i\ (v_i)$ and node $j\ (v_j)$. A useful way to represent a graph is through an adjacency matrix

$\left(A \in \mathbb{R}^{|\mathcal{V}|\times|\mathcal{V}|}\right)$, where the presence of an edge is encoded as an entry with $A_{ij} = 1$ if there is an edge between (v_i) and (v_j) and $A_{ij} = 0$ otherwise. Additionally, each node i has associated feature information or embeddings denoted as $h_i^{(0)}$.

GNNs are machine learning methods that receive data in the form of graphs and use neural message passing to generate embeddings for graphs and subgraphs. In [2], the author provides an overview of neural message passing, which can be expressed as follows:

$$h_u^{(k+1)} = update^{(k)}\left(h_u^{(k)}, agg^{(k)}(\{h_v, \forall\, v \in N(u)\})\right). \tag{1}$$

In this equation, $h_u^{(k)}$ is the current embedding of node u where the embeddings (h_v) of neighboring nodes will be sent; $N(u)$, the neighborhood of node u; and $update^{(k)}$ and $agg^{(k)}$, permutation-invariant functions.

There exist various GNN models that differ in their approach to the *aggregation* or *update* function expressed in Equation (1) and in their ability to perform prediction tasks at node, edge, or network level [36]. The theory of the three GNN models used in this study is presented below.

2.1. Graph Convolutional Networks (GCNs)

The goal of GCNs is to generalize the convolution operation to graph data by aggregating both self-features and neighbors' features [37]. Following the update rule given by Equation (2), GCNs enforce self-connections by making $\tilde{A} = A + I$ and stack multiple convolutional layers followed by nonlinear activation functions.

$$H^{(k+1)} = \sigma\left(\tilde{D}^{-\frac{1}{2}}\tilde{A}\tilde{D}^{-\frac{1}{2}}H^{(k)}W^{(k)}\right) \tag{2}$$

In this equation, H is the feature matrix containing the embeddings of the nodes as rows, and $\tilde{D}$ denotes the degree matrix of the graph, which is computed as $\tilde{D}_{ij} = \sum_j \tilde{A}_{ij}$. Moreover, $\sigma(\cdot)$ is an activation function, and W is a trainable weight matrix.

2.2. Graph SAmple and aggreGatE (GraphSAGE)

GraphSAGE, a framework built on top of the original GCN model [38], updates each node's embedding information by sampling the number of neighbors at different hop values and aggregating their respective embedding information. This iterative process allows nodes to increasingly gain information from different parts of the graph.

The main difference between the GCN model and GraphSAGE lies in the aggregation function. Where GCNs use an average aggregator, GraphSAGE employs a generalized aggregation function. Also, in GraphSAGE, self-features are not aggregated at each layer. Instead, the aggregated neighborhood features are concatenated with self-features, as shown in Equation (3).

$$h_u^{(k+1)} = \sigma\left(\left[W^{(k)} agg\left(\{h_v^{(k)}, \forall\, v \in N(u)\}\right), B^{(k)} h_u^{(k)}\right]\right) \tag{3}$$

In this equation, B is a trainable weight matrix, and agg denotes a generalized aggregation function, such as mean, pooling, or LSTM.

2.3. Graph Attention Networks (GATs)

In GCNs (Equation (2)), graph node features are averaged at each layer, with weights determined by coefficients obtained from the degree matrix $(\tilde{D})$. This implies that the outcomes of GCNs are highly dependent on the graph structure. GATs [39], for their part, seek to reduce this dependency by implicitly calculating these coefficients, taking into account the importance assigned to each node's features using the attention mechanism [40]. The purpose of this is to increase the model's representational capacity.

The expression for GATs is presented in Equation (4).

$$h_u^{(k+1)} = \sigma\left(\sum_{v \in N(u)} \alpha_{uv} W^{(k)} h_v^{(k)}\right) \quad (4)$$

In this equation, α_{uv} represents the attention coefficients of the neighbors of node u, $v \in N(u)$, regarding the aggregation feature aggregation at this node. These coefficients are computed as

$$\alpha_{uv} = \frac{exp(a^\top LeakyReLU(W[h_u, h_v]))}{\sum_{j \in N(u)} exp(a^\top LeakyReLU(W[h_u, h_j]))}, \quad (5)$$

with a denoting a trainable attention vector [41].

3. Materials

3.1. *UrbanSound8K*

UrbanSound8K is an audio dataset [42] that contains 8732 labeled audio files in WAV format and lasts four seconds or less. Each audio file belongs to one of the following ten classes: *air conditioner*, *car horn*, *children playing*, *dog bark*, *drilling*, *engine idling*, *gun shot*, *jackhammer*, *siren*, and *street music*.

The audio files are originally pre-distributed across ten folds, as depicted in Figure 1. To avoid errors that could invalidate the results and enable fair comparisons with existing literature, it is advised to perform cross-validation using the ten predefined folds.

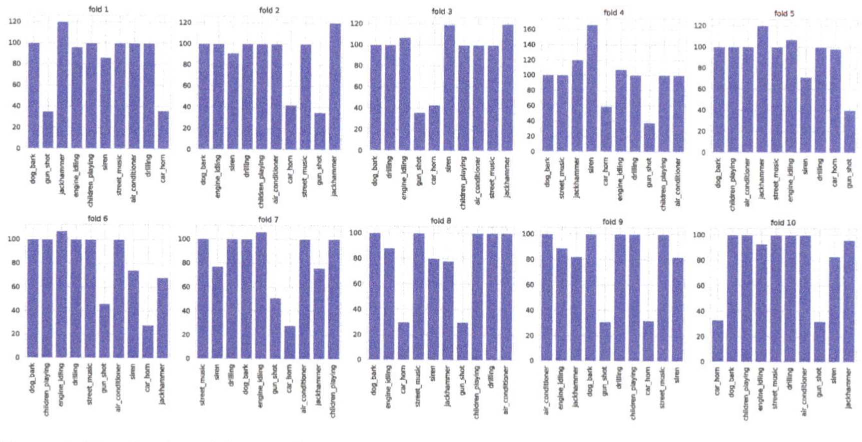

Figure 1. Distribution of the ten classes across the predefined folds.

3.2. *Rey Zamuro Reserve*

This dataset arises from a passive acoustic monitoring study conducted at Rey Zamuro and Matarredonda Private Reserves (3°31′02.5″ N, 73°23′43.8″ W), located in the municipality of San Martín in the Department of Meta, Colombia. The reserve covers an expanse of 6000 hectares, predominantly characterized by natural savanna constituting around 60%, interspersed with introduced pasture areas. The remaining 40% is covered by forests. This region falls within the tropical humid biome of the Meta foothills, showcasing an average temperature of 25.6 °C.

Data were acoustically recorded in September of 2022. A 13×8 grid was installed with 94 AudioMoth automatic acoustic devices placed 400 m from each other; of these recorders, one was not used due to deteriorated audio. The recording was made every fourteen minutes for seven consecutive days. The recordings were captured in mono format at a sampling rate of 192,000 Hz. The study encompassed various habitats, such as forest interiors, edges, and adjacent areas, each with distinct characteristics, including undergrowth. The recording heights were standardized at 1.5 m above the ground.

Depending on the kind of land cover, each acoustic recording of Rey Zamuro soundscapes was classified as forest, savanna, or pasture. These labels were given based on the placement of each automated recording unit. A total of 71,497 recordings were obtained, of which 14,546 correspond to forest class, 14,994 to savanna class, and 41,957 to pasture class. In all, 80% of the dataset is used as the training set, and the remaining 20% as the test set.

3.3. Pre-Trained Models for Audio Feature Extraction

We use pre-trained deep learning audio models to extract deep features from each audio file, which will be used as node information in the constructed graphs, i.e., as the values $h_i^{(0)}$. Specifically, we employed the following three models: VGGish, YAMNet, and PANNs.

3.3.1. VGGish

VGGish is a pre-trained neural network architecture particularly designed to generate compact and informative representations, or deep embeddings, for audio signals [43]. It is inspired by the Visual Geometry Group (VGG) network architecture originally developed for image classification [44]. The deep embeddings generated by VGGish effectively capture relevant acoustic features and serve as a foundation for various audio processing tasks, such as audio classification, content-based retrieval, and acoustic scene understanding [45,46]. VGGish was trained on AudioSet [47], a publicly available and widely used large-scale audio dataset comprising millions of annotated audio clips and 527 classes, including animal sounds, musical instruments, human activities, environmental sounds, and more.

The architecture of VGGish consists of several layers, including convolutional, max-pooling, and fully connected layers. In this model, the processed audio is segmented into 0.96-second clips, and a log-Mel spectrogram is calculated for each clip, serving as the input to the neural network. Then, the convolutional layers apply a set of learnable filters to the input audio spectrogram, aiming to detect local patterns and extract low-level features. Following each convolutional layer, max-pooling layers are employed to reduce the spatial dimensions of the obtained feature maps while retaining the most important information. This process helps capture and preserve relevant patterns at different scales and further abstract the representations. Lastly, the final layers of VGGish, i.e., the fully connected layers, take the flattened output of the preceding convolutional and max-pooling layers and map it to a 128-dimensional representation. This mapping aims to capture global and high-level dependencies, resulting in deep embeddings that encode meaningful information about the audio signal and can serve as input for subsequent shallow or deep learning methods.

3.3.2. PANNs

Large-scale Pretrained Audio Neural Networks (PANNs) are pre-trained models specifically developed for audio pattern recognition [48]. Their architecture is built upon CNNs, which are well-suited for analyzing audio mel-spectrograms. PANNs have multiple layers, including convolutional, pooling, and fully connected layers. These layers work together to learn hierarchical representations of audio patterns at various levels of abstraction.

The training process of PANNs involves pre-training the model on the large-scale AudioSet dataset. By being trained on this dataset, PANNs learn to capture a wide range of audio patterns, making them strong audio feature extractors. These audio patterns are then mapped to a 2048-dimensional output space.

3.3.3. YAMNet

Yet another Audio Mobilenet Network (YAMNet) is a pre-trained neural network architecture that utilizes the power of deep CNNs and transfer learning to perform accurate and efficient audio analysis [49].

YAMNet is a mobilenet-based architecture consisting of a stack of convolutional layers, followed by global average pooling and a final fully connected layer with softmax activation. The convolutional layers extract local features by convolving small filters over the input audio spectrogram, thereby capturing different levels of temporal and spectral patterns. Then, the global average pooling operation condenses the extracted features into a fixed-length representation. Finally, the fully connected layer produces the classification probabilities for each sound class.

YAMNet's primary objective is to accurately classify audio signals into a wide range of sound categories. However, the embeddings obtained after the global average pooling operation can also be useful.

To process audio, YAMNet divides the audio into segments of 0.96 s with a hop length of 0.48 s. For each segment, a feature output comprising 1024 dimensions is generated.

3.4. Graph Construction

A popular way to determine the edges of a graph is to define whether two points are neighbors through the *k*-nearest neighbors (*k*-NN) algorithm. According to this method, the neighbors of node v_i are those *k*-nearest neighbors in the feature space [50]. Thus, the *k*-NN algorithm assigns edges between v_i and its neighbors.

4. Experimental Framework

The proposed methodology of this study to assess the effectiveness of using graphs to represent audio data by leveraging pre-trained audio models to generate node information is depicted in Figure 2, and involves the following stages: (i) VGGish, YAMNet, and PANNs pre-trained audio models are used to extract features from both datasets, (ii) those deep features are used independently to construct graphs where each node represents an audio file, and edges are determined based on the *k*-NN algorithm, and (iii) the constructed graphs are used to train and optimize certain hyperparameters on GCN, GraphSAGE, and GAT models to perform node classification.

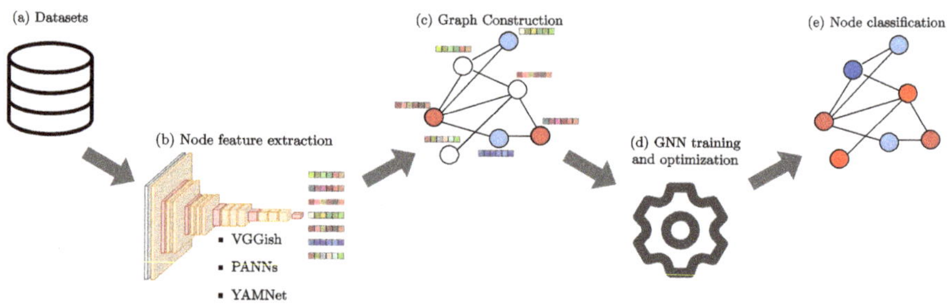

Figure 2. The workflow diagram proposed in this study illustrates that for each audio of a dataset (**a**) deep features are extracted with pre-trained audio models (**b**), then graphs are constructed by including those features as node information and setting edges with *k*-NN (**c**). For test data, the nodes present information but no labels (in the diagram the nodes unfilled are the test nodes). Subsequently, some GNN models are trained and optimized (**d**). Finally, trained models allow discriminating test nodes between classes (red or blue in the diagram) through transductive learning (**e**).

As a first step, we employed the VGGish, PANNs, and YAMNet pre-trained models to extract features from the audio files in both datasets to be used as node embedding vectors. In the UrbanSound8K dataset, fold information was preserved for the extracted features, as shown in Figure 3. VGGish model generates a 128-dimensional deep feature vector for every 0.96 s of an audio clip, and YAMNet produces a 1024-dimensional deep feature vector for every 0.48 s. Since the audio files have a maximum duration of four seconds for UrbanSound8K and 60 s for Rey Zamuro, to obtain node embeddings of the same length, we averaged those 128-dimensional VGGish-based and 1024-dimensional YAMNet-based deep features.

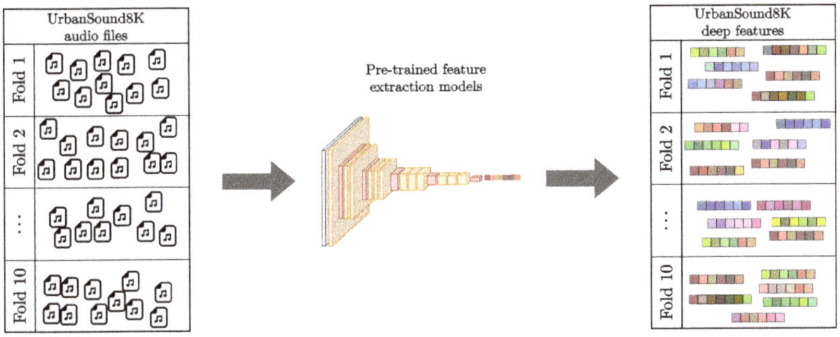

Figure 3. Feature extraction scheme. The audio files from each fold of the UrbanSound8K dataset were characterized using pre-trained models.

Subsequently, for each dataset characterized using the pre-trained models, we constructed a graph where the nodes represented the audio embeddings, and the edges were defined by applying the k-NN algorithm, where each node is connected with its k nearest neighbors. The value k was optimized for each architecture using Optuna [51]. Then, we implemented the GCN, GraphSAGE, and GAT architectures using PyTorch Geometric [52]. For the GCN and GraphSAGE models, we employed a two-layer architecture with a hidden dimension optimized by Optuna and an output dimension equal to the number of classes, i.e., three for the Rey Zamuro dataset and ten for UrbanSound8K. For the GAT model, we used a two-layer architecture, with the first layer having a value for hidden dimension optimized by Optuna and 10 heads, followed by a second layer with an output dimension corresponding to the number of classes and one head.

To compute the attention coefficients, we employed a slope of 0.2 on the LeakyReLU activation function in Equation (5). For all trained GNNs, we used the ReLU activation function and a dropout with a probability of 0.5. All models were trained to minimize cross-entropy loss using the Adam optimizer (with a learning rate of 0.001 and weight decay of 5×10^{-4}) for 300 and 1300 epochs for UrbanSound8K and Rey Zamuro dataset, respectively.

Finally, for UrbanSound8K, we evaluated the performance of the models in terms of accuracy using ten-fold cross-validation, i.e., following the dataset's distribution across the ten predefined folds. Alternatively, due to the large amount of data and the associated computational cost for training use, the performance of the models for the Rey Zamuro dataset was evaluated with the test set.

5. Results and Discussion

Tables 1 and 2 present the accuracy results of the three GNN models (GCN, GraphSAGE, and GAT) trained for audio file classification, with nodes representing the audio data in a graph. These nodes are characterized by three distinct feature sets derived from pre-trained models (VGGish, PANNs, and YAMNet) applied to UrbanSound8K and Rey Zamuro datasets. Additionally, the tables display the optimal hyperparameters determined by Optuna for each GNN model and node characterization combination. For the UrbanSound8K dataset, where fold distribution is predefined, accuracy results are presented as mean values accompanied by their corresponding standard deviations. Conversely, accuracy results for the Rey Zamuro dataset focus solely on the test set.

Table 1. UrbanSound8K accuracies.

Feature Model	GNN Architecture	Best Hyperparameters	Accuracy
VGGish	GCN	$k = 10$ n_hidden = 55	0.77 ± 0.04
	GraphSAGE	$k = 12$ n_hidden = 57	0.76 ± 0.05
	GAT	$k = 9$ n_hidden = 52	0.79 ± 0.05
YAMNet	GCN	$k = 5$ n_hidden = 196	0.81 ± 0.04
	GraphSAGE	$k = 11$ n_hidden = 55	0.8 ± 0.03
	GAT	$k = 6$ n_hidden = 252	0.82 ± 0.04
PANNs	GCN	$k = 4$ n_hidden = 40	0.83 ± 0.03
	GraphSAGE	$k = 5$ n_hidden = 183	0.82 ± 0.03
	GAT	$k = 10$ n_hidden = 206	0.83 ± 0.03

Table 2. Rey Zamuro accuracies.

Feature Model	GNN Architecture	Best Hyperparameters	Accuracy
VGGish	GCN	$k = 5$ n_hidden = 48	0.63
	GraphSAGE	$k = 10$ n_hidden = 63	0.63
	GAT	$k = 6$ n_hidden = 49	0.63
YAMNet	GCN	$k = 5$ n_hidden = 62	0.76
	GraphSAGE	$k = 6$ n_hidden = 56	0.74
	GAT	$k = 6$ n_hidden = 53	0.78
PANNs	GCN	$k = 6$ n_hidden = 64	0.87
	GraphSAGE	$k = 7$ n_hidden = 63	0.85
	GAT	$k = 5$ n_hidden = 51	0.91

The results reveal the consistent superiority of PANNs across both datasets and all three trained GNN models. In particular, on the Rey Zamuro dataset, PANNs show a significant improvement of up to 18% in accuracy. The higher performance can be attributed to the larger dimensional feature space produced by PANNs, with 2048 dimensions, compared to VGGish and YAMNet, which have dimensions of 128 and 1024, respectively. This larger feature space of PANNs is more suitable for capturing detailed information from audio data.

Furthermore, among the compared GNN models, GAT emerges as the top performer, demonstrating sustained superiority across both datasets. This underscores the effectiveness of the attention mechanism in exploiting graph information and optimizing aggregation strategies. Tables 3 and 4 present the computational costs of the experiments conducted, measured in terms of time and the number of trainable parameters of the networks for the UrbanSound8K and Rey Zamuro datasets, respectively. It is important to note that each model possesses a different number of neurons in the hidden layer due to the optimization performed with Optuna. The GAT model has the highest number of parameters for both datasets and the feature sets generated with the pre-trained models. Specifically, the largest GAT model for the UrbanSound8K dataset has 8M parameters when using PANNs' deep features. Regarding training time, the GAT model for this dataset can take up to 35 times longer than training GCN and GraphSAGE models. Concerning the Rey Zamuro dataset, we also calculate the time for each model under test. Once again, the GAT model demonstrates the largest number of parameters, as well as longer training and testing times. However, during testing, the times are closer to those of the other two models. Although training time can indeed be long, it is worth considering that a trained network can be scalable regardless of the amount of data. However, it is crucial to consider the computational requirements for building and storing the graph.

Our results show that representing audio datasets through graphs and using deep features extracted from pre-trained models as node features enables sound classification. However, it is important to acknowledge an ongoing research challenge in the graph-building step, particularly in setting its node feature information and edges. To the best of our knowledge, only one study has employed GNNs for sound classification on UrbanSound8K dataset [34]. In one such study, the overall classification accuracy obtained using GNNs was 63.5%, which improved to 73% when GNNs were used in combination with features learned from a CNN. However, our results surpass this, even in the case of GraphSAGE, whose lowest accuracy is 76% for VGGish features. Moreover, our findings are comparable to those reported in other studies employing 1D CNN models. For example, in [18], RawNet CNN was presented, which worked with the raw waveform and achieved an accuracy of 87.7 ± 0.2. Additionally, in [19], a CNN called EnvNet-v2 obtained an accuracy of 78.3%, in [20] with very deep 1D convolutional networks a maximum accuracy of 71.68% only for the 10th fold used as the test set, while in [21], a proposed end-to-end 1D CNN achieved 89% accuracy. In addition, 2D CNN models have also been used on the UrbanSound8K dataset, reaching 79% [22], 70% [23], 83.7% [24], and 97% [25]. It should be noted that although other studies used the UrbanSound8K dataset to train 1D or 2D CNNs, they often employ unofficial random splits of the dataset, conducting their own cross-validations or training-test splits. This causes them to use different training and validation data than published papers that follow the official distribution, making comparison unfair.

Table 3. Computational cost for UrbanSound8K dataset tests.

Feature Model	GNN Architecture	# Parameters	Training Time [s]
VGGish	GCN	7655	13.3
	GraphSAGE	15,799	10.9
	GAT	145,640	86.7
YAMNet	GCN	202,870	18.6
	GraphSAGE	113,805	33.0
	GAT	5,221,480	370.9
PANNs	GCN	82,370	12.7
	GraphSAGE	753,421	37.1
	GAT	8,487,240	453.9

Table 4. Computational cost for Rey Zamuro dataset tests.

Feature Model	GNN Architecture	# Parameters	Training Time [s]	Test Time [ms]
VGGish	GCN	6682	14.3	4.9
	GraphSAGE	17,461	25.8	12.9
	GAT	137,240	232.9	84.0
YAMNet	GCN	64,180	20.9	6.1
	GraphSAGE	115,874	68.4	80.4
	GAT	1,098,200	295.4	99.9
PANNs	GCN	131,786	27.2	7.8
	GraphSAGE	259,381	136.9	190.6
	GAT	2,101,240	325.7	120.3

6. Conclusions

In this paper, we explored using graphs as a suitable representation of acoustic data for sound classification tasks, focusing on the UrbanSound8K dataset and a passive acoustic monitoring study. Particularly, this study offers novel insights into the potential of graph representation learning methods for analyzing audio data.

First, we utilized pre-trained audio models, namely VGGish, PANNs, and YAMNet, to compute node embeddings and extract informative features. Then, we trained GCNs, GraphSAGE, and GATs and evaluated their performance. For the UrbanSound8K dataset, we employed a ten-fold cross-validation approach with the dataset's predefined folds for performance evaluation. Additionally, we partitioned the Rey Zamuro Dataset into train and test sets to validate its results. Moreover, during the training stage, we conducted hyperparameter optimization to attain the best possible model for the built graphs.

Our findings demonstrate the effectiveness of using graphs to represent audio data. In addition, they show that GNNs can achieve a competitive performance in sound classification tasks. Most notably, it is shown that it is possible to identify ecosystem states through audio and GNNs. Notably, the best results were obtained when employing PANNs-based deep features with the three GNN models. Among the GNN models, the GAT model outperforms the others. This advantage stems from its attention-based operation, enabling it to aggregate node information by assigning weights to its neighbors based on relevance.

To further our research, we plan to explore the feasibility of using temporal GNNs for sound classification tasks to leverage graphs constructed using deep features based on temporal segments of the audio signal, such as those obtained with VGGish and YAMNet. Additionally, the proposed methodology will be applied to the area of soundscape ecology, seeking to generate acoustic heterogeneity maps from the treatment of large volumes of data with GNN techniques that allow exploiting the acoustic relationships between different recording sites.

Author Contributions: Conceptualization, A.E.C.-O.; Formal analysis, A.E.C.-O., M.A.S.-S. and J.D.M.-V.; Methodology, A.E.C.-O., M.A.S.-S. and L.S.V.-E.; Project administration, C.I.; Supervision, J.D.M.-V.; Validation, L.S.V.-E. and C.I.; Writing—original draft, A.E.C.-O. and M.A.S.-S.; Writing—review and editing, L.S.V.-E., C.I. and J.D.M.-V. All authors have read and agreed to the published version of the manuscript.

Funding: This work was supported by Universidad de Antioquia, Instituto Tecnológico Metropolitano de Medellín, Alexander von Humboldt Institute for Research on Biological Resources, and Colombian National Fund for Science, Technology and Innovation, Francisco Jose de Caldas—MINCIENCIAS (Colombia). [Program No. 111585269779].

Institutional Review Board Statement: Not applicable.

Informed Consent Statement: Not applicable.

Data Availability Statement: The data presented in this study are available under request.

Conflicts of Interest: The authors declare no conflicts of interest.

References

1. Veličković, P. Everything is connected: Graph neural networks. *Curr. Opin. Struct. Biol.* **2023**, *79*, 102538. [CrossRef] [PubMed]
2. Hamilton, W.L. Graph representation learning. In *Synthesis Lectures on Artifical Intelligence and Machine Learning*; Morgan and Claypool: San Rafael, CA, USA, 2020; Volume 14, pp. 1–159.
3. Angles, R.; Gutierrez, C. Survey of graph database models. *ACM Comput. Surv. (CSUR)* **2008**, *40*, 1–39. [CrossRef]
4. Goyal, P.; Ferrara, E. Graph embedding techniques, applications, and performance: A survey. *Knowl.-Based Syst.* **2018**, *151*, 78–94. [CrossRef]
5. Dong, G.; Tang, M.; Wang, Z.; Gao, J.; Guo, S.; Cai, L.; Gutierrez, R.; Campbel, B.; Barnes, L.E.; Boukhechba, M. Graph neural networks in IoT: A survey. *ACM Trans. Sens. Netw.* **2023**, *19*, 1–50. [CrossRef]
6. Su, X.; Xue, S.; Liu, F.; Wu, J.; Yang, J.; Zhou, C.; Hu, W.; Paris, C.; Nepal, S.; Jin, D.; et al. A comprehensive survey on community detection with deep learning. *IEEE Trans. Neural Netw. Learn. Syst.* **2022**. [CrossRef] [PubMed]
7. Chen, F.; Wang, Y.C.; Wang, B.; Kuo, C.C.J. Graph representation learning: A survey. *APSIPA Trans. Signal Inf. Process.* **2020**, *9*, e15. [CrossRef]
8. Wu, Z.; Pan, S.; Chen, F.; Long, G.; Zhang, C.; Philip, S.Y. A comprehensive survey on graph neural networks. *IEEE Trans. Neural Netw. Learn. Syst.* **2020**, *32*, 4–24. [CrossRef]
9. Bansal, A.; Garg, N.K. Environmental Sound Classification: A descriptive review of the literature. *Intell. Syst. Appl.* **2022**, *16*, 200115. [CrossRef]
10. Passricha, V.; Aggarwal, R.K. A hybrid of deep CNN and bidirectional LSTM for automatic speech recognition. *J. Intell. Syst.* **2019**, *29*, 1261–1274. [CrossRef]
11. Mustaqeem; Kwon, S. A CNN-assisted enhanced audio signal processing for speech emotion recognition. *Sensors* **2019**, *20*, 183. [CrossRef]
12. Dias, F.F.; Ponti, M.A.; Minghim, R. A classification and quantification approach to generate features in soundscape ecology using neural networks. *Neural Comput. Appl.* **2022**, *34*, 1923–1937. [CrossRef]
13. Quinn, C.A.; Burns, P.; Gill, G.; Baligar, S.; Snyder, R.L.; Salas, L.; Goetz, S.J.; Clark, M.L. Soundscape classification with convolutional neural networks reveals temporal and geographic patterns in ecoacoustic data. *Ecol. Indic.* **2022**, *138*, 108831. [CrossRef]
14. Kostrzewa, D.; Brzeski, R.; Kubanski, M. The classification of music by the genre using the KNN classifier. In *Beyond Databases, Architectures and Structures. Facing the Challenges of Data Proliferation and Growing Variety, Proceedings of the 14th International Conference, BDAS 2018, Held at the 24th IFIP World Computer Congress, WCC 2018, Poznan, Poland, 18–20 September 2018*; Springer: Berlin/Heidelberg, Germany, 2018; pp. 233–242.
15. Prabavathy, S.; Rathikarani, V.; Dhanalakshmi, P. Classification of Musical Instruments using SVM and KNN. *Int. J. Innov. Technol. Explor. Eng.* **2020**, *9*, 1186–1190. [CrossRef]
16. Tsalera, E.; Papadakis, A.; Samarakou, M. Monitoring, profiling and classification of urban environmental noise using sound characteristics and the KNN algorithm. *Energy Rep.* **2020**, *6*, 223–230. [CrossRef]
17. Malik, H.; Bashir, U.; Ahmad, A. Multi-classification neural network model for detection of abnormal heartbeat audio signals. *Biomed. Eng. Adv.* **2022**, *4*, 100048. [CrossRef]
18. Li, S.; Yao, Y.; Hu, J.; Liu, G.; Yao, X.; Hu, J. An ensemble stacked convolutional neural network model for environmental event sound recognition. *Appl. Sci.* **2018**, *8*, 1152. [CrossRef]
19. Tokozume, Y.; Ushiku, Y.; Harada, T. Learning from between-class examples for deep sound recognition. *arXiv* **2017**, arXiv:1711.10282.
20. Dai, W.; Dai, C.; Qu, S.; Li, J.; Das, S. Very deep convolutional neural networks for raw waveforms. In Proceedings of the 2017 IEEE International Conference on Acoustics, Speech and Signal Processing (ICASSP), New Orleans, LA, USA, 5–9 March 2017; pp. 421–425.
21. Abdoli, S.; Cardinal, P.; Koerich, A.L. End-to-end environmental sound classification using a 1D convolutional neural network. *Expert Syst. Appl.* **2019**, *136*, 252–263. [CrossRef]
22. Salamon, J.; Bello, J.P. Deep convolutional neural networks and data augmentation for environmental sound classification. *IEEE Signal Process. Lett.* **2017**, *24*, 279–283. [CrossRef]
23. Pons, J.; Serra, X. Randomly weighted cnns for (music) audio classification. In Proceedings of the ICASSP 2019-2019 IEEE International Conference on Acoustics, Speech and Signal Processing (ICASSP), Brighton, UK, 12–17 May 2019; pp. 336–340.
24. Zhang, Z.; Xu, S.; Cao, S.; Zhang, S. Deep convolutional neural network with mixup for environmental sound classification. In Proceedings of the Chinese Conference on Pattern Recognition and Computer Vision (PRCV), Guangzhou, China, 23–26 November 2018; pp. 356–367.
25. Su, Y.; Zhang, K.; Wang, J.; Madani, K. Environment sound classification using a two-stream CNN based on decision-level fusion. *Sensors* **2019**, *19*, 1733. [CrossRef]
26. Gong, Y.; Chung, Y.A.; Glass, J. Ast: Audio spectrogram transformer. *arXiv* **2021**, arXiv:2104.01778.

27. Gan, J. Music feature classification based on recurrent neural networks with channel attention mechanism. *Mob. Inf. Syst.* **2021**, *2021*, 7629994. [CrossRef]
28. Banuroopa, K.; Shanmuga Priyaa, D. MFCC based hybrid fingerprinting method for audio classification through LSTM. *Int. J. Nonlinear Anal. Appl.* **2021**, *12*, 2125–2136.
29. Zhuang, Y.; Chen, Y.; Zheng, J. Music genre classification with transformer classifier. In Proceedings of the 2020 4th International Conference on Digital Signal Processing, Chengdu, China, 19–21 June 2020; pp. 155–159.
30. Nogueira, A.F.R.; Oliveira, H.S.; Machado, J.J.; Tavares, J.M.R. Transformers for urban sound classification—A comprehensive performance evaluation. *Sensors* **2022**, *22*, 8874. [CrossRef]
31. Zhang, Y.; Li, B.; Fang, H.; Meng, Q. Spectrogram transformers for audio classification. In Proceedings of the 2022 IEEE International Conference on Imaging Systems and Techniques (IST), Kaohsiung, Taiwan, 21–23 June 2022; pp. 1–6.
32. Zhu, W.; Omar, M. Multiscale audio spectrogram transformer for efficient audio classification. In Proceedings of the ICASSP 2023—2023 IEEE International Conference on Acoustics, Speech and Signal Processing (ICASSP), Rhodes Island, Greece, 4–10 June 2023; pp. 1–5.
33. Zhang, S.; Qin, Y.; Sun, K.; Lin, Y. Few-Shot Audio Classification with Attentional Graph Neural Networks. In Proceedings of the Interspeech 2019, Graz, Austria, 15–19 September 2019; pp. 3649–3653.
34. Aironi, C.; Cornell, S.; Principi, E.; Squartini, S. Graph-based representation of audio signals for sound event classification. In Proceedings of the 2021 29th European Signal Processing Conference (EUSIPCO), Dublin, Ireland, 23–27 August 2021; pp. 566–570.
35. Hou, Y.; Song, S.; Yu, C.; Wang, W.; Botteldooren, D. Audio event-relational graph representation learning for acoustic scene classification. *IEEE Signal Process. Lett.* **2023**, *30*, 1382–1386. [CrossRef]
36. Bishop, C.M.; Bishop, H. Graph Neural Networks. In *Deep Learning: Foundations and Concepts*; Springer: Berlin/Heidelberg, Germany, 2023; pp. 407–427.
37. Kipf, T.N.; Welling, M. Semi-supervised classification with graph convolutional networks. *arXiv* **2016**, arXiv:1609.02907.
38. Hamilton, W.; Ying, Z.; Leskovec, J. Inductive representation learning on large graphs. *Adv. Neural Inf. Process. Syst.* **2017**, *30*, 1–11.
39. Veličković, P.; Cucurull, G.; Casanova, A.; Romero, A.; Lio, P.; Bengio, Y. Graph attention networks. *arXiv* **2017**, arXiv:1710.10903.
40. Bahdanau, D.; Cho, K.; Bengio, Y. Neural machine translation by jointly learning to align and translate. *arXiv* **2014**, arXiv:1409.0473.
41. Brody, S.; Alon, U.; Yahav, E. How attentive are graph attention networks? *arXiv* **2021**, arXiv:2105.14491.
42. Salamon, J.; Jacoby, C.; Bello, J.P. A dataset and taxonomy for urban sound research. In Proceedings of the 22nd ACM international conference on Multimedia, Orlando, FL, USA, 3–7 November 2014; pp. 1041–1044.
43. Hershey, S.; Chaudhuri, S.; Ellis, D.P.; Gemmeke, J.F.; Jansen, A.; Moore, R.C.; Plakal, M.; Platt, D.; Saurous, R.A.; Seybold, B.; et al. CNN architectures for large-scale audio classification. In Proceedings of the 2017 IEEE International Conference on Acoustics, Speech and Signal Processing (ICASSP), New Orleans, LA, USA, 5–9 March 2017; pp. 131–135.
44. Simonyan, K.; Zisserman, A. Very deep convolutional networks for large-scale image recognition. *arXiv* **2014**, arXiv:1409.1556.
45. Kim, B.; Pardo, B. Improving content-based audio retrieval by vocal imitation feedback. In Proceedings of the ICASSP 2019—2019 IEEE International Conference on Acoustics, Speech and Signal Processing (ICASSP), Brighton, UK, 12–17 May 2019; pp. 4100–4104.
46. Tsalera, E.; Papadakis, A.; Samarakou, M. Comparison of pre-trained cnns for audio classification using transfer learning. *J. Sens. Actuator Netw.* **2021**, *10*, 72. [CrossRef]
47. Gemmeke, J.F.; Ellis, D.P.; Freedman, D.; Jansen, A.; Lawrence, W.; Moore, R.C.; Plakal, M.; Ritter, M. Audio set: An ontology and human-labeled dataset for audio events. In Proceedings of the 2017 IEEE International Conference on Acoustics, Speech and Signal Processing (ICASSP), New Orleans, LA, USA, 5–9 March 2017; pp. 776–780.
48. Kong, Q.; Cao, Y.; Iqbal, T.; Wang, Y.; Wang, W.; Plumbley, M.D. Panns: Large-scale pretrained audio neural networks for audio pattern recognition. *IEEE/ACM Trans. Audio Speech Lang. Process.* **2020**, *28*, 2880–2894. [CrossRef]
49. Models/Research/Audioset/Yamnet at Master · Tensorflow/Models—github.com. Available online: https://github.com/tensorflow/models/tree/master/research/audioset/yamnet (accessed on 18 April 2023).
50. Maier, M.; Luxburg, U.; Hein, M. Influence of graph construction on graph-based clustering measures. *Adv. Neural Inf. Process. Syst.* **2008**, *21*, 1–8.
51. Akiba, T.; Sano, S.; Yanase, T.; Ohta, T.; Koyama, M. Optuna: A next-generation hyperparameter optimization framework. In Proceedings of the 25th ACM SIGKDD International Conference on Knowledge Discovery & Data Mining, Anchorage, AK, USA, 4–8 August 2019; pp. 2623–2631.
52. Fey, M.; Lenssen, J.E. Fast graph representation learning with PyTorch Geometric. *arXiv* **2019**, arXiv:1903.02428.

sensors

MDPI

Article

A Quantitative Evaluation of the Performance of the Low-Cost AudioMoth Acoustic Recording Unit

Sam Lapp *, Nickolus Stahlman and Justin Kitzes

Department of Biological Sciences, University of Pittsburgh, 103 Clapp Hall, Fifth and Ruskin Avenues, Pittsburgh, PA 15260, USA; nickstahlman@gmail.com (N.S.); justin.kitzes@pitt.edu (J.K.)
* Correspondence: sam.lapp@pitt.edu

Abstract: The AudioMoth is a popular autonomous recording unit (ARU) that is widely used to record vocalizing species in the field. Despite its growing use, there have been few quantitative tests on the performance of this recorder. Such information is needed to design effective field surveys and to appropriately analyze recordings made by this device. Here, we report the results of two tests designed to evaluate the performance characteristics of the AudioMoth recorder. First, we performed indoor and outdoor pink noise playback experiments to evaluate how different device settings, orientations, mounting conditions, and housing options affect frequency response patterns. We found little variation in acoustic performance between devices and relatively little effect of placing recorders in a plastic bag for weather protection. The AudioMoth has a mostly flat on-axis response with a boost above 3 kHz, with a generally omnidirectional response that suffers from attenuation behind the recorder, an effect that is accentuated when it is mounted on a tree. Second, we performed battery life tests under a variety of recording frequencies, gain settings, environmental temperatures, and battery types. We found that standard alkaline batteries last for an average of 189 h at room temperature using a 32 kHz sample rate, and that lithium batteries can last for twice as long at freezing temperatures compared to alkaline batteries. This information will aid researchers in both collecting and analyzing recordings generated by the AudioMoth recorder.

Keywords: ARU; automated recording unit; AudioMoth; acoustic monitoring

Citation: Lapp, S.; Stahlman, N.; Kitzes, J. A Quantitative Evaluation of the Performance of the Low-Cost AudioMoth Acoustic Recording Unit. *Sensors* **2023**, *23*, 5254. https://doi.org/10.3390/s23115254

Academic Editor: Hector Eduardo Roman

Received: 5 May 2023
Revised: 25 May 2023
Accepted: 28 May 2023
Published: 1 June 2023

1. Introduction

Acoustic monitoring is an increasingly widespread technique for surveying populations of sound-producing species in the field [1–4]. The growth in acoustic field surveys in recent years has been supported in large part by the development and availability of a variety of inexpensive autonomous recording units (ARUs), devices that are designed to record audio in the field passively without the need for a human surveyor to be present. ARUs generally use battery power and store recordings locally on the device and can be programmed to record ambient sound at predetermined dates and times [5]. Commercial ARUs have been sold for many years by companies including Titley Scientific and Wildlife Acoustics [6,7]. More recently, inexpensive open-source designs based on Raspberry Pi's [8–10] demonstrated that ARUs could potentially be produced more widely and at a far lower cost than previously possible.

In 2017, Open Acoustic Devices released the first version of the AudioMoth [11], which has rapidly become one of the most widely used ARUs, with over 30,000 units produced and sold in the first four years after its release [12]. AudioMoths have been fabricated for as little as USD 50 and have most recently been sold through group purchases for USD 80 [13]. These devices combine many of the user-friendly features of commercial devices at a fraction of the cost.

Despite its popularity, little information is currently available about the performance characteristics of the AudioMoth recorder. From the perspective of audio quality, sensitivity, directionality, and frequency response (relative sensitivity across frequencies) are fundamental characteristics of audio recording equipment that must be measured to interpret audio data captured by a device. Although the frequency response of the AudioMoth's isolated microphone component is available online [14], the device's end-to-end on-axis and polar frequency response and sensitivity across gain settings have not been published. This is particularly important as deployment instructions have suggested attaching the AudioMoth to the trunk of a 200–400 mm tree [15], but the effect of this mounting option on audio recordings is not known. These quantitative measurements of recording equipment are distinct from measuring the maximum detection distance of biotic sounds, which depends on a combination of the properties of the recording equipment, the strength and characteristics of the sound source, and environmental noise. Other studies have investigated the maximum detection distance of wildlife using AudioMoth recorders [16,17], and previous literature has described the importance of considering maximum detection distance during the design of acoustic monitoring studies and the analysis of bioacoustics data [18,19].

The recording lifespan of the AudioMoth on one set of batteries has previously been measured only for a subset of the possible configuration settings. Hill et al. reported the battery life of the AudioMoth using 3000 mAh lithium batteries for some common configurations [15], reporting that an AudioMoth could record for 115 days recording at 8 kHz, the lowest sample rate, for 30 s every 5 min. The developers also reported the AudioMoth lasted 9 days recording nonstop at a 48 kHz sample rate. While the AudioMoth configuration app provides estimates of battery life for any chosen configuration settings, these estimates have not to our knowledge been validated empirically.

Here, we report the results of two sets of tests that we conducted to empirically characterize the performance of the AudioMoth ARU. First, to characterize the acoustic properties of the device and its onboard microphone, we conducted controlled playback tests alongside a test microphone across a variety of device orientations and mounting options. Second, to characterize the longevity of the device in the field, we investigated the expected battery life for the device across a variety of available settings, battery types, and ambient temperatures. This information will assist investigators in designing field experiments using AudioMoth ARUs as well as analyzing the resulting recordings.

2. Materials and Methods

Both tests described below were performed using AudioMoth 1.1.0 recorders from 2021 through 2023. Schematics, printed circuit board layout, and components for AudioMoth 1.1.0 are publicly available [15]. Firmware versions varied between the tests and are described below. A full list of all equipment used is compiled in Appendix A. Additional information on these tests and results can be found in [20,21].

2.1. Acoustic Performance Test

Our first test had the goal of characterizing the end-to-end sensitivity of the AudioMoth recorder from the microphone through analog-to-digital conversion. To do this, we examined five specific assessments, described below. In all testing and reporting of the results, we follow the guidelines and recommendations of Eargle [22] wherever feasible. All uses of the word "decibels" refer to a logarithmic value and are reported with respect to a reference value. For sound pressure level (a physical measurement of sound in air), dBA is used: the "A-weighted" average of sound pressure level across frequencies, with a reference point of 20 micro-Pascals = 0 dBA [23]. For digital values, dBFS for "decibels full scale" is used, where 0 dBFS is the highest measurement possible in the digital system. In many cases, frequency responses are reported relative to a reference, for instance, relative to the value of the frequency response at 1 kHz, or relative to the on-axis frequency response.

Unless noted otherwise, the gain setting is 0 (low) for all assessments, which allows the greatest difference between the test signal and environmental noise without clipping. The sampling rate was set to 48 kHz and all AudioMoths were using 1.3.0 firmware. The pink noise test signal used throughout the assessments was generated with a Mackie SRM-450 loudspeaker at a level of 86 dBA at one meter in front of the speaker. The AudioMoth was placed 1 m in front of the speaker, except where noted. All frequency responses and analyses use a frequency range of 100 Hz to 17 kHz, as the speaker was unable to reliably reproduce sounds above this range. This range covers the vocalizing range of virtually all birds and frogs and some insects, but not bats [24–27]. All plots of frequency response show frequency on the x-axis on a logarithmic scale from 100 Hz to 17 kHz and decibels (dB) on the y-axis with a range of 55 dB so that figures can be compared easily.

The first assessment compared variations in the frequency response patterns of five new AudioMoths. Comparisons occurred in a 4×6 m room in a residential area with acoustically absorbent panels surrounding the microphone to reduce ambient noise levels and reflections. The ambient noise level was measured to be 45.5 ± 0.5 dBA using an Enviro meter EM80 A-weighted Sound Level Meter (Sealed Unit Parts Co., Inc., Allenwood, NJ, USA).

Second, we assessed the effect of the gain setting on frequency response. The AudioMoth has five gain settings, numbered from 0 (low) to 4 (high). For this assessment, pink noise was played at 69 dBA at 1 m to avoid clipping on the highest gain setting of 4. Comparisons occurred in the same room as the first test.

Third, we evaluated on-axis frequency response in a grassland environment using three different AudioMoth protective housing conditions: no case, a Ziploc bag, and a sealed vacuum bag with air. The grassland environment was located at the University of Pittsburgh's Pymatuning Lab of Ecology Wood Lab site. This environment minimized reflections from buildings and other objects, as the nearest building was over 30 m from the testing location and the speaker faced away from that building. Ambient noise levels recorded before, during, and after testing were 49 ± 3 dBA.

Fourth, we evaluated the polar pattern, which is the sensitivity of a microphone when the sound arrives from different angles relative to the device. The microphone on the AudioMoth is omnidirectional, meaning that it has equal sensitivity to sound arriving from any direction. However, the device itself is unlikely to be truly omnidirectional, especially for high frequencies, because the device will reflect or absorb some sounds before they reach the microphone. We tested the polar pattern at various frequencies by incrementally rotating the AudioMoth 360 degrees in the horizontal or vertical plane while keeping the source at a fixed position. These tests were performed in the same grassland environment as the on-axis tests. Housing was not found to substantially affect response; therefore, results in different housings are not reported here but can be found in [20].

Fifth, we assessed the acoustic effect of mounting AudioMoth on trees, a common means of deploying devices in forested environments. The forest environment used for testing was a mixed coniferous and deciduous second-growth forest located at the Pymatuning Lab of Ecology housing site. Ambient noise levels recorded before, during, and after testing were 47 ± 1.5 dBA. AudioMoths were strapped to the front, back, and sides of trees from the perspective of the sound source, with the microphone facing away from the tree. We examined three trees with circumferences of 41 cm, 97 cm, and 170 cm. Housing was not found to affect response; therefore, results in different housings are not reported here but can be found in [20].

2.2. Battery Life Test

Our second test had the goal of estimating the total number of hours that an AudioMoth can record under a variety of configurations and conditions expected to be encountered in the field. We specifically evaluated the influence of sample rate, device gain, device temperature, battery type, and in some cases their interactions on AudioMoth recording time. All tests were performed in a controlled indoor environment using AudioMoths with 1.5.0 firmware and 64 GB SanDisk Ultra microSDXC cards. For all tests, AudioMoths were programmed to record continuously, and battery life was defined as the number of hours of recordings made successfully before the batteries were unable to continue powering the AudioMoths.

First, we evaluated the effect of the sample rate. A total of twenty-one new AudioMoths were configured to record using 2 (medium) gain at room temperature (21.5 °C) with Procell 2100 mAh alkaline batteries. Three AudioMoths were set to the same sample rate across each available setting (8, 16, 32, 48, 96, 192, 250, and 384 kHz). For comparison purposes, the expected battery life was determined while using the AudioMoth Configuration desktop app by dividing the Procell battery capacity by the daily estimated power consumption. Sample rates above 48 kHz filled the SD card before the batteries died, and cards were replaced as needed to obtain the total estimates of battery life.

Second, we evaluated the effect of gain setting. A total of twelve AudioMoths were configured to record at a 32 kHz sample rate at room temperature with Procell batteries. Three AudioMoths were set to record at the same gain across each available setting (low, low–medium, medium, medium–high, and high).

Third, we evaluated the interacting effects of battery type and temperature on recording time. Procell 2100 mAh alkaline batteries were compared against Energizer Ultimate Lithium 35,000 mAh batteries. Devices with both battery types were placed into three temperature conditions in a consumer-grade refrigerator/freezer combination. We tested battery life using three temperature treatments: room (20.5 °C), fridge (3.4 °C), and freezer (−16.1 °C). The average temperature for each category was determined using an infrared thermometer to obtain readings of the surface next to the AudioMoths during the morning, midday, and evening for each condition and across three separate days. A total of 18 AudioMoths were set to record at a sample rate of 32 kHz with a medium gain. For each battery type (alkaline and lithium), three devices were tested in each temperature condition (room, fridge, and freezer).

3. Results

3.1. *Acoustic Performance Tests*

3.1.1. Frequency Response Variation

AudioMoth devices appear to have consistent frequency response patterns with minimal variation. The maximum variation at any frequency between devices was +4.0/−3.5 dBFS, and the overall recording levels were indistinguishable. Small increases or decreases in measured sensitivity at specific frequencies may have been caused by imperfect measurement conditions rather than differences in the devices themselves. Figure S1 shows the frequency response variation of each device relative to one reference device.

3.1.2. Effect of Gain on Frequency Response Variation

The gain settings do not affect the frequency response, besides an overall change in the recorded level. The average levels for gain settings 0, 1, 2, and 3 relative to the highest settings are −14.2, −12.0, −5.7, and −2.6 dB, respectively. Figure S2 plots the frequency response of one device set to each gain setting, relative to the highest gain setting.

3.1.3. On-Axis Frequency Response

There was wind during on-axis frequency response testing, which registers as low-frequency noise in the recordings. For this reason, the frequency responses reported here may be less accurate below 1 kHz. Figure 1 shows the frequency response of AudioMoth in various protective housings. The 0 dB reference for all lines in this plot is the level of the control, no housing, at 1 kHz. The vacuum bag has the least effect, with the only substantial impact being a 5–10 dB loss above 10 kHz. The frequency response is mostly flat but has sharp dips at around 12 kHz and 15 kHz. The Ziploc bag has slight losses overall, with a bump from 5–10 kHz and a drop-off above 10 kHz. The frequency response of the two housings relative to the control is shown numerically in Table S1.

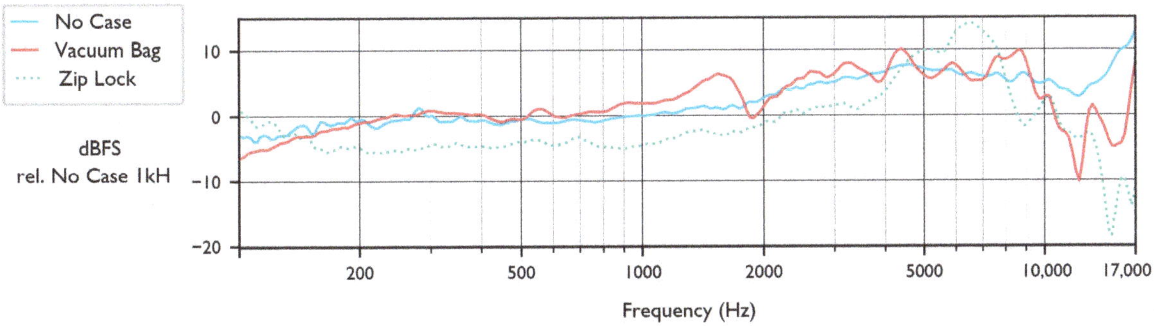

Figure 1. On-axis frequency response of three 1.1.0 AudioMoths recording pink noise playback from 1 m away. One AudioMoth was placed in a closed Ziploc bag, one was placed in a sealed, but not vacuumed, bag, and the control AudioMoth (No Case) was not placed in any housing.

3.1.4. Polar Response

As with on-axis testing, there was wind during polar response testing, and the frequency responses reported here may be less accurate below 1 kHz. Figure 2 shows the horizontal polar response of the AudioMoth with no case for several frequencies. Each frequency is plotted in 30-degree increments, relative to that frequency's level at 0 degrees (on-axis). Lines that dip towards the center of the plot indicate a reduction in sensitivity to that frequency. The polar response is relatively omnidirectional for low frequencies, losing about 5 dB for off-axis sounds. This is expected because sound waves below 2 kHz have wavelengths significantly longer than the dimensions of the device (a 2 kHz sound wave is approximately 17 cm) and are minimally attenuated or reflected by the device. Higher frequencies show significant attenuation directly behind the device, especially at 5 kHz (over 25 dB reduction). The largest effects are losses of 28 dB at 5 kHz and 20 dB at 10 kHz when sound arrives from behind the device. Figure 2 also shows the vertical polar response where the sound was coming from "above" and "below" the AudioMoth. The vertical rotation of the AudioMoth has little effect on low frequencies other than a surprising boost of 500 Hz arriving from slightly below the device. This could be due to a resonance mode in the device or may be a measurement error. In the high frequencies, 15 kHz is severely reduced when it arrives from behind and slightly above the device. This may be a result of cancellation or absorption by the battery pack (15 kHz has a wavelength of about 2 cm).

3.1.5. Impact of Trees on Frequency Response

Figure 3 shows the frequency spectrums of pink noise playback tests recorded by an AudioMoth with no case strapped to the front, back, or side of each tree from the perspective of the sound source. When the AudioMoth is on the same side as the tree as the sound source (0 degrees), a narrow notch filter (reduction in sensitivity) appears at 2.3 kHz, indicating cancellation of 15 cm sound waves. One possible cause of this is a reflection of incoming sound off of the cambium of the tree: if sound travels 3.75 cm to the reflective surface and 3.75 cm back to the receiver, it will travel half a wavelength (7.5 cm) round trip and cause destructive interference of 2.3 kHz sound at the microphone. Interestingly, if this is true, this "dead spot" will only have a strong effect directly in front of the microphone and at a specific frequency, as varying the angle of incidence will change the center frequency of the notch filter. Besides the dramatic notch filters, the effects of the trees on frequency response match our expectations. When sound arrives from behind the tree, attenuation increases with frequency and with the size of the tree.

For the smallest tree (41 cm circumference), high frequencies are substantially reduced while low frequencies are unaffected. With increasing tree radius, overall attenuation increases while the pattern of more attenuation at higher frequencies remains.

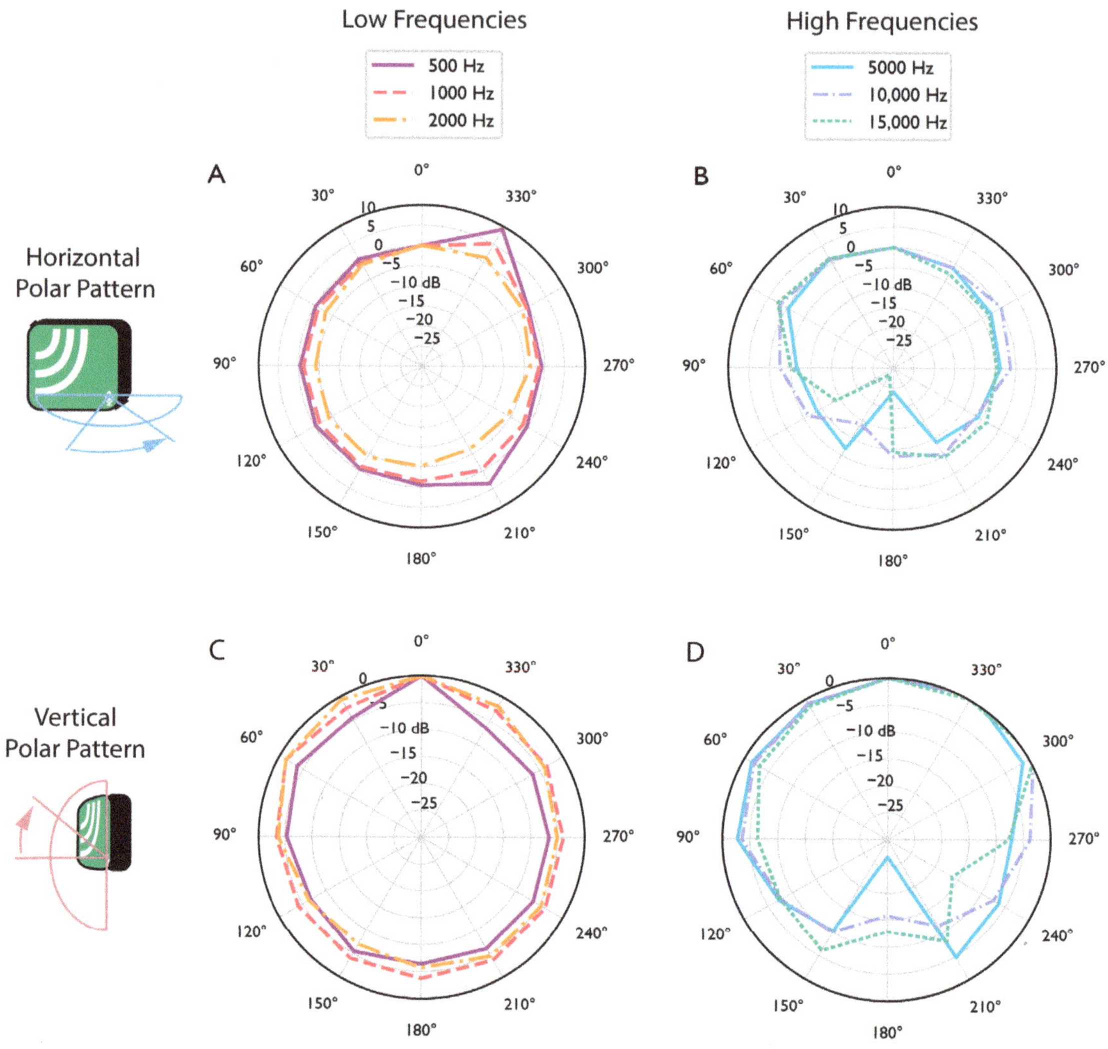

Figure 2. The horizontal polar response of an AudioMoth 1.1.0 with no case is shown at (**A**) 500–2000 Hz and (**B**) 5000–15,000 Hz (90 degrees is to the left and 270 degrees is to the right of the device). The vertical polar response for the same AudioMoth is shown at (**C**) 500–2000 Hz and (**D**) 5000–15,000 Hz (90 degrees is above and 270 degrees is below the device).

3.2. Battery Life Tests

3.2.1. Effect of Sample Rate

The average hours of audio recorded by each AudioMoth at varying sample rates is summarized in Table 1. There was little variation in recording performance between devices, and replicate performance is shown in Table S2. Under what could be considered baseline conditions, an AudioMoth records an average of 189 h of audio at 20.5 °C with Procell batteries at a 32 kHz sample rate, very close to the 187 h estimated by the AudioMoth configuration app. As expected, there is a negative relationship between the sample rate and the hours of audio the AudioMoth will record. In general, we found that the AudioMoth Configuration App underestimates recording time at low frequencies and overestimates recording time at high frequencies.

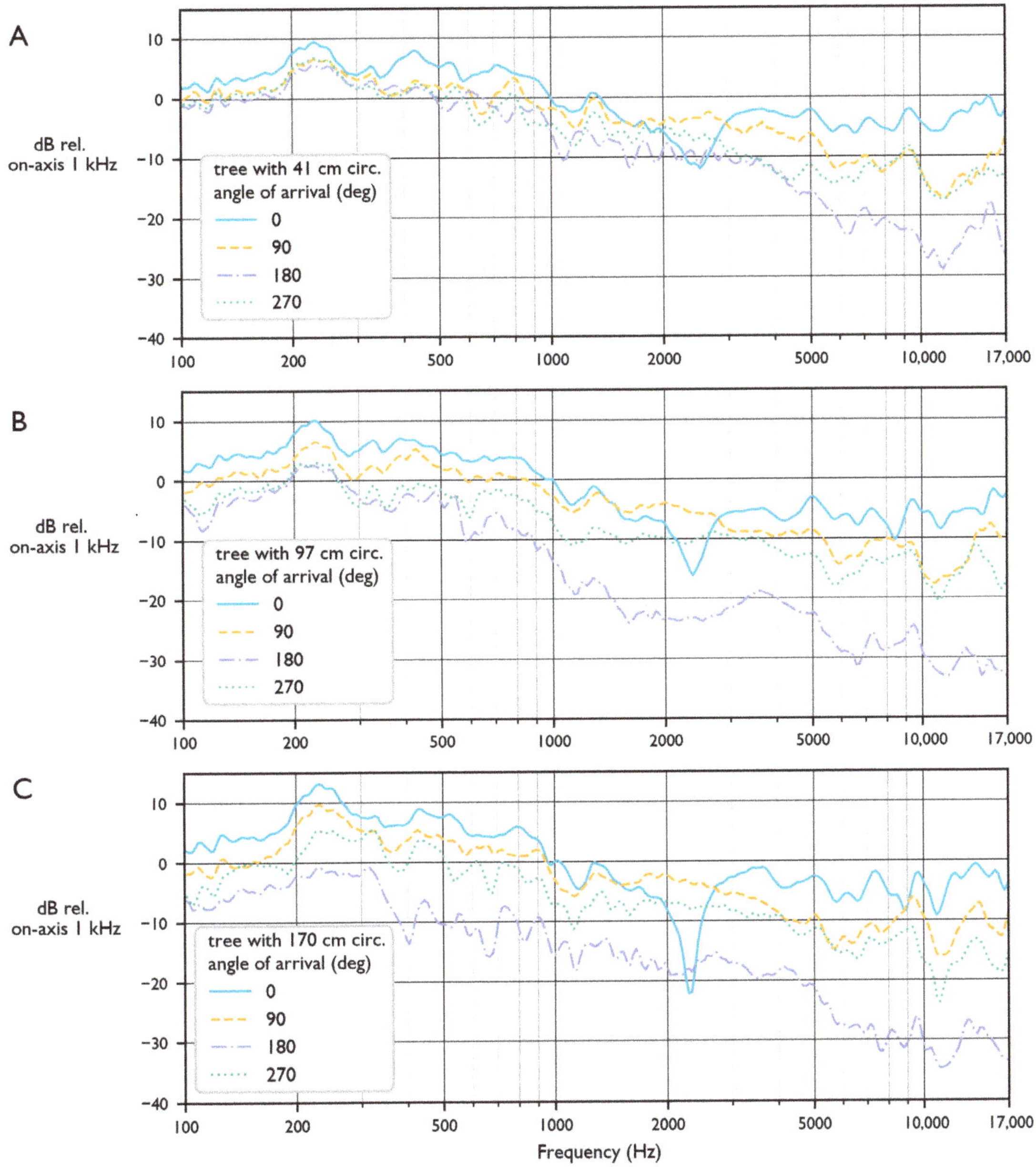

Figure 3. Pink noise recorded by three 1.1.0 AudioMoths that were strapped 1 m away from the playback speaker to the front (0°), back (180°), and sides (90° and 270°) of trees with various circumferences: (**A**) 41 cm, (**B**) 97 cm, and (**C**) 170 cm.

3.2.2. Effect of Gain

The gain setting had a negligible effect on the total hours of audio recorded. The average across settings varied by a maximum of 2%. Individual performance and the average is shown in Table S3.

Table 1. The average, minimum, and maximum hours of audio recorded across three 1.1.0 AudioMoths at each sample rate and the average amount of data written to the SD cards compared against the AudioMoth configuration app estimations.

Sample Rate (kHz)	Hours Recorded			Configuration App Estimate (h)
	Mean	Min	Max	
8	249	247	251	229
16	224	224	225	210
32	189	187	191	187
48	161	160	163	168
96 [1]	91	90	91	133
192 [1]	60	59	61	87
250 [1]	47	47	47	80
384 [1]	43	43	43	55

[1] Audio files at these sample rates filled the SD card before the batteries depleted, and required a second SD card to be inserted once the red and green LEDs began flashing.

3.2.3. Effect of Battery Type and Temperature

The Procell and lithium temperature trials are summarized in Table 2. For Procell batteries, there is a small 3% decrease in hours recorded between the control devices at room temperature (20.5 °C) and the devices at 3.4 °C. In contrast, devices at −16.1 °C recorded just more than half as long as those at room temperature. The lithium batteries had relatively consistent recording times across all temperatures, all of which were higher than the recording times for the Procell batteries. With lithium batteries, the AudioMoths recorded for a slightly longer time at colder temperatures compared to room temperature. Results were relatively consistent across devices as shown in Tables S4 and S5.

Table 2. A comparison of the average, minimum, and maximum hours of audio recorded by 1.1.0 AudioMoths set to record with a 32 kHz sample rate at different temperature ranges using three Procell alkaline or Energizer lithium AA batteries.

Temperature	Procell			Lithium		
	Avg	Min	Max	Avg	Min	Max
Room (20.5 °C)	189	187	191	234	228	239
Fridge (3.4 °C)	183	181	185	241	239	244
Freezer (−16.1 °C)	103	99	105	238	236	241

4. Discussion

4.1. Acoustic Performance Tests

In summary, our results show that the AudioMoth has favorable acoustic recording properties and battery life. The frequency response was found to be generally flat, with a slight increase in sensitivity above 3 kHz, peaking between 5 kHz and 10 kHz (Figure 1). The AudioMoth's frequency response does not vary significantly between devices (Figure S2). Changing the gain settings results in an overall change in recorded signal level, providing a total range of adjustment of 14.2 dB, and changing the gain setting does not affect the relative sensitivity across frequencies. Plastic bags used as weatherproof housing were found to attenuate frequencies above 10 kHz (Figure 1). This result is congruent with physical acoustics, which predicts that sound transmission loss through a membrane will be low for thin, flexible membranes, but will increase with frequency [28]. The polar response pattern of the AudioMoth showed that the device has lower sensitivity behind the device than in front, an effect accentuated when the recorder was mounted on a tree. This lack of omnidirectionality should be considered during data analysis, especially when estimating the area surveyed by a recorder or estimating the distance to a sound source. During deployment, attenuation of sounds arriving from behind the device can be minimized by mounting the AudioMoths on trees or stakes with small diameters (3–7 cm). We note that all tests were completed using new devices that had not previously been deployed in the field, and that acoustic performance may be affected by microphone damage after devices are left in the field for extended periods [29].

In practice, the AudioMoth is rarely used without protective housing and is often attached to a tree. When planning deployments of AudioMoths for field data collection, the effects of such choices should be considered. If a deployment strategy will cause substantial reductions in certain frequency bands, the downstream effects of this lost information should be carefully considered. For instance, strapping an AudioMoth to a 30 cm diameter tree will result in a sensitivity to high frequencies that is about 25 decibels higher in front of the device than behind the device. Considering that sound decays approximately 6 dB per doubling of distance ignoring absorption and attenuation, the maximum recording distance of an event behind the device would effectively be four times smaller than the front.

Our results show that even when the AudioMoth is not in a case and not strapped to a tree, it is not truly omnidirectional. When sound arrives from behind the microphone, there is an overall reduction of at least 5 dB for all frequencies, and certain frequencies (around 5 kHz and 10 kHz) are sharply reduced. A reduction of level in a frequency response plot can be thought of as a loss of information for each frequency. While an overall loss of level simply results in quieter files (and effectively a smaller sampling radius), the loss of specific frequencies more than others "distorts" the data by recording some sounds quieter than others of equivalent volume. This would be problematic if, for instance, a study was to attempt to compare the presence of two species with vocalizations in two specific frequency ranges, one of which was recorded with 15 dB less sensitivity than the other. During analysis, it is important to account for any frequency-dependent sensitivity of the deployment strategy.

Cases made of solid materials will have different effects on frequency response than the thin, flexible bags used for housing in this report. Sound travels easily through flexible membranes but is absorbed or reflected by thick solid membranes [28]. Higher frequencies will be attenuated more than lower frequencies when traveling across a solid membrane, and overall attenuation will be correlated with the thickness of the membrane [28]. Because the precise effects of a protective housing on the recorded audio are difficult to predict, the effect of alternative housings on frequency response should be measured before deployment.

When choosing a gain setting for an AudioMoth deployment, the goal is to choose the highest gain setting that does not cause clipping. Clipping occurs when the signal level exceeds the maximum levels that can be recorded, and causes harmonic distortion of the audio signal, which results in the introduction of harmonics not present in the real-world signal. Because field sites and organisms can differ greatly in noise levels, experimenting with different gain settings at a specific field site is the best way to determine an appropriate gain setting that maximizes recording levels while avoiding clipping.

4.2. Battery Life Tests

In summary, our results show that recording times decreased with an increasing sample rate, as expected. Expected recording times predicted by the AudioMoth configuration app for alkaline batteries were consistent with our results for sampling rates of 32 kHz. The configuration app underestimated recording times at lower sampling rates and overestimated recording times at higher sampling rates. When planning field deployments around AudioMoth battery life, we recommend using the results of our empirical experiments rather than the configuration app's estimates. As expected, cold temperatures decreased recording times with alkaline batteries, although this effect was not substantial until all temperatures fell below 0 °C. Lithium batteries were recorded for similar amounts of time at all three examined temperatures.

5. Conclusions

Our tests show that the AudioMoth's long battery life, directional pattern, and flat frequency response make it an effective recording hardware choice for bioacoustic monitoring. The recorder has a relatively flat frequency response and records sound from all directions effectively, although we note that the device is not fully omnidirectional. We found that plastic bags have little effect on sensitivity to frequencies below 10 kHz, making them ideal housings in environments where they provide sufficient protection. When using AudioMoths or any other automated recording unit, the characteristics of the recording hardware and housing should be considered during data analysis. In particular, the frequency response and directionality of the recorder, and the effect of mounting the recorder on solid objects such as trees, should be considered when evaluating recordings. The frequency response measurements and battery life tests provided in this report can be used to inform the survey design and data analysis of acoustic recordings collected by AudioMoths.

Supplementary Materials: The following supporting information can be downloaded at: https://www.mdpi.com/article/10.3390/s23115254/s1, Figure S1: Frequency response variation between five AudioMoth devices without cases relative to Device 1017; Figure S2: Frequency response for pink noise recorded at the 5 gain settings, relative to the highest gain setting; Table S1: On-axis frequency response in each housing treatment; Table S2: Individual results and overall average for the effect of sample rate on battery life.; Table S3: Individual results and overall average for the effect of gain setting on battery life.; Table S4: Individual results and overall average for the effect of temperature on Procell alkaline battery life; Table S5: Individual results and overall average for the effect of temperature on Energizer lithium battery life.

Author Contributions: Conceptualization, S.L., N.S. and J.K.; methodology, S.L., N.S. and J.K.; formal analysis, S.L. and N.S.; investigation, S.L. and N.S.; resources, S.L. and J.K.; data curation, S.L. and N.S.; writing—original draft preparation, S.L., N.S. and J.K.; writing—review and editing, S.L., N.S. and J.K.; visualization, S.L. and N.S.; supervision, J.K.; project administration, J.K.; funding acquisition, J.K. All authors have read and agreed to the published version of the manuscript.

Funding: This research was funded by the National Science Foundation, grant number 1935507, the Gordon and Betty Moore Foundation, grant number 10540, and the Department of Biological Sciences at the University of Pittsburgh.

Institutional Review Board Statement: Not applicable.

Informed Consent Statement: Not applicable.

Data Availability Statement: All data and results are available in GitHub repositories at https://github.com/kitzeslab/audiomoth-performance (accessed on 27 May 2023) and https://github.com/kitzeslab/ARU_battery_longevity (accessed on 27 May 2023).

Acknowledgments: We thank Open Acoustic Devices for productive conversations regarding the AudioMoth recorder that assisted us in carrying out these tests and Joe Begley in the Electronics Shop in the Dietrich School of Arts and Sciences at the University of Pittsburgh for assistance in fabricating AudioMoth recorders. We thank R. P. Lyon and L. M. Chronister for their assistance in performing acoustic testing. We also thank the Pymatuning Lab of Ecology at the University of Pittsburgh for providing space for acoustic testing.

Conflicts of Interest: The authors declare no conflict of interest.

Appendix A

Acoustic Performance Test Equipment List

Enviro meters A-weighted EM80 Sound Level (Sealed Unit Parts Co., Inc., Allenwood, NJ, USA)

- AudioMoth version 1.1.0 with 1.3.0 firmware (LABmaker, Berlin, Germany)
- 64 GB SanDisk Ultra microSDXC UHS-I Card (SanDisk, Milpitas, CA, USA)
- dbx RTA-M Reference Microphone (HARMAN International, Stamford, CT, USA)

- Scarlett 2i2 Audio Interface (Focusrite, High Wycombe, United Kingdom)
- Macbook Pro 2013 with Logic Pro X and Audacity 2.2.2 (Apple, Cupertino, CA, USA)
- Mackie SRM-450 Loudspeaker (Mackie, Woodinville, WA, USA)

Battery Life Test Equipment List

- AudioMoth version 1.1.0 with 1.5.0 firmware (LABmaker, Berlin, Germany)
- 64 GB SanDisk Ultra microSDXC UHS-I Card (SanDisk, Milpitas, CA, USA)
- Procell PC1500 Alkaline AA Batteries (Duracell Inc., Chicago, IL, USA)
- Energizer Ultimate Lithium AA Batteries (Energizer Holdings, Inc., St. Louis, MO, USA)
- Frigidaire Top Freezer Refrigerator Model: FFET1022UV (Frigidaire, Charlotte, NC, USA)

References

1. Sugai, L.S.M.; Silva, T.S.F.; Ribeiro, J.W.; Llusia, D. Terrestrial Passive Acoustic Monitoring: Review and Perspectives. *BioScience* **2019**, *69*, 15–25. [CrossRef]
2. Rhinehart, T.A.; Chronister, L.M.; Devlin, T.; Kitzes, J. Acoustic Localization of Terrestrial Wildlife: Current Practices and Future Opportunities. *Ecol. Evol.* **2020**, *10*, 6794–6818. [CrossRef] [PubMed]
3. Laiolo, P. The Emerging Significance of Bioacoustics in Animal Species Conservation. *Biol. Conserv.* **2010**, *143*, 1635–1645. [CrossRef]
4. Marques, T.A.; Thomas, L.; Martin, S.W.; Mellinger, D.K.; Ward, J.A.; Moretti, D.J.; Harris, D.; Tyack, P.L. Estimating Animal Population Density Using Passive Acoustics. *Biol. Rev.* **2013**, *88*, 287–309. [CrossRef] [PubMed]
5. Johnson, E.; Campos-Cerqueira, M.; Jumail, A.; Yusni, A.S.A.; Salgado-Lynn, M.; Fornace, K. Applications and Advances in Acoustic Monitoring for Infectious Disease Epidemiology. *Trends Parasitol.* **2023**, *39*, 386–399. [CrossRef] [PubMed]
6. Titley Scientific Acoustic Monitoring Products. Available online: https://www.titley-scientific.com/us/products/anabat-systems (accessed on 24 April 2023).
7. Wildlife Acoustics Recorders/Software Products. Available online: https://www.wildlifeacoustics.com/products (accessed on 24 April 2023).
8. Sethi, S.S.; Ewers, R.M.; Jones, N.S.; Orme, C.D.L.; Picinali, L. Robust, Real-time and Autonomous Monitoring of Ecosystems with an Open, Low-cost, Networked Device. *Methods Ecol. Evol.* **2018**, *9*, 2383–2387. [CrossRef]
9. Caldas-Morgan, M.; Alvarez-Rosario, A.; Rodrigues Padovese, L. An Autonomous Underwater Recorder Based on a Single Board Computer. *PLoS ONE* **2015**, *10*, e0130297. [CrossRef] [PubMed]
10. Whytock, R.C.; Christie, J. Solo: An Open Source, Customizable and Inexpensive Audio Recorder for Bioacoustic Research. *Methods Ecol. Evol.* **2017**, *8*, 308–312. [CrossRef]
11. Hill, A.P.; Prince, P.; Piña Covarrubias, E.; Doncaster, C.P.; Snaddon, J.L.; Rogers, A. AudioMoth: Evaluation of a Smart Open Acoustic Device for Monitoring Biodiversity and the Environment. *Methods Ecol. Evol.* **2018**, *9*, 1199–1211. [CrossRef]
12. Roedel, K. Reno Startup Helps Fund Global Production of Acoustic Recording Device. Northern Nevada Business Weekly. 2021. Available online: https://www.nnbw.com/news/2021/sep/21/reno-startup-helps-fund-global-production-acoustic (accessed on 24 April 2023).
13. GroupGets AudioMoth by Open Acoustic Devices. Available online: https://groupgets.com/manufacturers/open-acoustic-devices/products/audiomoth (accessed on 24 April 2023).
14. Knowles Product Data Sheet—SPM0408LE5H-TB Amplified Zero-Height SiSonic Microphone with Enhanced RF Protection. Available online: https://media.digikey.com/pdf/Data%20Sheets/Knowles%20Acoustics%20PDFs/SPM0408LE5H-TB.pdf (accessed on 26 April 2023).
15. Hill, A.P.; Prince, P.; Snaddon, J.L.; Doncaster, C.P.; Rogers, A. AudioMoth: A Low-Cost Acoustic Device for Monitoring Biodiversity and the Environment. *HardwareX* **2019**, *6*, e00073. [CrossRef]
16. Barber-Meyer, S.M.; Palacios, V.; Marti-Domken, B.; Schmidt, L.J. Testing a New Passive Acoustic Recording Unit to Monitor Wolves. *Wildl. Soc. Bull.* **2020**, *44*, 590–598. [CrossRef]
17. Manzano-Rubio, R.; Bota, G.; Brotons, L.; Soto-Largo, E.; Pérez-Granados, C. Low-Cost Open-Source Recorders and Ready-to-Use Machine Learning Approaches Provide Effective Monitoring of Threatened Species. *Ecol. Inform.* **2022**, *72*, 101910. [CrossRef]
18. Pérez-Granados, C.; Traba, J. Estimating Bird Density Using Passive Acoustic Monitoring: A Review of Methods and Suggestions for Further Research. *Ibis* **2021**, *163*, 765–783. [CrossRef]
19. Darras, K.; Batáry, P.; Furnas, B.; Celis-Murillo, A.; Van Wilgenburg, S.L.; Mulyani, Y.A.; Tscharntke, T. Comparing the Sampling Performance of Sound Recorders versus Point Counts in Bird Surveys: A Meta-Analysis. *J. Appl. Ecol.* **2018**, *55*, 2575–2586. [CrossRef]
20. GitHub: KitzesLab—A Quantitative Report of Audio Recording Quality for the AudioMoth. Available online: https://github.com/kitzeslab/audiomoth-performance (accessed on 12 April 2023).
21. GitHub: KitzesLab—ARU Battery Longevity Report. Available online: https://github.com/kitzeslab/ARU_battery_longevity (accessed on 12 April 2023).

22. Rayburn, R.A.; Eargle, J. *Eargle's Microphone Book: From Mono to Stereo to Surround: A Guide to Microphone Design and Application*, 3rd ed.; Focal Press/Elsevier: Waltham, MA, USA, 2012; ISBN 978-0-240-82075-0.
23. Decibels. Available online: https://www.dsprelated.com/freebooks/mdft/Decibels.html (accessed on 12 April 2023).
24. Bat Echolocation. Available online: https://dnr.maryland.gov/wildlife/Pages/plants_wildlife/bats/batelocu.aspx (accessed on 12 April 2023).
25. Dooling, R.J.; Lohr, B.; Dent, M.L. Hearing in Birds and Reptiles. In *Comparative Hearing: Birds and Reptiles*; Dooling, R.J., Fay, R.R., Popper, A.N., Eds.; Springer Handbook of Auditory Research; Springer: New York, NY, USA, 2000; Volume 13, pp. 308–359. [CrossRef]
26. Heffner, H.E.; Heffner, R.S. Hearing. In *Comparative Psychology*, 1st ed.; Greenberg, G., Haraway, M.M., Eds.; Routledge: New York, NY, USA, 1998; pp. 290–303. [CrossRef]
27. Sarria-S, F.A.; Morris, G.K.; Windmill, J.F.C.; Jackson, J.; Montealegre-Z, F. Shrinking Wings for Ultrasonic Pitch Production: Hyperintense Ultra-Short-Wavelength Calls in a New Genus of Neotropical Katydids (Orthoptera: Tettigoniidae). *PLoS ONE* **2014**, *9*, e98708. [CrossRef]
28. Long, M. Sound Transmission Loss. In *Architectural Acoustics*; Elsevier: Amsterdam, The Netherlands, 2014; pp. 345–382; ISBN 978-0-12-398258-2.
29. Turgeon, P.J.; Van Wilgenburg, S.L.; Drake, K.L. Microphone Variability and Degradation: Implications for Monitoring Programs Employing Autonomous Recording Units. *Avian Conserv. Ecol.* **2017**, *12*, 9. [CrossRef]

sensors

MDPI

Article

Efficient Speech Detection in Environmental Audio Using Acoustic Recognition and Knowledge Distillation

Drew Priebe [1], Burooj Ghani [2] and Dan Stowell [1,2,*]

1 Department of Cognitive Science and Artificial Intelligence, Tilburg University, 5037 Tilburg, The Netherlands
2 Naturalis Biodiversity Center, 2333 Leiden, The Netherlands; burooj.ghani@naturalis.nl
* Correspondence: dan.stowell@naturalis.nl

Abstract: The ongoing biodiversity crisis, driven by factors such as land-use change and global warming, emphasizes the need for effective ecological monitoring methods. Acoustic monitoring of biodiversity has emerged as an important monitoring tool. Detecting human voices in soundscape monitoring projects is useful both for analyzing human disturbance and for privacy filtering. Despite significant strides in deep learning in recent years, the deployment of large neural networks on compact devices poses challenges due to memory and latency constraints. Our approach focuses on leveraging knowledge distillation techniques to design efficient, lightweight student models for speech detection in bioacoustics. In particular, we employed the MobileNetV3-Small-Pi model to create compact yet effective student architectures to compare against the larger EcoVAD teacher model, a well-regarded voice detection architecture in eco-acoustic monitoring. The comparative analysis included examining various configurations of the MobileNetV3-Small-Pi-derived student models to identify optimal performance. Additionally, a thorough evaluation of different distillation techniques was conducted to ascertain the most effective method for model selection. Our findings revealed that the distilled models exhibited comparable performance to the EcoVAD teacher model, indicating a promising approach to overcoming computational barriers for real-time ecological monitoring.

Keywords: passive acoustic monitoring; eco-acoustics; deep learning; knowledge distillation; bioacoustics; classification; transfer learning; speech detection

Citation: Priebe, D.; Ghani, B.; Stowell, D. Efficient Speech Detection in Environmental Audio Using Acoustic Recognition and Knowledge Distillation. *Sensors* **2024**, *24*, 2046. https://doi.org/10.3390/s24072046

Academic Editor: Hector Eduardo Roman

Received: 27 January 2024
Revised: 1 March 2024
Accepted: 7 March 2024
Published: 22 March 2024

1. Introduction

Bioacoustics is the scientific discipline that focuses on sounds generated by animals [1]. The field offers insight into the behaviors, communication, and migration patterns of different species. Recent advances in computational bioacoustics, such as data storage and digital recording costs, have enabled the application of more advanced analytical approaches like deep learning [1]. While early deep learning methods focused on neural networks such as the multilayer perceptron (MLP), Convolutional Neural Network (CNN) and Recurrent Neural Network (RNN) models currently surpass and exceed MLP models in the field [1]. More recently, a convolution-free Audio Spectrogram Transformer (AST), an attention-based model for audio classification, was designed [2]. However, due to the quadratic complexity of self-attention, transformer-based models such as AST are known to be computationally expensive, resulting in increased latency and model size when compared with lightweight CNNs [3].

Despite the recent progress in computational bioacoustics, some practical and theoretical obstacles remain that prevent deep learning methods from broad usage in the field. A notable obstacle arises from the intricacies of dealing with human speech recordings in wildlife settings. Although these recordings serve as a useful proxy for quantifying human disturbance in ecosystems, they also allow for a more precise assessment of human presence [4]. This increased precision, while beneficial in one aspect, could lead to significant data privacy concerns as acoustic monitoring equipment becomes more advanced and more widely implemented [1]. The implications of this obstacle extend even further given the documented impact of human activity on the temporal dynamics of animal activity patterns, which include an increase in nocturnality and potential consequences for ecological interactions [5–7]. Moreover, noise pollution levels in protected areas have doubled, affecting critical habitat areas for endangered species [8]. In response to these challenges, Cretois, Rosten, and Sethi [4] developed a voice activity detection (VAD) model, EcoVAD, aimed at addressing both the need for precise measurement of human presence and privacy preservation in eco-acoustic data.

In addition to the above-mentioned theoretical challenges, there are practical challenges that prevent deep models, such as EcoVAD, from being deployed in eco-acoustic environments. The deployment of such models has high latency costs [1]. The current state-of-the-art acoustic monitoring tool, AudioMoth [9], is a low-cost, low-power solution to certain technical challenges in bioacoustics. However, AudioMoth is not efficient enough to execute deep neural networks (DNNs) in real time [10]. The challenges that DNNs bring for deploying models on small devices have led to a series of model compression and acceleration techniques, one of which is knowledge distillation [11]. The main idea behind knowledge distillation is that a student model is trained to emulate the processing performed by a larger teacher model in order to distill refined knowledge and obtain a competitive performance versus the teacher [11]. This technique allows efficient DNNs to be trained from large DNNs without a substantial drop in accuracy [11].

While knowledge distillation addresses the compression framework required for deployment on edge devices, architectural efficiency remains another critical aspect for real-time inference [1,12]. In an attempt to design a more efficient architecture, Howard et al. [13] introduced the MobileNetV1 architecture, which replaced the convolutional layer of CNNs with depth-wise separable convolutions. Specifically, the utilization of factorized convolutions through the combination of depth-wise and point-wise convolution reduced the computation required by the convolutional block by a factor of eight [13]. While the introduction of MobileNetV1 allowed for a reduction in parameters without a significant loss in accuracy, it was not effective at efficiently extracting the manifold of interest (MOI) [14]. This issue was in part due to the application of the nonlinear functions (RELU) on low-dimensional activations, which lead to information loss in the MOI. To confront this problem within the MobileNetV1 architecture, ref. [14] introduced MobileNetV2, which incorporated inverted residuals with a linear bottleneck. In order to improve the representational power of the CNN architecture, Hu, Shen, and Sun [15] implemented a Squeeze-and-Excitation block (SE), which allows the weighting of interdependencies between channels for feature selection. In light of this development, researchers then attempted to augment MobileNetV2 and introduced the SE block in the MobileNetV3 architecture. As a result, this led to an improvement in both the latency and parameter size of the model [16].

Despite the design of MobileNet architectures addressing the model complexity and latency costs for deployment on small mobile devices, these architectures are not optimized for other edge devices, such as Raspberry Pi, NVIDIA Jetson Nano, or Google Coral, which contain different hardware specifications [17,18]. In an attempt to improve the MobileNetV3 design for Raspberry Pi devices, MobileNetV3-Small-Pi was developed [18]. This architecture replaced the 5x5 filter with a 3x3 filter in the convolution block and changed the hard-swish activation function to RELU. The modifications made to the MobileNetV3 led to improvements in both latency and accuracy in MobileNetV3-Small-Pi [18].

Silva et al. [19] built a CNN-based VAD model using audio spectrograms to detect speech in audio signals. Using the LeNet 5 CNN and the Half Total Error Rate metric, the proposed method outperformed several baseline VAD models in low-, medium-, and high-noise conditions. In an effort to further optimize VAD models in noisy conditions, ref. [20] integrated a two-layer bottleneck Denoising Autoencoder (DAE) with a CNN. The researchers carried out experiments using two different feature sets, MFCCs (Mel-Frequency

Cepstral Coefficients) and filterbanks, and compared their performance in various Signal-to-Noise Ratio (SNR) conditions. The results demonstrate an improvement in classifying speech in high-noise environments. In an attempt to measure human disturbance in ecological settings, ref. [4] proposed an alternative approach for acoustic VAD models. Researchers trained CNN models on synthetic datasets containing human voices mixed with typical background noises encountered in eco-acoustic data. By proposing a specialized preprocessing pipeline for audio augmentation and synthetic dataset building, the results indicate the performance of a custom VGG11 model established a new state-of-the-art benchmark for VAD models in ecological settings. Despite the advances demonstrated in the aforementioned studies with respect to the accuracy of VAD models, the challenge of designing models that are suitable for real-time inference and deployment on edge devices remains a significant challenge. Ref. [21] proposed a lightweight CNN with data augmentation and regularization techniques to improve the generalization ability of the model. Utilizing the PreAct ResNet-18 architecture as a teacher and log-scaled Mel Spectrogram as feature inputs, researchers trained a student model using response-based distillation resulting in a lower equal error rate and latency from the distilled model. In a similar piece of research, ref. [22] proposed a response-based knowledge distillation approach, where the teacher estimates the frame probability for each sound event and provides frame-level supervision to the student model, which was trained to then discriminate ground truth speech from non-speech-labeled events. With the aim of deployment on embedded devices such as Raspberry Pi, the results indicate a 98% reduction in parameters while outperforming the teacher model.

This study addresses the challenge of deploying deep learning models for ecological speech detection within the computational constraints of small, edge devices. These cost-effective and low-power devices struggle to efficiently run complex neural networks like EcoVAD, hampering real-time bioacoustic monitoring. To circumvent these challenges, our research focuses on applying knowledge distillation to create streamlined student models that parallel the larger EcoVAD teacher model's performance. This approach is intended to overcome the inherent memory, latency, and computational limitations of such devices while facilitating a more robust model capable of effective ecological monitoring.

2. Materials and Methods

In the current study, we build on the previous research discussed above, which has proven instrumental in developing efficient, compact deep learning models suitable for deployment. We design and execute experiments to optimize deep neural networks for real-time speech detection. To achieve this objective, we investigate the suitability of MobileNetV3-Small-Pi [18] model as a student architecture for EcoVAD [4]. The aforementioned studies also highlight the significance of specialized preprocessing, efficient lightweight architectures, and distillation techniques for optimizing VAD models for such deployment. Consequently, we employ different knowledge distillation techniques while incorporating variations in the MobileNetV3-Small-Pi architecture to achieve optimal performance. Finally, we examine how reductions in parameters, floating-point operations per second (FLOPs), multiplications, and memory utilization in student VAD models influence the performance of the resulting architectures.

2.1. Knowledge Distillation Techniques

Hinton, Vinyals, and Dean [23] first popularized the knowledge distillation method by training a smaller student network, using a teacher for distilled knowledge transfer. The method, known as response -based distillation, trains the student to optimize the loss function based on the student and teacher's softened outputs. While response-based distillation allowed for "dark knowledge" to be distilled, depth is a critical aspect of feature representation learning [11,24].

In an attempt to distill intermediate representations, ref. [24] introduced feature-based distillation, which trained a student network to optimize the loss function based on the student's outputs and ground truth labels, along with the feature maps from an intermediary layer within the student and teacher, respectively. This method, which selects a teacher hidden layer as a "hint" layer and student hidden layer as a "guide", improved the generalization and accuracy of the student when compared with the teacher [11].

While featured distillation allowed for deeper representation learning, the knowledge distilled is independent of outside data examples. Thus, Park et al. [25] introduced relational knowledge distillation, a method relying upon the relations between learned representations. This method trained the student network to optimize the loss function based on the angle-wise and distance-wise relations between different data points, allowing the teacher to distill refined instance relations between the layers and outputs of the model [25].

2.2. Model Architectures

The teacher architecture used for knowledge distillation is based on a customized VGG11 architecture [4], adapted to process 128 × 128 single-color channel images, in contrast to the standard VGG11's handling of 224 × 224 RGB images. Significant modifications included the reconfiguration of input and output neurons, the introduction of batch normalization after each convolutional layer, and the implementation of a dropout strategy in fully connected layers to enhance the model's specificity for binary speech detection. Additionally, a Fast Fourier Transform (FFT) window duration of 64 milliseconds (equivalent to 1024 samples at a sampling rate of 16 kHz) with a 50% overlap (hop size of 512 samples) was selected for its proven effectiveness in audio classification tasks, as detailed in [4]. This approach is further validated by the findings of [26], particularly highlighting the significant role of normalizing the Mel Spectrograms along each frequency bin in enhancing classifier performance. By compressing the frequency into 128 Mel scale bands and implementing this normalization, the model's input is finely tuned, thereby improving its capability to accurately differentiate between speech and nonspeech elements. All student architectures were based on MobileNetV3-Small-Pi (MSP) [18]. The student architectures maintained the differences implemented in [18] with respect to MobileNetV3, more specifically, the adjustment from a 5 × 5 filter with a 3 × 3 filter in the later convolution blocks and an adjustment from the hard-swish activation function to RELU. However, the architectural differences in the students differ from MSP in a number of ways.

With the goal of analyzing the efficiency of student architectures, four different student designs were trained to measure the tradeoffs in accuracy and efficiency. The primary differences between these four architectures lie within the number of channels used in the convolutional and bottleneck layers, as well as the overall depth of the architecture, allowing for an exploration of established principles [27] to find an optimal balance between model complexity and computational efficiency. Student 1 starts with an initial 3 × 3 convolutional layer with 16 output channels, followed by a series of bottleneck layers with channels ranging from 16 to 512. This design leverages concepts from residual learning to reduce the computational cost and enhance feature extraction capability compared with prior CNNs by using depth and channel expansion to capture complex patterns within the data [14]. Student 2, while similar to Student 1, has a reduction in the number of bottleneck layers and a difference in the input channels prior to the Adaptive Average Pooling layer. The input channels are changed from 256 to 512 in this case. The reduction in bottleneck layers allows for a continuation of the reduction in the depth of the network while maintaining a higher learning capacity for feature extraction in the later stages of the network.

Student 3 was initiated with a smaller number of channels compared with the aforementioned student architectures, starting at only 4 output channels in a 3 × 3 convolutional layer and progressing through a series of more compact bottleneck layers that scale from 4 to 128 channels. This architecture emphasizes an experimental divergence from its predecessors to examine efficiency with an inherent reduction in model capacity. The decrease in initial channels and compact bottleneck design was to ensure a reduction in calculations performed for each convolutional operation while ensuring computations performed within the bottleneck layers were reduced due to smaller feature maps. The final student, Student 4, maintains a similar foundational structure to Student 3, with a 3 × 3 convolutional layer with 4 output channels in the initial bottleneck. However, the number of bottleneck layers is reduced in this case, leading to a more compact architecture with fewer layers. The channel sizes range from 4 to 64. These design changes reflect our efforts to prioritize a reduction in depth and complexity in order to assess the generalization capabilities of a simplified network.

Each student architecture maintains a similar final layer construction, which consists of an Adaptive Average Pooling layer, two 1 × 1 convolutional layers, and a flatten layer. The models also maintain the presence or absence of the SE block, as in [18]. Furthermore, each student's architecture maintains the same expansion ratio pattern, with the exception of Student 4. The differences in teacher and student architectures, which are highlighted in Table 1, influence each respective model's capacity for feature extraction and performance on the voice activity detection task.

Table 1. Summary of different teacher and student model characteristics. Avg. inference time is defined as the average time taken by the model to make a prediction on a single input instance, measured over 100 trials.

Model	Parameters	Layers	FLOPS	Multiplications	Memory (MB)	Avg. Inference Time (s)
Teacher	59,568,769	20	2,485,390,000	1,242,700,000	227	0.17
Student 1	4,662,017	215	388,459,000	194,230,000	17	0.038
Student 2	2,930,177	179	337,257,000	168,628,000	11	0.042
Student 3	502,793	179	27,353,400	13,676,700	1.91	0.0087
Student 4	52,253	114	8,648,350	4,324,170	0.19	0.0050

2.3. Dataset and Preprocessing

The current study used three distinct datasets for the EcoVAD preprocessing pipeline:

The Soundscape Dataset [4], collected from the Bymarka forest near Trondheim, Norway, contains a total of 10 days of acoustic data recorded in files of 55 s at a sampling frequency of 44.1 kHz. From the initial 10 days of recordings, a subset of data were used for the EcoVAD preprocessing pipeline, consisting of 9037 raw audio signals from a continuous 5-day forest recording sampled with the same rate and intervals.

The Libri-Speech Dataset [28], a corpus containing 1000 h of 16kHz of read English speech with a 1:1 male-to-female ratio was used for voice active detection. The data used for the EcoVAD preprocessing pipeline were a subset from the corpus containing 360 h, of which 200 h of English reading speech with a 1:1 male-to-female ratio was extracted.

The Background Noise Dataset is a combination of the ESC-50 dataset [29] and BirdClef 2017 dataset [30]. The ESC-50 dataset, used for environmental sound classification, contains 2000 environmental recordings organized in 50 classes. For training, we subsetted the data to only include 1600 recordings organized into 40 classes at 5 s intervals, removing human-related sounds. The BirdClef 2017 dataset, which includes audio recordings of various bird species, contains 36,496 audio recordings with 1500 species classes. Due to storage limitations, a subset of the dataset was used, accounting for 11,889 audio recordings belonging to 501 species. The three datasets, namely Soundscape, Libri-Speech, and Background Noise, were collectively utilized as inputs for the EcoVAD preprocessing pipeline.

The EcoVAD preprocessing pipeline [4] was used to generate a synthetic dataset consisting of 20,000 audio files, with a 1:1 distribution between speech and nonspeech audio files. The pipeline augments raw soundscape audio into processed 3 s soundscape audio clips, which are accompanied by ground truth labels denoting the presence or absence of speech. These processed 3 s soundscape audio clips were augmented with speech, background, and bird species audio recordings to build an accurate representation of the ecological soundscape. To refine the raw audio signals into features for the speech detection task, the signals were converted into 128 × 128 Mel Spectrograms containing a single color channel, as in [4]. The Mel Spectrograms were then used as input into the student and teacher architectures for training.

The Evaluation Playback Dataset [4] is an extensive collection of audio recordings designed to simulate diverse environmental conditions for the purpose of testing voice activity detection (VAD) models. This unique curated collection of three-second audio clips is derived from 48 two-minute recordings within forest and seminatural grassland environments. This dataset, consisting of 5140 audio files, incorporates audio recordings of male, female, and child voices, both in speech and nonspeech contexts, captured at distances of 1, 5, 10, and 20 m. The playback dataset allows for the final evaluation and verification of the robustness of the various student models across distinct landscapes and at varying distances.

2.4. Training and Evaluation

The synthetic dataset generated for training each student model was broken down into training, evaluation, and test sets with ratios of 60%, 20%, and 20%, respectively. All models utilized in this study were subjected to a training process that involved a maximum of 50 epochs, employing batch sizes of 32. The number of inputs for each model was set to the Mel Spectrograms' feature dimensions, where the outputs for each model were set to one. Given that the task is binary classification, this allows for the model to produce values between 0 and 1 in order to represent a prediction for speech detection. Furthermore, we use binary cross entropy with logits loss for the student losses and binary cross-entropy for the teacher loss function to accurately predict the binary classification task and replicate the training procedure implemented in [4].

Moreover, after initial hyperparameter testing, we found that the Adam optimizer [31] was best suited for the optimization algorithm. Additionally, in each distillation experiment, we employed a learning rate of 0.001. The temperature parameter, which is used to soften the probability distribution of the logits, was set to 5. The alpha parameter, which controls the balance between the distillation loss and student loss in the total loss function, was set to 0.2. Finally, an early stopping method was used to prevent overfitting. The method involved comparing the present validation loss with the best validation loss. Furthermore, a patience parameter was introduced and set to 3 in order to ensure that if the loss failed to improve over a predetermined number of epochs specified by the patience parameter, the training of the model would be completed.

The evaluation metrics used to measure student model performance include the F1 score and the Area Under the Receiver Operating Characteristic Curve (AUC) score. The F1 score is a statistical measure used to evaluate the accuracy of a binary classifier, which can be seen as the harmonic mean of precision and recall. It provides a single performance measurement that balances both the false positives and false negatives [32] . On the other hand, the AUC score represents the likelihood that the classifier will rank a randomly chosen positive instance higher than a randomly chosen negative one. It measures the area under a curve that plots the true positive rate (TPR) against the false positive rate (FPR), offering an aggregate measure of performance across all possible classification thresholds [4]. Both the F1 score and AUC score were chosen to evaluate the student models based on the metrics employed in the training of the teacher model.

2.5. Software

The python programming language (3.10.11) was used throughout the study. The preprocessing pipeline was developed using EcoVAD [4], which utilizes Librosa v.0.8.1 [33] and Pydub v.0.25.1 [34] as the audio processing libraries. The data visualizations were performed using matplotlib [35]. The pandas [36] and NumPy [37] libraries were used for data loading and preprocessing. PyTorch (2.0.0) [38] was used for developing the deep learning models. The Scikit-learn [39] library was used for the evaluation of the models. The Google Colaboratory Environment [40] was used for training the models.

3. Results

3.1. Refinement of Knowledge Distillation Techniques

The performance of student models employing various knowledge distillation techniques was assessed using median Area Under the Curve (AUC) and F1 scores, providing robust central tendency measures appropriate for our data's non-normal distribution (see Table 2). In multiple experiment runs without fixed seeds, soft target distillation yielded a median AUC of 0.98625 with a confidence interval of 0.9704 to 0.989, and a median F1 score of 0.95395 with a confidence interval of 0.9216 to 0.9596. Feature-based distillation exhibited a median AUC of 0.98795 with a confidence interval of 0.98755 to 0.99015 and a median F1 score of 0.95460 with a confidence interval of 0.95015 to 0.9583. Relational-based distillation demonstrated a median AUC of 0.98905 with a confidence interval of 0.9879 to 0.9898 and a median F1 score of 0.95900, with a confidence interval of 0.9538 to 0.96125.

Table 2. Table of results for different student models employing distinct distillation techniques. We report median AUC and F1 scores of the student models and distillation methods with bootstrap confidence intervals given in brackets.

Model	Soft Target Distillation	Feature-Based Distillation	Relational-Based Distillation
Student 1	AUC: 0.9892 F1: 0.9599	AUC: 0.9880 F1: 0.9520	AUC: 0.9899 F1: 0.9595
Student 2	AUC: 0.9908 F1: 0.9593	AUC: 0.9899 F1: 0.9594	AUC: 0.9897 F1: 0.9619
Student 3	AUC: 0.9850 F1: 0.9492	AUC: 0.9874 F1: 0.9483	AUC: 0.9880 F1: 0.9502
Student 4	AUC: 0.9870 F1: 0.9528	AUC: 0.9877 F1: 0.9542	AUC: 0.9878 F1: 0.9552
Overall	AUC: 0.98625 [0.9704–0.9895] F1: 0.95395 [0.9216–0.9596]	AUC: 0.98795 [0.98755–0.99015] F1: 0.95460 [0.95015–0.9583]	AUC: 0.98905 [0.9879–0.9898] F1: 0.95900 [0.9538–0.96125]

A pairwise comparison of the different distillation methods, assessed by the Mann–Whitney U test, did not reveal statistically significant differences in median AUC or F1 scores between the distillation methods (all p-values > 0.05). This indicates that the performance of student models is consistent across different distillation methods, suggesting that while the refinement of knowledge distillation techniques did not improve the performance of the resulting models, no substantial reduction in performance was observed either.

3.2. Impact of Parameter Reduction and Efficiency on Model Accuracy

The reduction in parameters, FLOPs, multiplications, and memory utilization had varied accuracies across different distillation techniques (Figure 1). Despite these reductions, the F1 scores of the student models did not decrease when compared with the teacher-replica model. For instance, Student 1, with only 4,662,017 parameters and 388,459,000 FLOPs, achieved a median F1 score of 0.9552 in the relational distillation method, which was higher than the teacher-replica model's F1 score of 0.9376.

The results indicate that the models are not in alignment with the assumption that a direct linear relationship exists between reductions in model parameters—inclusive of floating-point operations per second (FLOPs), multiplications, and memory utilization—and model accuracy, as Student 2 and Student 4 outperformed Student 1 and Student 3, respectively.

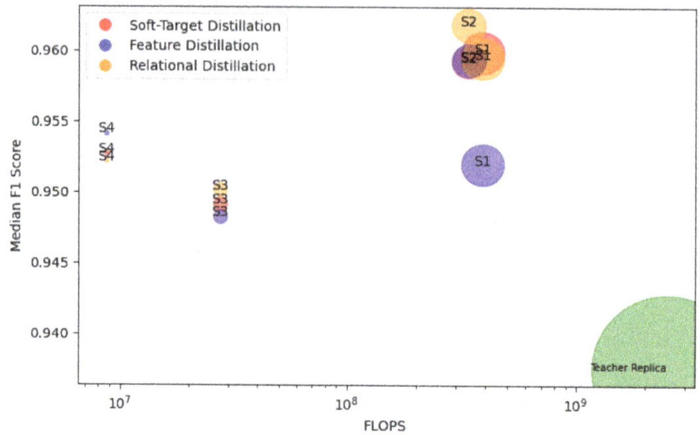

Figure 1. Varied distillation technique results per student with respect to FLOPS and Size. S1–S4 corresponds to the four student models, while the teacher replica is the EcoVAD model. The size of the circles corresponds to the number of parameters.

3.3. *Performance of Lightweight Student Models On Playback Dataset*

In terms of performance, the student models demonstrated comparable, and in one instance superior, performance relative to the EcoVAD teacher model (Figure 2). For instance, Student 1 achieved average F1 scores of 0.94595, 0.93945, 0.93875, and 0.79895 at 1, 5, 10, and 20 m, respectively, compared with the EcoVAD teacher model's average F1 scores of 0.93500, 0.94000, 0.96500, and 0.83200 at the same distances.

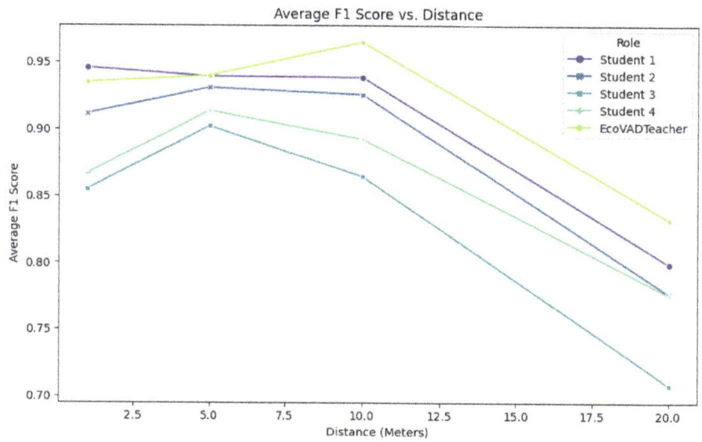

Figure 2. Avg. F1 scores based on distance on the playback evaluation data set for relational-based models. Please note: In this figure, we report mean rather than median scores to facilitate comparison with [4].

Furthermore, while the avg. F1 score across all distances for the EcoVAD model was 0.917, the student averages were 0.905, 0.886, 0.832, and 0.862 for Students 1–4, respectively. These results indicate that efficient, lightweight student models can achieve comparable performance relative to the more complex EcoVAD teacher model.

These results highlight the potential of using knowledge distillation techniques for generating efficient, lightweight models for VAD tasks. Furthermore, these models maintain their accuracy despite significant reductions in parameters, FLOPs, multiplications, and memory utilization.

4. Discussion

The goal of this study was to build an efficient general-purpose algorithm for voice detection in environmental audio by comparing different knowledge distillation techniques and student model architectures, more specifically, using the EcoVAD model [4] as a teacher and variations in MobileNetV3-Small-Pi [18] as student models to compare knowledge distillation techniques for designing an efficient EcoVAD model. The results of this study were compared with the EcoVAD model on a playback dataset to evaluate the robustness of the efficient EcoVAD models on different landscapes with varying distances.

This study demonstrates that efficient, lightweight student models can indeed achieve comparable performance relative to the EcoVAD teacher architecture using knowledge distillation and efficient student architectures. Students 1 and 2 maintained similar avg. F1 scores on the playbacks dataset using the relational distillation models when compared with the EcoVAD teacher model. This outcome supports the findings of previous research that distillation techniques can be used to create smaller, more efficient models without a significant reduction in accuracy [11]. Furthermore, the statistical analyses conducted across various distillation techniques revealed no significant effect attributable to the refinement of knowledge distillation processes on enhancing the performance of student models. Interestingly, in both feature-based and relational distillation experiments, Student 2's architecture outperformed Student 1's in the VAD task on the test dataset; however, Student 1 outperformed Student 2 on the evaluation playback dataset. This could be the result of Student 2 being overfitted on the test set; however, further testing would need to be conducted in order to determine if this is the case.

Given their reduced computational demands, these student models are well-suited for deployment in edge devices, where efficiency is paramount. This study also demonstrates that reductions in parameters, FLOPs, multiplications, and memory utilization do not necessarily result in a significant decrease in model accuracy. However, the results are not linear. The student models' performances on the evaluation dataset demonstrated that while Student 1 and Student 2 outperformed the smaller models, Student 4 consistently outperformed Student 3 on both the test set and playback dataset. This could be the result of certain architectural design features between student models, such as a difference in the expansion ratio and SE block implementation; however, further testing would need to be carried out in order to validate these claims.

5. Conclusions

This study demonstrates that efficient student models can achieve comparable performance to EcoVAD. The findings indicate that MobileNetV 3-Small-Pi [18] can serve as a backbone for building efficient EcoVad models capable of achieving results comparable to the EcoVAD teacher model [4]. This study suggests that Student 1 illustrates the feasibility of deploying effective lightweight EcoVAD models on small-edge devices for real-time ecological monitoring. These advancements are crucial for the field of ecological monitoring, offering a scalable solution for biodiversity assessment and the monitoring of human impacts on natural habitats.

The results of the current study are subject to certain limitations. This study incorporated a limited range of distillation techniques; therefore, other distillation methods could serve to improve upon the current results. Additionally, the experiments ran were nondeterministic; therefore, the implementation of a fixed seed could potentially enhance the reproducibility of these experiments. Moreover, portions of the data used to generate the synthetic dataset are proprietary and therefore restricted to research purposes only. Future research could explore the use of other distillation techniques while further investigating different variations in the student EcoVAD models presented. Additionally, research could investigate the performance of these models in real-world settings.

Our work contributes to ongoing efforts to expand eco-acoustic monitoring technologies. Our focus on efficiency and the deployment feasibility of VAD models paves the way for such algorithms to be deployed on small embedded devices, such as Raspberry Pi, to detect and remove human voices where privacy is a strong constraint or equally well to monitor patterns of human disturbance. The present study indicates that the optimization and design of efficient lightweight student models can lead to results comparable to the larger EcoVAD model. While the current study is in no way a thorough investigation into efficient VAD model design, it can be considered a contribution toward the design of an efficient general-purpose algorithm for voice detection in ecological settings.

Author Contributions: Conceptualization, D.S. and D.P.; methodology, D.S. and D.P.; software, D.P.; validation, D.P., B.G. and D.S.; formal analysis, D.P.; investigation, D.P., B.G. and D.S.; resources, D.S.; data curation, D.P.; writing—original draft preparation, D.S., B.G. and D.S.; writing—review and editing, D.S., B.G. and D.S.; visualization, D.P. and B.G.; supervision, B.G. and D.S.; project administration, B.G. and D.S. All authors have read and agreed to the published version of the manuscript.

Funding: B. Ghani is supported by the EU-funded projects Mambo and Guarden. Specifically, Mambo is funded under grant agreement No. 101060639, and Guarden is supported by grant agreement No. 101060693.

Institutional Review Board Statement: Not applicable.

Informed Consent Statement: Not applicable.

Data Availability Statement: The data presented in this study are available on request from [4].

Conflicts of Interest: The authors declare no conflict of interest.

Abbreviations

The following abbreviations are used in this manuscript:

CNN	Convolutional Neural Networks
AST	Audio Spectrogram Transformer
DNN	Deep neural network

References

1. Stowell, D. Computational bioacoustics with deep learning: A review and roadmap. *PeerJ* **2022**, *10*, e13152. [CrossRef]
2. Gong, Y.; Chung, Y.A.; Glass, J. Ast: Audio spectrogram transformer. *arXiv* **2021**, arXiv:2104.01778.
3. Pan, J.; Bulat, A.; Tan, F.; Zhu, X.; Dudziak, L.; Li, H.; Tzimiropoulos, G.; Martinez, B. Edgevits: Competing light-weight cnns on mobile devices with vision transformers. In Proceedings of the Computer Vision–ECCV 2022: 17th European Conference, Tel Aviv, Israel, 23–27 October 2022; pp. 294–311.
4. Cretois, B.; Rosten, C.M.; Sethi, S.S. Voice activity detection in eco-acoustic data enables privacy protection and is a proxy for human disturbance. *Methods Ecol. Evol.* **2022**, *13*, 2865–2874. [CrossRef]
5. Gaynor, K.M.; Hojnowski, C.E.; Carter, N.H.; Brashares, J.S. The influence of human disturbance on wildlife nocturnality. *Science* **2018**, *360*, 1232–1235. [CrossRef]
6. Lewis, J.S.; Spaulding, S.; Swanson, H.; Keeley, W.; Gramza, A.R.; VandeWoude, S.; Crooks, K.R. Human activity influences wildlife populations and activity patterns: Implications for spatial and temporal refuges. *Ecosphere* **2021**, *12*, e03487. [CrossRef]
7. Hoke, K.L.; Hensley, N.; Kanwal, J.K.; Wasserman, S.; Morehouse, N.I. Spatio-temporal Dynamics in Animal Communication: A Special Issue Arising from a Unique Workshop-Symposium Model. *Integr. Comp. Biol.* **2021**, *61*, 783–786. [CrossRef] [PubMed]
8. Buxton, R.T.; McKenna, M.F.; Mennitt, D.; Fristrup, K.; Crooks, K.; Angeloni, L.; Wittemyer, G. Noise pollution is pervasive in US protected areas. *Science* **2017**, *356*, 531–533. [CrossRef]
9. Hill, A.P.; Prince, P.; Snaddon, J.L.; Doncaster, C.P.; Rogers, A. AudioMoth: A low-cost acoustic device for monitoring biodiversity and the environment. *HardwareX* **2019**, *6*, e00073. [CrossRef]
10. Solomes, A.M.; Stowell, D. Efficient bird sound detection on the bela embedded system. In Proceedings of the ICASSP 2020–2020 IEEE International Conference on Acoustics, Speech and Signal Processing (ICASSP), Barcelona, Spain, 4–8 May 2020; pp. 746–750.
11. Gou, J.; Yu, B.; Maybank, S.J.; Tao, D. Knowledge distillation: A survey. *Int. J. Comput. Vis.* **2021**, *129*, 1789–1819. [CrossRef]
12. Hershey, S.; Chaudhuri, S.; Ellis, D.P.; Gemmeke, J.F.; Jansen, A.; Moore, R.C.; Plakal, M.; Platt, D.; Saurous, R.A.; Seybold, B.; et al. CNN architectures for large-scale audio classification. In Proceedings of the 2017 IEEE International Conference on Acoustics, Speech and Signal Processing (ICASSP), New Orleans, LA, USA, 5–9 March 2017; pp. 131–135.
13. Howard, A.G.; Zhu, M.; Chen, B.; Kalenichenko, D.; Wang, W.; Weyand, T.; Andreetto, M.; Adam, H. Mobilenets: Efficient convolutional neural networks for mobile vision applications. *arXiv* **2017**, arXiv:1704.04861.

14. Sandler, M.; Howard, A.; Zhu, M.; Zhmoginov, A.; Chen, L.C. Mobilenetv2: Inverted residuals and linear bottlenecks. In Proceedings of the IEEE Conference on Computer Vision and Pattern Recognition, Salt Lake City, UT, USA, 18–22 June 2018; pp. 4510–4520.
15. Hu, J.; Shen, L.; Sun, G. Squeeze-and-excitation networks. In Proceedings of the IEEE Conference on Computer Vision and Pattern Recognition, Salt Lake City, UT, USA, 18–22 June 2018; pp. 7132–7141.
16. Howard, A.; Sandler, M.; Chu, G.; Chen, L.C.; Chen, B.; Tan, M.; Wang, W.; Zhu, Y.; Pang, R.; Vasudevan, V.; et al. Searching for mobilenetv3. In Proceedings of the IEEE/CVF International Conference on Computer Vision, Seoul, Republic of Korea, 27 October–2 November 2019; pp. 1314–1324.
17. Kang, P.; Somtham, A. An Evaluation of Modern Accelerator-Based Edge Devices for Object Detection Applications. *Mathematics* **2022**, *10*, 4299. [CrossRef]
18. Glegoła, W.; Karpus, A.; Przybyłek, A. MobileNet family tailored for Raspberry Pi. *Procedia Comput. Sci.* **2021**, *192*, 2249–2258. [CrossRef]
19. Silva, D.A.; Stuchi, J.A.; Violato, R.P.V.; Cuozzo, L.G.D. Exploring convolutional neural networks for voice activity detection. In *Cognitive Technologies*; Springer: Cham, Switzerland, 2017; pp. 37–47.
20. Lin, R.; Costello, C.; Jankowski, C.; Mruthyunjaya, V. Optimizing Voice Activity Detection for Noisy Conditions. In Proceedings of the Interspeech 2019, 20th Annual Conference of the International Speech Communication Association, Graz, Austria, 15–19 September 2019; pp. 2030–2034.
21. Alam, T.; Khan, A. Lightweight CNN for Robust Voice Activity Detection. In Proceedings of the Speech and Computer: 22nd International Conference, SPECOM 2020, St. Petersburg, Russia, 7–9 October 2020; pp. 1–12.
22. Dinkel, H.; Wang, S.; Xu, X.; Wu, M.; Yu, K. Voice activity detection in the wild: A data-driven approach using teacher-student training. *IEEE/ACM Trans. Audio Speech Lang. Process.* **2021**, *29*, 1542–1555. [CrossRef]
23. Hinton, G.; Vinyals, O.; Dean, J. Distilling the knowledge in a neural network. *arXiv* **2015**, arXiv:1503.02531.
24. Romero, A.; Ballas, N.; Kahou, S.E.; Chassang, A.; Gatta, C.; Bengio, Y. Fitnets: Hints for thin deep nets. *arXiv* **2014**, arXiv:1412.6550.
25. Park, W.; Kim, D.; Lu, Y.; Cho, M. Relational knowledge distillation. In Proceedings of the IEEE/CVF Conference on Computer Vision and Pattern Recognition, Long Beach, CA, USA, 15–20 June 2019; pp. 3967–3976.
26. Nagrani, A.; Chung, J.S.; Zisserman, A. VoxCeleb: A Large-Scale Speaker Identification Dataset. In Proceedings of the Interspeech 2017, 18th Annual Conference of the International Speech Communication Association, Stockholm, Sweden, 20–24 August 2017; pp. 2616–2620.
27. Hu, X.; Chu, L.; Pei, J.; Liu, W.; Bian, J. Model Complexity of Deep Learning: A Survey. *arXiv* **2021**, arXiv:2103.05127.
28. Panayotov, V.; Chen, G.; Povey, D.; Khudanpur, S. Librispeech: An asr corpus based on public domain audio books. In Proceedings of the 2015 IEEE International Conference on Acoustics, Speech and Signal Processing (ICASSP), South Brisbane, Australia, 19–24 April 2015; pp. 5206–5210.
29. Piczak, K.J. ESC: Dataset for environmental sound classification. In Proceedings of the 23rd ACM International Conference on Multimedia, Brisbane, Australia, 26–30 October 2015; pp. 1015–1018.
30. Kahl, S.; Wilhelm-Stein, T.; Hussein, H.; Klinck, H.; Kowerko, D.; Ritter, M.; Eibl, M. Large-Scale Bird Sound Classification using Convolutional Neural Networks. In Proceedings of the 8th CLEF Conference, Dublin, Ireland, 11–14 September 2017.
31. Kingma, D.P.; Ba, J. Adam: A method for stochastic optimization. *arXiv* **2014**, arXiv:1412.6980.
32. Lipton, Z.C.; Elkan, C.; Naryanaswamy, B. Optimal thresholding of classifiers to maximize F1 measure. In Proceedings of the Machine Learning and Knowledge Discovery in Databases: European Conference, ECML PKDD 2014, Nancy, France, 15–19 September 2014; pp. 225–239.
33. McFee, B.; Raffel, C.; Liang, D.; Ellis, D.P.; McVicar, M.; Battenberg, E.; Nieto, O. librosa: Audio and music signal analysis in python. In Proceedings of the 14th Python in Science Conference, Austin, TX, USA, 6–12 July 2015; Volume 8, pp. 18–25.
34. Robert, J. Pydub: Manipulate Audio with a Simple and Easy High Level Interface. 2018. Available online: https://pypi.org/project/pydub/ (accessed on 26 January 2024).
35. Hunter, J.D. Matplotlib: A 2D graphics environment. *Comput. Sci. Eng.* **2007**, *9*, 90–95. [CrossRef]
36. McKinney, W. Data structures for statistical computing in python. In Proceedings of the 9th Python in Science Conference, Austin, TX, USA, 28 June–3 July 2010; Volume 445, pp. 51–56.
37. Harris, C.R.; Millman, K.J.; Van Der Walt, S.J.; Gommers, R.; Virtanen, P.; Cournapeau, D.; Wieser, E.; Taylor, J.; Berg, S.; Smith, N.J.; et al. Array programming with NumPy. *Nature* **2020**, *585*, 357–362. [CrossRef] [PubMed]
38. Paszke, A.; Gross, S.; Massa, F.; Lerer, A.; Bradbury, J.; Chanan, G.; Killeen, T.; Lin, Z.; Gimelshein, N.; Antiga, L.; et al. Pytorch: An imperative style, high-performance deep learning library. *Adv. Neural Inf. Process. Syst.* **2019**, *32*, 8026–8037
39. Pedregosa, F.; Varoquaux, G.; Gramfort, A.; Michel, V.; Thirion, B.; Grisel, O.; Blondel, M.; Prettenhofer, P.; Weiss, R.; Dubourg, V.; et al. Scikit-learn: Machine learning in Python. *J. Mach. Learn. Res.* **2011**, *12*, 2825–2830.
40. Bisong, E. *Building Machine Learning and Deep Learning Models on Google Cloud Platform*; Springer: Berlin/Heidelberg, Germany, 2019.

sensors

MDPI

Article

Directional Resonant MEMS Acoustic Sensor and Associated Acoustic Vector Sensor

Justin Ivancic, Gamani Karunasiri and Fabio Alves *

Department of Physics, Naval Postgraduate School, Monterey, CA 93943, USA; justin.ivancic@nps.edu (J.I.); gkarunas@nps.edu (G.K.)
* Correspondence: fdalves@nps.edu

Abstract: This paper reports on the design, modeling, analysis, and evaluation of a micro-electromechanical systems acoustic sensor and the novel design of an acoustic vector sensor array (AVS) which utilized this acoustic sensor. This research builds upon previous work conducted to develop a small, lightweight, portable system for the detection and location of quiet or distant acoustic sources of interest. This study also reports on the underwater operation of this sensor and AVS. Studies were conducted in the lab and in the field utilizing multiple acoustic sources (e.g., generated tones, gun shots, drones). The sensor operates at resonance, providing for high acoustic sensitivity and a high signal-to-noise ratio (SNR). The sensor demonstrated a maximum SNR of 88 dB with an associated sensitivity of −84.6 dB re 1 V/μPa (59 V/Pa). The sensor design can be adjusted to set a specified resonant frequency to align with a known acoustic signature of interest. The AVS demonstrated an unambiguous, 360-degree, in-plane, azimuthal coverage and was able to provide an acoustic direction of arrival to an average error of within 3.5° during field experiments. The results of this research demonstrate the potential usefulness of this sensor and AVS design for specific applications.

Keywords: MEMS acoustic sensor; MEMS acoustic vector sensor; resonant sensor

Citation: Ivancic, J.; Karunasiri, G.; Alves, F. Directional Resonant MEMS Acoustic Sensor and Associated Acoustic Vector Sensor. *Sensors* **2023**, *23*, 8217. https://doi.org/10.3390/s23198217

Academic Editor: Hector Eduardo Roman

Received: 25 August 2023
Revised: 26 September 2023
Accepted: 29 September 2023
Published: 1 October 2023

1. Introduction

The design, modeling, analysis, and evaluation of a micro-electromechanical systems (MEMS) directional acoustic sensor operating at resonance and an associated acoustic vector sensor (AVS) is presented. Determining the direction of arrival (DOA) of sound has long been an active field of study in acoustics. The DOA can be determined via many different devices and techniques.

1.1. Biologically Inspired Sensors

In the field of MEMS vector sensors, significant inspiration has been drawn from biology. Many MEMS acoustic sensors have been designed based on the hearing organs of humans as well as certain lizards, mosquitos, locusts, and flies [1]. The acoustic device presented here draws inspiration from the hearing organ of a fly.

In 1995 Miles et al. [2] described the hearing organ of the fly, *Ormia ochracea*, and how the fly is able to determine the DOA of sound. This hearing organ consists of two mechanically coupled tympana (eardrums). In 2006 Arthur and Hoy [3] demonstrated in lab conditions that the fly could reliably navigate towards sounds of interest. In 2008 Akcakaya and Nehorai [4] analyzed the physical performance of the ear with respect to DOA estimation and since then, various MEMS acoustic sensors, inspired by *Ormia ochracea*, have been investigated [5–28]. In general, these designs consisted of mechanically coupled vibrating membranes and a method to sense the vibration. The membrane vibration was typically sensed with capacitive circuits, piezoelectric arms, or optical sensors.

1.2. Directional Sensors

The most important factor to *Ormia ochracea* inspired sensors is that these sensors typically exhibited a cosine-like directionality, where a maximum response is detected with the DOA normal to the sensor surface. This response drops sinusoidally to zero as the DOA is rotated by ninety degrees [29].

Ishfaque et al. [12], in 2019, described a circular membrane MEMS sensor with a piezoelectric sensing system. This research investigated methods to minimize noise levels and to maximize the signal-to-noise ratio (SNR). They reported a sensitivity of −167 dB re 1 V/μPa (4.36 mV/Pa) at 1 kHz and an SNR of 66.77 dB. In 2019 Rahaman and Kim [13] demonstrated an AVS which consisted of two MEMS sensors aligned orthogonally. The DOA of incoming sound (limited to a single quadrant) was calculated using an arctangent algorithm. While a graph comparing measured and actual DOA was presented, no explicit DOA accuracy was discussed. Rahaman and Kim [15] presented a different AVS design in 2020 consisting of two coupled wing-like diaphragms. They employed an arccosine function to determine DOA with 180-degree azimuthal coverage and average error of 2.6 degrees. The SNR of the sensors was reported to be about 68.5 dB. In 2021, Rahaman et al. [16] demonstrated a double-wing-designed sensor utilizing a piezoelectric readout with a sensitivity of −139 dB re 1 V/μPa (110.5 mV/Pa) measured at 1 kHz. This sensitivity measurement was taken at a frequency significantly below the sensor's primary eigenmodes. Ren and Qi [17], in 2021, reported on a double-winged sensor that utilized a laser to measure the wing deflection, and showed a low noise floor and highly repeatable sound pressure measurements. Also in 2021, Shen et al. [18] described an *Ormia ochracea*-inspired sensor utilizing an intermembrane bridge. The sensor consisted of two separate circular membranes mechanically coupled by a structure that pivoted between the two membranes. They reported a theoretical acoustic DOA resolution of two degrees with an optimal frequency range of 300 Hz to 1500 Hz. In 2022 Rahaman and Kim [19] described an array of three double-wing sensors collocated and arranged at 120-degree angles to each other. They reported the angular resolution of the sound source localization to be ±2 degrees in the horizontal azimuth and elevation; however, localization required a priori knowledge of one or the other. The reported SNR of the sensor was 66.77 dB with a sensitivity of −167 dB re 1 V/μPa (4.36 mV/Pa).

1.3. Resonant Sensors

Another key aspect of the sensor presented in this paper is that it is designed to operate at, or near, resonance. Typically, microphones are designed to have a relatively constant sensitivity over a large frequency range. Sensors that are designed to operate near resonance trade increased performance at the cost of a reduced frequency range. In applications where tonal detection or signature-based harmonic detection is desired, resonant acoustic sensors can be designed with the advantage of mechanically filtering noise outside of the frequency range of interest [30–36].

In 1998 Schoess et al. [31] demonstrated a simple, resonant, integrated microbeam sensor for detecting pending mechanical failure in aircraft components. The reported SNR of the sensor was 6:1 (15.6 dB) and operated at a resonant frequency of 312 kHz.

In 2017, Kusano et al. [37] demonstrated a small-sized yet low-frequency (430 Hz), 3D-printed, spiral-shaped resonator based on the human cochlea. This resonator was coupled with a commercial MEMS microphone to create a low-power system tuned to detect acoustic frequencies of interest. They demonstrated a 9 dB amplification in sound pressure at the fundamental mode of the resonator with an acoustic sensitivity of up to −154.4 dB re 1 V/μPa (19 mV/Pa).

Although not a MEMS sensor, in 2020 Lee et al. [35] presented a resonant acoustic sensor for DOA determination using acoustically coupled Helmholtz resonators. In 2020 Li et al. [36] presented a 210-beam sensor. The research explored the optimization of a silicone-layer thickness for acoustic sensitivity enhancement, reporting multi-cantilever sensitivity of 72 V/m/s (sensitivity measured in terms of output voltage to unit velocity).

1.4. Combined Resonant and Directional Sensors

The sensor presented in this paper follows years of research at the Naval Postgraduate School. Touse et al. [20,21], in 2010, demonstrated a double-wing sensor design. This design featured interdigitated comb fingers between the ends of the wings and a substrate enabling a capacitive readout. The research investigated the cosine-like directionality of the sensor as well as geometric design options to emphasize particular vibration modes. In 2014, Downey and Karunasiri [22] investigated device-layer thickness and comb finger design effects on the sensors' acoustic sensitivity and overall wing displacement. In 2016, Wilmott et al. [23] developed an AVS using two double-wing sensors canted with a 30-degree offset. This design moved towards reducing the azimuthal ambiguity of an AVS. At a resonance frequency of 1.69 kHz, a sensitivity of −92.0 dB re 1 V/μPa (25 V/Pa) was measured. The AVS was able to determine the DOA to within a 3.4-degree accuracy over a ±60-degree arc. In 2020, Espinoza et al. [24] presented an investigation of a similar sensor in an underwater environment. Two sensors were presented, a double-wing and single-wing design. The sensors were placed in a housing filled with silicone oil. Underwater sensitivity for both sensors was approximately −165 dB re 1 V/μPa (6 mV/Pa) measured at resonance (125 Hz single wing, 242 Hz double-wing). Both sensors demonstrated a cosine-like directivity pattern.

Rabelo et al. [25], in 2020, demonstrated how separating a sensor's output signal into a superposition of rocking and bending modes, and determining the phase shift between them, could be used to calculate the DOA over a 180-degree arc. In 2022, Alves et al. [26] presented another underwater version of the sensor. The sensor was enclosed in a near neutrally buoyant, air-filled housing. At resonance (1.6 kHz), the sensitivity was −149 dB re 1 V/μPa (37 mV/Pa) with an SNR of about 38 dB. The sensor displayed a dipole directionality pattern that was more cosine-like than previous underwater designs. Again in 2022, Alves et al. [27] demonstrated double-wing sensors where each wing had a different resonant frequency (e.g., left wing: 718 Hz, right wing: 658 Hz). Combining the outputs of each wing was shown to broaden the frequency peak of the sensor. A sensitivity of −97.7 dB re 1 V/μPa (13 V/Pa) was measured with an average SNR of 91 dB for a band pass of 120 Hz. The sensor displayed a cosine-like directionality. Crooker et al. [28], in 2023, presented algorithms for calculating unambiguous 360-degree DOA. These algorithms account for minor differences in the individual acoustic sensors, used by the AVS, which would otherwise introduce DOA errors.

A common problem with the AVSs discussed above is the limitation in azimuthal range. Single sensors or a combination of them were not able to resolve 360 degrees of acoustic DOA without ambiguity or a priori knowledge of some parameters. The AVS design presented in this paper addresses these shortfalls. In this paper, we describe the design and characterization of a double-wing MEMS sensor, similar to the ones demonstrated in [23,27]. Using these MEMS sensors, we demonstrate a novel AVS design capable of non-ambiguous DOA determination over 360 degrees with a significantly higher SNR than similar sensors.

2. Design and Modeling

2.1. Design Requirements

Distant and quiet acoustic sources can be difficult to detect and track with conventional acoustic detectors. High SNR, directionality, small size, light weight, and low power consumption are sensor characteristics often required by many applications, particularly those associated with national defense. AVSs meeting these requirements can be made using two MEMS directional acoustic sensors operating at resonance [20–28] coupled with one omnidirectional microphone.

While the sensor described in this paper is inspired by the *Ormia ochracea*, there are significant differences between the fly and this sensor. The fly utilizes multiple vibration modes of its hearing organ to determine acoustic DOA with a single pair of mechanically coupled membranes. The sensor presented here utilizes only one vibration mode and calculates acoustic DOA using two MEMS sensors and an omnidirectional microphone.

2.2. Sensor Construction

The directional sensor presented in this paper is a MEMS device consisting of a pair of identical wings connected by a bridge, as shown in Figure 1. The wings vibrate out of the plane of the sensor substrate when exposed to acoustic waves. The bridge is anchored to the substrate by torsional legs that are aligned perpendicular to the bridge. In this sensor design, the torsional legs are located along the center of the bridge, making the sensor symmetrical. Interdigitated comb fingers are located between the ends of the wings and the substrate, allowing the displacement of the wings to be measured via a capacitance-sensing circuit.

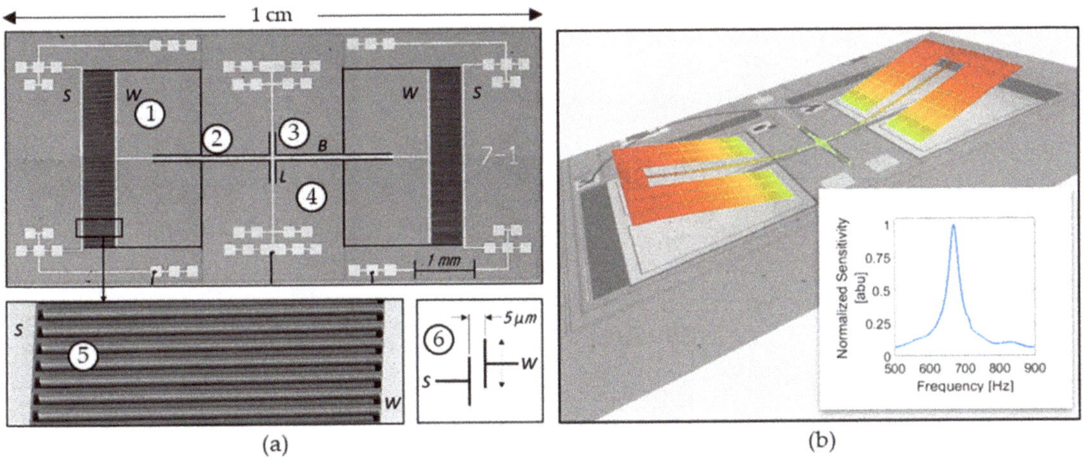

Figure 1. Image of MEMS sensor: (**a**) microscope image of MEMS sensor: (1) wing, (2) bridge, (3) torsional legs, (4) substrate, (5) SEM image of comb fingers (for capacitive sensing), (6) diagram showing the 5 μm gap between comb fingers; and (**b**) laser vibrometry image of MEMS sensor vibrating in bending mode. Inset: normalized sensitivity response of sensor showing a resonance at about 680 Hz.

When exposed to acoustic waves, the sensor is subject to a rocking mode (the wings move opposite to each other) and a bending mode (wings move with each other), as shown in Figure 1b. In this design, the back of the sensor is open to the environment, enhancing the bending mode and diminishing the rocking mode. The bending mode exhibits a cosine-like response to acoustic DOA. The sensor is designed to detect sounds with component frequencies near the resonance (~700 Hz), allowing for increased SNR in a narrow frequency band (FWHM ~25 Hz). This also reduces the effects of noise outside of the frequency band. The resonance and FWHM can be tailored by design by changing the size and shape of the sensor components. The impact of most of these design parameters translates into the stiffness of the beam, effective mass, characteristic length, and fluid density. Detailed parametric simulations are beyond the scope of this paper. Minor physical differences (due to microfabrication tolerances) in each sensor lead to observable differences in sensor responses. A detailed analysis of the fabrication imperfections is beyond the scope of this paper; however, differences in the mass of the wings due to under-etch or over-etch can cause changes in the resonant peak. Thinner or thicker comb fingers due to under- or over-etch can cause differences in the mass of the wings (changes in the resonant peak) and damping (changes in the quality factor). Microscope inspection reveals that minor differences exist from sensor to sensor (e.g., missing comb fingers, inconsistent gap etching between wing and substrate comb fingers, device-layer thickness). Laser vibrometry measurements of multiple sensors show an average resonant frequency of 671 Hz and average quality factor of 27.

The sensor is microfabricated by a commercial foundry (MEMSCAP) [38] on a 400 µm thick silicon-on-insulator (SOI) wafer with a 25 µm silicon device layer deposited on top of the wafer. The device layer includes gold contact pads which allow for the sensor to be wire-bonded to a printed circuit board (PCB) which contains a capacitance readout circuit, as shown in Figure 2.

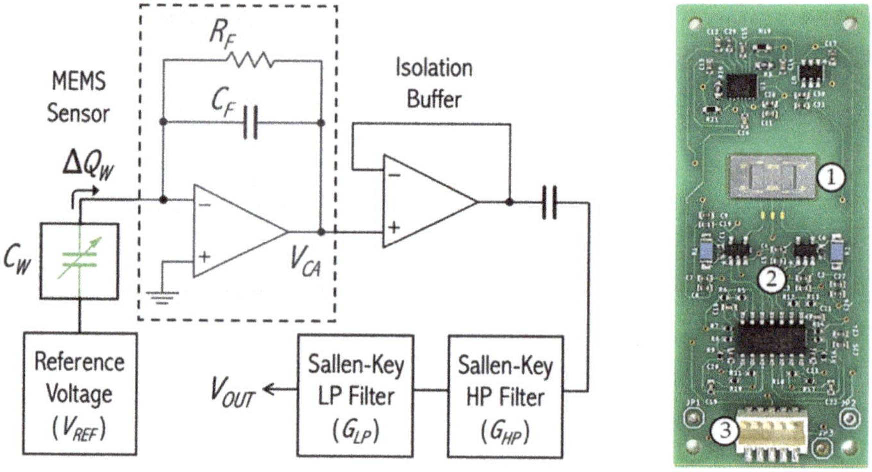

(a) (b)

Figure 2. (**a**) MEMS sensor capacitive readout circuit diagram; and (**b**) MEMS sensor mounted in PCB: (1) MEMS sensor, (2) capacitive readout circuitry, (3) wired connection to supply power and to read sensor output.

Figure 2a shows a block diagram of the electronic readout. The charge amplifier circuit (dashed rectangle) includes an operational amplifier, which connects sensor and feedback capacitors to a high-impedance virtual ground at its inverting input. The variable capacitance C_W represents the interdigitated comb finger capacitors of the MEMS directional sound sensor. Assuming that C_W is biased under a constant voltage V_{REF}, when the wings are at rest, an equilibrium position of the comb fingers' overlap is achieved and the capacitance between them is constant. Consequently, the accumulated charge on its plates, Q_W, is constant as well. Under these conditions, no electric current (ideally) flows to, or from, the op amp nor does it flow through the feedback network; the output voltage V_{CA} follows the inverting input voltage, which in this case is virtual ground. When an acoustic pressure wave impinges on the sensor's wings, their equilibrium position at rest is changed along with the overlap of comb fingers, resulting in a change in the wing capacitance to a new value, C'_W. If the biasing voltage is kept constant, the accumulated charge in the comb finger capacitors must readjust to a new value, Q'_W. The charge difference, ΔQ_W, is pushed to, or pulled from, the remainder of the circuit. Since the op amp input impedance is much larger than that of the feedback capacitor, C_F, all the charge variations from the sense capacitor, C_W, move through C_F, developing an output voltage, V_{CA}, given by

$$V_{CA} = -\frac{\Delta Q_W}{C_F} = -\frac{\Delta C_W}{C_F} V_{REF}. \tag{1}$$

Two Sallen–Key filter stages are added to limit the passband between 100 and 3500 Hz.

The sensor is intended to be used both in air and underwater. For underwater operation, the sensor is installed in a smaller PCB with a similar readout circuit. The sensor and PCB are enclosed in an air-filled, watertight housing, shown in Figure 3. The housing is a

3-D printed cylinder constructed of plastic (Rigur, VeroWhitePlus) and rubber (AGILUS30 FLX935) [39]. This material was found to be acoustically opaque (~3.5% transmissivity) for the frequency range of the sensor. For underwater usage, the sensor is in a near neutrally buoyant housing. In this configuration, the housing vibrates with the same velocity as the particle velocity of the fluid that would result if the housing was removed [40]. When the sensor housing is exposed to an acoustic wave, the MEMS sensor behaves like an inertial sensor. The wings act as a proof mass while the substrate vibrates with the rest of the sensor housing.

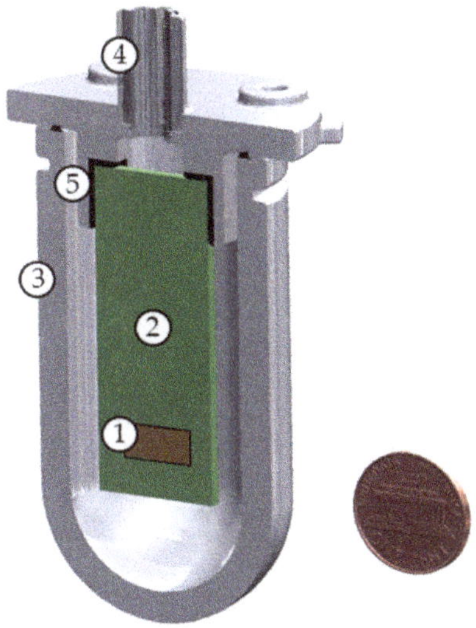

Figure 3. Planar sliced diagram of 3-D printed underwater housing with mounted PCB (penny included for sense of scale): (1) MEMS sensor, (2) PCB, (3) watertight sensor housing, (4) watertight wire penetration, and (5) bracket to mount PCB to housing.

2.3. *Analytical Modeling*

The sensor operation can be approximated as a driven, damped harmonic oscillator. The complexities in the shape of the sensor and various sources of damping make this design appropriate for finite element modeling (FEM). However, some simple assumptions and approximations allow for analytical methods to be used for the sensor design.

In the bending mode, the wings bend up and down together, causing no torque to be applied to the torsional arms. Analytical models of the sensor can be simplified to a single wing in a clamped-free configuration. The stiffness of the micro-scale sensor bridge matches that of larger-scale beams [41].

$$k_{bridge} = \frac{Ewt^3}{4L^3}, \tag{2}$$

where E is Young's modulus, w is the width of the beam, t is the thickness of the beam, and L is the length of the beam. The stiffness of the wing is much greater than the stiffness of the bridge, therefore the wing does not bend significantly compared to the bridge. Therefore, the sensor can be further simplified to a simple mass-loaded beam. In this model, the wing is treated as a point mass (m_{eff}) located at the end of the bridge with an equivalent moment of inertia in the wing. The natural resonant frequency can then be modeled by

$$\omega_0 = \sqrt{\frac{k_{bridge}}{m_{eff}}}. \tag{3}$$

Resonating MEMS devices are subject to multiple damping effects. The significance of these damping sources depends strongly on the design of the device [42]. The damping caused by the fluid surrounding our sensor is modeled by accounting for two primary effects: the damping that exists due to a beam vibrating alone in a fluid and damping caused by the fluid flow between the sensor and the substrate. Sader [43] describes a method for modeling a cantilever beam of an arbitrary but uniform cross-section that vibrates while immersed in a viscous fluid. Our sensor does not have a uniform cross-section; however, Sader's method was adapted to model it. Sader's method calculates a Reynolds number using a characteristic length based on the width of the beam. Our model uses the width of the wing when calculating this Reynold's number, given by

$$Re = \frac{\rho \omega b^2}{4\mu}, \tag{4}$$

where ω is the vibrational frequency of the beam, ρ is the fluid density, μ is the fluid dynamic viscosity, and b is the characteristic length (width of the wing for our purposes). The Reynold's number is used to calculate a hydrodynamic function (Γ), which is used to determine the resonant frequency and quality factor. Multiple equations are needed to calculate Γ, and are beyond the scope of this paper. Suffice to say, we used the hydrodynamic function to calculate the resonant frequency and quality factor using

$$\frac{\omega_1}{\omega_0} = \left(1 + \frac{\pi \rho b^2}{4\mu}\Gamma_r\right)^{\frac{-1}{2}}, \tag{5}$$

and

$$Q_1 = \frac{\frac{4\mu}{\pi \rho b^2} + \Gamma_r}{\Gamma_i}, \tag{6}$$

where ω_1 and Q_1 are the resonant frequency and quality factor modeled by Sader's method. Γ_r and Γ_i are the real and imaginary parts of the hydrodynamic function detailed in [43]. When the sensor operates in air, Sader's method produces estimated resonant frequencies within 6% of the average measured values. However, when the sensor is placed in a more dense, less viscous fluid, Sader's model by itself is insufficient.

Sader's model assumes that the beam vibrates alone in the fluid. Our sensor is surrounded by a substrate. There is a narrow gap between the edges of the wing and substrate, and an even more narrow gap between the interdigitated comb fingers of the wing and substrate. Couette flow in these gaps was considered. The drag force from the Couette flow can be related to the mechanical resistance of a simple harmonic oscillator system [44,45] as

$$R_m = \mu\left(\frac{A_w}{g_w} + \frac{A_c N}{g_c}\right), \tag{7}$$

where R_m is the mechanical resistance. A_w, A_c, g_w, g_c, and N are the surface area on the sides of the wing, surface area on the side of a single comb finger, the gap distance between the wing and substrate, gap distance between comb fingers, and number of comb fingers on the wing, respectively. The Couette flow contribution to the quality factor of the sensor is then

$$Q_2 = \frac{m_{eff}\omega_0}{R_m}. \tag{8}$$

These two separate damping sources are accounted for in our analytical model by calculating a total quality factor, Q_t, via

$$Q_t = \left(\frac{1}{Q_1} + \frac{1}{Q_2}\right)^{-1}. \tag{9}$$

A best fit curve for Q as a function of frequency is calculated based on $Q \to 0$ as $\omega \to 0$, $Q \to \infty$ as $\omega \to \omega_0$, and $Q(\omega_1) = Q_1$. Q_t is evaluated against this best fit curve to determine the resonant frequency, ω_t.

To date, the sensor described in this paper (model 7-1) has been operated only in air (the sensor has been used in underwater applications while encased in an air-filled housing). However, a similar sensor (model 7-3) has been operated in air and in a low-viscosity silicone oil. Table 1 shows a comparison of the analytically modeled resonance frequency and quality factor against average values measured in laboratory conditions.

Table 1. Modeled sensor behavior compared against sensor behavior measured in laboratory. The error of the 7-1 analytical model is notably larger than the 7-3 error. A contributing factor is the characteristic length, b, used to calculate the Reynolds number in (3). The analytical model assumes that wing width is sufficient to use as the characteristic length, b. However, the ratio of the wing width to bridge width for the 7-1 design is an order of magnitude greater than it is for the 7-3 design.

Model	Key Design Parameters [μm]	Resonant Freq Model [Hz]	Q_1	Q_2	Quality Factor Model	Resonant Freq Measured [Hz]	Quality Factor Measured
7-1 Air	Wing Width = 3000 Wing Length = 1450 Bridge Width = 80 Bridge Length = 1960	607	26	162	22	664	27
7-3 Air 7-3 Oil	Wing Width = 2400 Wing Length = 1595 Bridge Width = 500 Bridge Length = 1400	2345 435	53 22	1017 48	50 15	2340 440	59 4

2.4. Finite Element Modeling

FEM analysis of the sensor designs was conducted using COMSOL Multiphysics software version 6.1. The FEM included the wings, substrate, and a sphere of surrounding fluid. To reduce computing requirements, the model was bisected along the center of the length of the sensor bridge with symmetry constraints applied along this bisection.

The material properties of the substrate and sensor components were modeled as anisotropic, single-crystal silicon based on the properties of silicon used in the sensor fabrication [46]. The elasticity matrix of the material was adjusted to match the orientation of the silicon in our sensor. The fluid surrounding the sensor can be modeled by a variety of substances. However, as this sensor is intended to operate in air, the fluid sphere was modeled as air at standard temperature and pressure from the COMSOL material library. The fluid sphere was surrounded by a shell that was a perfectly matched layer of air to prevent acoustic reflections.

The FEM included pressure acoustics and solid mechanics. Acoustic wave radiation conditions were applied to the outside edge of the sphere. Boundary loads were applied between the wings and substrate to simulate the resistance from the flow through the gaps between the sensor and substrate. Separate boundary loads were applied to the top surface of the wings to account for form friction. The acoustic wave was modeled as a background pressure field plane wave. The acoustic DOA was set using an adjustable parameter to control the wave direction. Figure 4a depicts the meshing scheme used for the sensor mode. Figure 4b shows the results of a sound-pressure-level study while depicting the sensor in the bending resonant mode. In both Figures, the fluid within the sphere is transparent for the sake of image clarity. Figure 4c depicts a detailed image of the sensor in the bending mode and highlights some of the key boundary conditions utilized in the model.

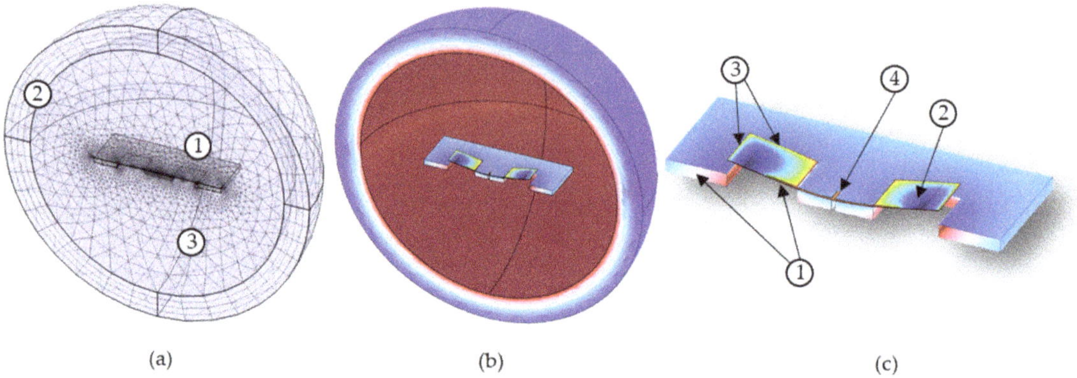

Figure 4. FEM study of MEMS acoustic sensor: (**a**) mesh scheme of sensor and fluid: (1) The sensor meshing utilizes a combination of free triangle mesh with swept mesh. (2) The boundary of the fluid sphere is a perfectly match layer that uses swept mesh. (3) The fluid inside the shell uses a free tetrahedral mesh; (**b**) results of a sound pressure level study showing the sensor in the bending mode; and (**c**) Detailed image of the sensor in the bending mode. Boundary conditions were applied to the model: (1) symmetry conditions along the bisected edge of the sensor and substrate, (2) boundary load conditions applied to top surfaces of wings, (3) separate boundary conditions applied to sides of wings and comb fingers along the gap between the sensor and substrate, and (4) fixed constraint attaching the torsional leg to the substrate.

Two studies were conducted as part of the FEM analysis: a frequency sweep and a directionality analysis, where the displacement of the wing and the associated phase shift from the pressure wave were modeled. The frequency sweep was performed for a constant angle of incidence of 45 degrees, as seen in Figure 5a. In the directionality analysis, the direction of propagation of the incident wave was rotated 360 degrees in 5-degree increments around the bisection while the frequency was kept constant. The FEM shows a cosine-like response to the DOA, as seen in Figure 5b.

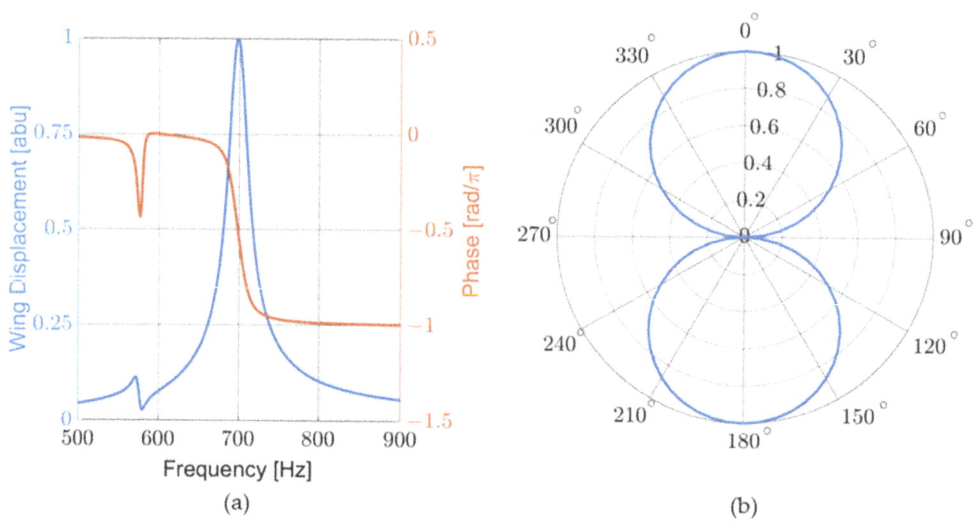

Figure 5. COMSOL FEM of sensor behavior: (**a**) normalized mechanical response and sensor phase response to frequency sweep. The large peak at 700 Hz corresponds to the bending mode. The phase response matches that of a simple harmonic oscillator; and (**b**) normalized mechanical response to acoustic propagation angle.

Electro-static effects were not modeled in the FEM analysis. Data collection on earlier generations of similar sensors showed that electrostatic forces do not appreciably affect the performance of the sensors. For this sensor, measurements were taken with the sensor electrically isolated and electrically connected to a powered PCB. No significant changes to the resonant frequency or quality factor were noted.

The FEM resonant frequency was 699 Hz with a quality factor of 33. Laser vibrometry measurements of multiple sensors showed an average bending mode frequency of 671 Hz and an average quality factor of 27. The sensor resonant frequencies ranged from 662 Hz to 679 Hz. These results indicate that there are physical effects which impact the sensor response that were not fully accounted for in the FEM, such as dimensional variation due to fabrication tolerances and actual material properties. However, the FEM, as it stands, is a useful tool for sensor design.

2.5. DOA Estimation

The MEMS acoustic sensors described in this paper were used to create an AVS array by placing two collocated MEMS sensors perpendicular to each other with an omnidirectional microphone (Knowles MEMS microphones model: SPM0687LR5H [47]) or hydrophone (Brüel & Kjær (B&K), Nærum, Denmark, Type 8103 [48]) placed between them. The DOA was measured from the normal of the reference MEMS sensor, which was designated here as the cosine sensor. The perpendicular MEMS sensor was designated as the sine sensor. A diagram of the AVS array design is shown in Figure 6a. A three-dimensional representation of how the individual sensors are arranged to form the AVS is shown in Figure 6b. The algorithm employed to calculate the DOA is explained in detail in [28] and it is given by

$$DOA = \operatorname{atan}\left(\frac{\sum|M_s V_s|\operatorname{sgn}(num)}{\sum|V_C|\operatorname{sgn}(den)}\right), \tag{10}$$

where

$$num = \sum Re\{M_S V_S V_O^*\} \tag{11}$$

and

$$den = \sum Re\{V_C V_O^*\} \tag{12}$$

where V_S, V_C, and V_O are the sine, cosine, and omnidirectional sensor voltage Fourier transforms. M_S is a correction function that accounts for the differences in frequency response of the two MEMS directional sensors. The term sgn(.) is the sign function. Since the omnidirectional microphone is not ideal, only the phase contribution was used (sgn(*num*) and sgn(*den*) terms) to determine the quadrant.

Many underwater acoustic localization applications require the knowledge of azimuth and elevation, which was not possible with the single AVS demonstrated in this manuscript. The use of two devices orthogonally placed would solve this problem. A more elegant option being studied by our group is to use the rocking vibrational mode of the MEMS sensors, which has a sine dependence on the incoming sound, in combination with the bending motion. This would provide 3D unambiguous coverage with the same arrangement demonstrated in this paper.

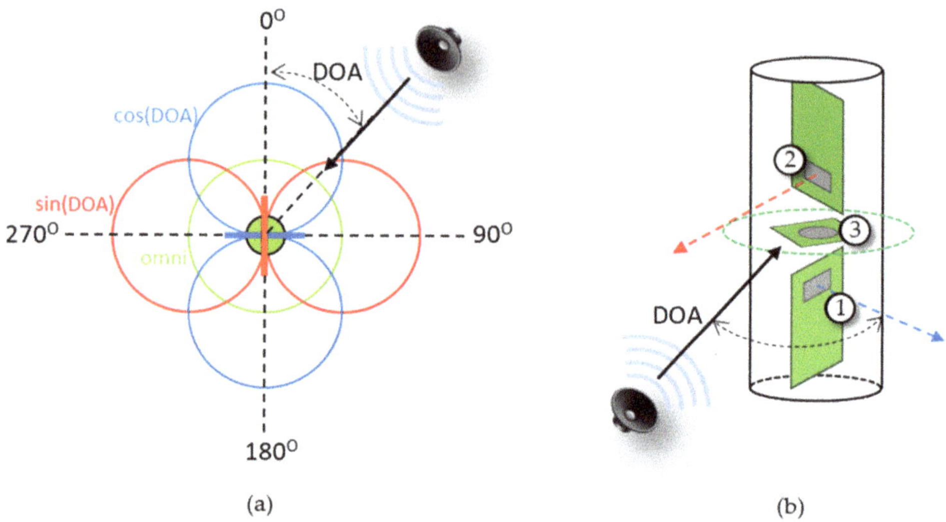

Figure 6. Diagram of AVS: (**a**) top-down view. The AVS can be used to determine the direction of incoming sound (DOA) and consists of a cosine sensor (blue), orthogonal sine sensor (red), and omnidirectional sensor (green); and (**b**) 3-D view: (1) cosine sensor, (2) sine sensor, and (3) omnidirectional sensor.

3. Methods

3.1. Mechanical Sensitivity Measurements

All in-air laboratory measurements were taken inside an anechoic chamber. The chamber is made with 12-inch-thick concrete walls and is mechanically and acoustically isolated from the rest of the building housing the chamber. It is surrounded on the walls, ceiling, and floor with fiberglass wedges which absorb 99% of incident sound for frequencies greater than 100 Hz [27].

Mechanical sensitivity measurements were taken using a laser vibrometer setup consisting of a Polytech data management system (DMS), OFV-534 laser unit, and OFV-5000 controller, as shown in Figure 7. An electrically isolated MEMS sensor was placed in a holder in the path of the laser beam. The deflection of the sensor's wings was measured at the edge of the wing just before the beginning of the comb fingers. A frequency sweep signal was generated from the DMS through a Techron 5507 amplifier to a JBL 2450H speaker with a 2380A cone pointed toward the sensor. A Piezotronics Model 378A21 calibrated reference microphone was positioned near the MEMS sensor. The microphone signal was sent through a Piezotronics Model 482C sensor signal conditioner to the DMS. The DMS software (Polytec Vibrometer version 4.7) averaged 5 consecutive frequency sweeps to calculate a mechanical sensitivity curve, as discussed in Section 4.

3.2. Electrical Experimental Setup in Air

Electrical sensitivity and directionality measurements were taken in an anechoic chamber with the sensor electrically connected to a PCB. The MEMS sensors (either individually or in an AVS configuration) were mounted on a B&K Model 5960 turntable, operated by a B&K Type 5997 turntable controller. A rubber dampening device was installed between the sensor and turntable to reduce mechanical coupling. An acoustic signal generated by a Zurich Instruments Multifunction Lock-in Amplifier (MFLI) was sent to a Techron 5507 amplifier to a JBL 2450H speaker with a 2380A cone, which was pointed toward the AVS. Parallel signals from each sensor were sent to individual MFLIs and to a microprocessor (which calculated the DOA). An Agilent 33220A waveform generator was used in conjunction with the MFLI to generate broadband white noise. The experimental setup is shown in Figure 8.

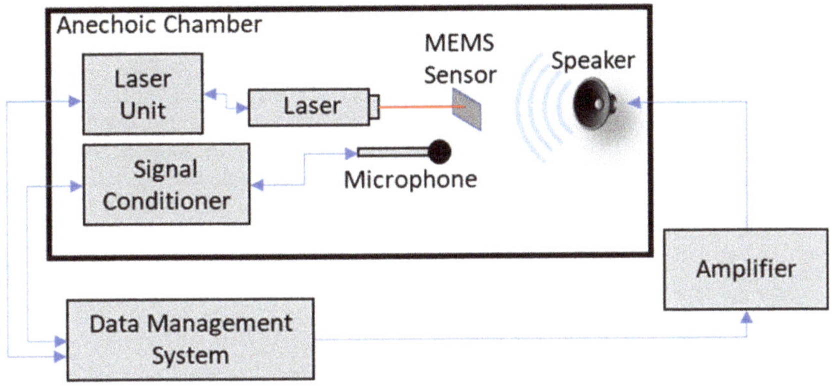

Figure 7. Experimental setup for mechanical sensitivity measurement.

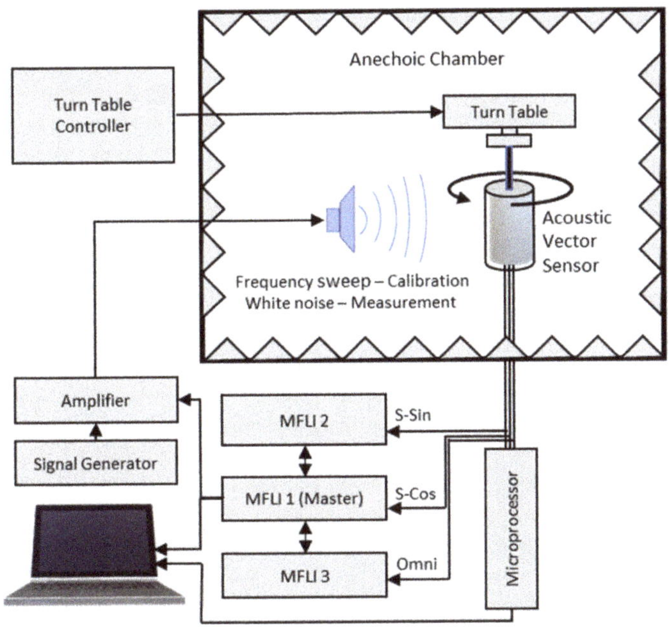

Figure 8. Schematic diagram of the experimental setup for AVS calibration and DOA estimation.

Electrical sensitivity measurements were made with the sensor facing the speaker (DOA = 0 degrees), while the speaker produced a frequency sweep generated by the MFLI. The output of the sensor was sent to the MFLI. A Piezotronics Model 378A21 calibrated reference microphone was positioned near the MEMS sensor during the frequency sweeps. The microphone signal was sent through a Piezotronics Model 482C sensor signal conditioner (with a gain applied such that the microphone signal corresponded to the pressure) to a separate MFLI. These signals were divided to generate the electrical acoustic sensitivity of the sensor, as discussed in Section 4.

Sensor directionality was measured with single tones (near sensor resonance) generated by the MFLI. The turntable was rotated at a constant angular velocity through a 360-degree circle. The sensor output was sent to the MFLI, which continuously recorded sensor output as a function of time during the rotation. The time corresponded to the DOA angle and a radial plot of sensor directionality was generated, as in Figure 11b.

Prior to an AVS being used to calculate DOA, it must be calibrated. To calibrate the AVS, the turntable was rotated such that the DOA was set to 45 degrees. The signal generator produced a frequency sweep. The outputs of each MEMS sensor and omnidirectional microphone were sent through a microprocessor control box to a computer. A computer program received the output from the three sensors and computed the correction function, M_S. This correction function was ingested in the arctangent estimator algorithm, which was run by a Teensy 4.0 microprocessor in the control box.

To characterize the AVS performance, the AVS was exposed to various acoustic signals (e.g., single tones, white noise, gunshot recordings, drone recordings). The turntable was adjusted to a known DOA value, and the microprocessor was triggered. The microprocessor calculated the DOA. The actual and calculated DOA were recorded. The turntable was then rotated to a new DOA value and the process was repeated.

3.3. Field Experimental Setup

AVSs were operated in the field to measure DOA accuracy when exposed to actual gunshots. A node consisting of the AVS (encased in a protective housing and dust cover) and microprocessor control box were placed on an Edelkrone Pan Pro motorized single-axis rotation device mounted on a tripod, as shown schematically in Figure 9. The rotation device allowed for precise DOA angles to be set and allowed for continuous rotation. The AVS was adjusted to point towards the location of the shooter (approximately 200 m away), the shooter would fire their weapon, the AVS would calculate the DOA to the gunshot. Both the actual and calculated DOAs were recorded. The AVS was rotated to a new DOA and the process was repeated. Various rifles and handguns were used for separate DOA measurements.

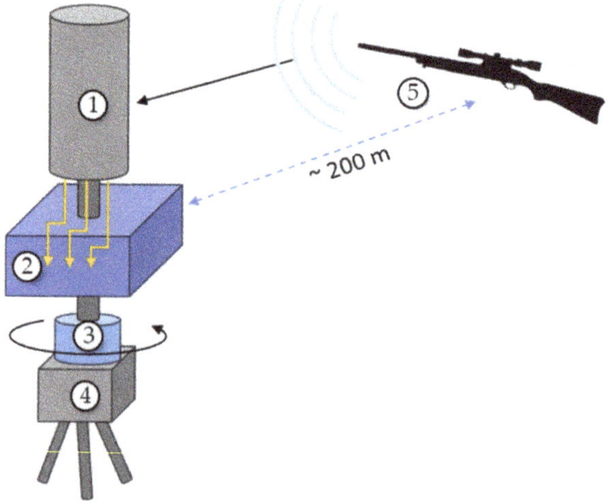

Figure 9. Schematic of field AVS DOA measurement setup: (1) AVS, (2) sensor signals connect to control box containing a microprocessor, (3) rotating mount, (4) tripod, and (5) rifle (sound source).

3.4. Experimental Setup under Water

Individual underwater MEMS sensors were measured in the SWT, which is an aluminum tube with an outer diameter of 30.5 cm (12 in), an inner diameter of 25.4 cm (10 in) and is 61.0 cm (24 in) tall. An Electro Voice UW30 underwater loudspeaker was placed at the bottom of the SWT on vibration-damping material. The SWT was filled with water such that the water surface was approximately 33 cm (13 in) above the top of the speaker. The sensor, contained in a watertight housing, was held on a suspended rod with a rotating mechanism, which allowed for the DOA of the sensor to be adjusted.

An acoustic signal was generated by an MFLI and sent to a Hewlett Packard 467A amplifier, which powered the speaker. Acoustic measurements were taken by a B&K Type 8103 hydrophone. The hydrophone output was sent to a Stanford Research Systems Model SR560 low-noise preamp and then to an MFLI. The MEMS sensor output was sent directly to an MFLI.

Electric acoustic sensitivity measurements were made with the sensor facing the speaker (DOA = 0 degrees). A frequency sweep was generated by the MFLI. Outputs from the hydrophone and MEMS sensor were sent to MFLIs. The output of the hydrophone was used to calculate the acoustic pressure, which was then used with the MEMS sensor output to determine sensitivity.

The directionality measurements for the underwater sensor were taken in the similar manner to the in-air directionality data. Directional measurements were made by connecting the Edelkrone Pan Pro rotation device to the rotator at the end of the suspension rod. The MFLI generated a single tone while the sensor was rotated at a constant angular velocity through 360 degrees.

Underwater AVS measurements were taken at the TRANSDEC anechoic pool. The AVS was suspended on a rotating pole 2 m away from a Lubell Labs VC2C underwater speaker. Both the sensor and speaker were 6 m from the surface and bottom of the pool. As with in-air AVS data collection, the MEMS sensor outputs were sent to the microprocessor control box. Instead of a microphone, the underwater AVS used a B&K Type 8103 hydrophone as the omnidirectional sensor. The hydrophone output was sent to a Stanford Research Systems Model SR560 low-noise preamp and then to the microprocessor. The rotating pole was adjusted to set the AVS to a known DOA while a constant tone was played by the underwater speaker. The microprocessor was triggered, and the DOA was calculated. The actual and calculated DOAs were recorded. Then, the pole was rotated to set a new DOA and the process was repeated.

4. Experimental Results

4.1. Operation in Air

Laser vibrometry measurements were taken for all MEMS sensors. Wing displacement was measured and compared against the acoustic pressure of a frequency sweep to determine the mechanical sensitivity (µm/Pa). These measurements were conducted in an anechoic chamber and the response of a typical pair of sensors is shown in Figure 10. Note the mismatch of the resonant peaks. This mismatch is compensated by the correction function, M_S, in (10) and (11).

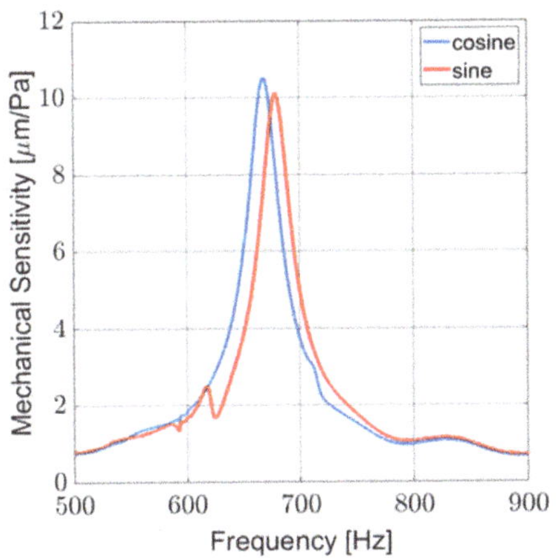

Figure 10. Laser vibrometry measurement of mechanical sensitivity of typical MEMS sensors used in AVS.

Electrical sensitivity (V/Pa), as well as directionality, were measured in an anechoic chamber. Figure 11a shows the response of a sensor in air for a frequency sweep conducted with the sensor facing the sound source (DOA ≈ 0 degrees). Figure 11b shows the cosine-like response of the sensor as it is rotated near an acoustic source producing a single tone near resonance. Although not pictured, additional directional response measurements were taken for a variety of single-tone frequencies and broadband stimuli with similar results.

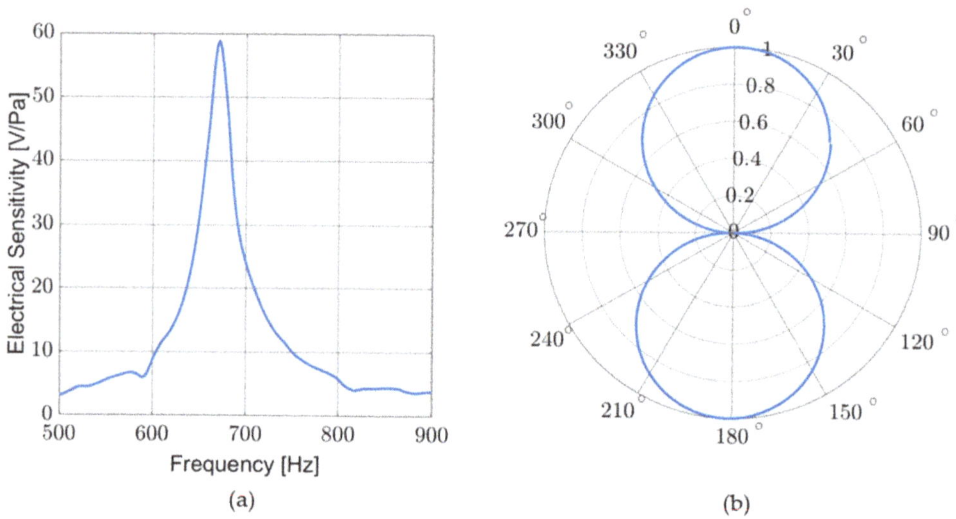

Figure 11. Sensor response in air: (**a**) electrical sensitivity measured for a frequency sweep. Electrical sensitivity corresponds to mechanical sensitivity; and (**b**) normalized sensor response to a single tone (near resonance) while rotating sensor to sweep DOA.

Noise measurements were performed in an anechoic chamber with the sensor readout electronically powered and with no acoustic stimuli in two different configurations. First, the sensor wings were free. Then, the sensor wings were cemented to the substrate to prevent their natural vibration. Figure 12 shows the noise spectral density (V/$\sqrt{\text{Hz}}$) over a 100 Hz to 10 kHz band. Since both curves are coincident, it is possible to infer that the electronic noise is predominant and that the natural vibrations of the sensor are not captured by the circuit. The SNR at resonance for an acoustic stimulus of 1 Pa over the 3 kHz band of the circuit is approximately 88 dB. Since the sensor is meant to operate at resonance, the Sallen–Key filter stages (Figure 2a) can be designed for a much narrower passband. With a 120 Hz bandwidth, the SNR at resonance becomes greater than 102 dB.

Prior to use in the field, the AVSs were calibrated in an anechoic chamber. This process is described in Section 3 and in [28]. The frequency response of the sine and cosine sensors are measured (V_C and V_S) with the DOA set to 45 degrees. Ideally, the frequency responses of both sensors at 45 degrees should be the same, however, since they are not, $M_S = V_C/V_S$ is applied to the sine signal. The signal processing electronics calculate the DOA, using (10), as the AVS is rotated.

Measurements were taken in an anechoic chamber and in the field for various acoustic sources, particularly gunfire and small multi-rotor aircraft (drones). Figure 13a shows the results for the DOA characterization measurements of an AVS. The blue squares represent data taken in an anechoic chamber with the AVS exposed to an audio recording of rifle fire. The red circles represent data taken in the field with the AVS exposed to actual fire from the same type of rifle. Figure 13b shows the detection error, which was calculated as the difference between the measured and actual DOA. Figure 14a,b shows the results for the similar data collection of an AVS operating in an anechoic chamber exposed to audio recordings of a typical four-rotor drone in flight.

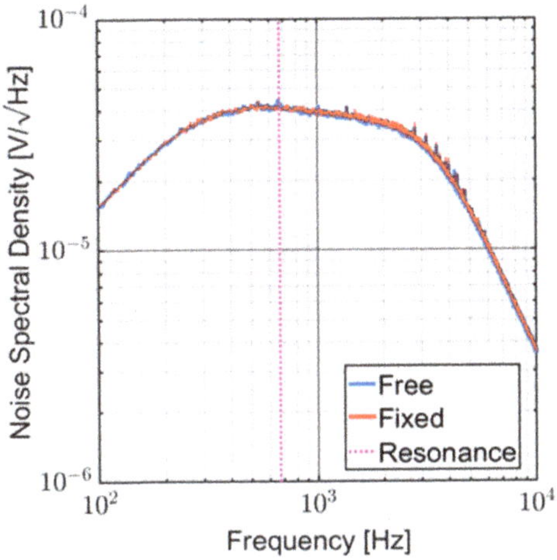

Figure 12. Noise spectral density of typical MEMs sensor. It is possible to notice that electronic noise is predominant, and the natural vibrations of the sensor are not captured by the circuit.

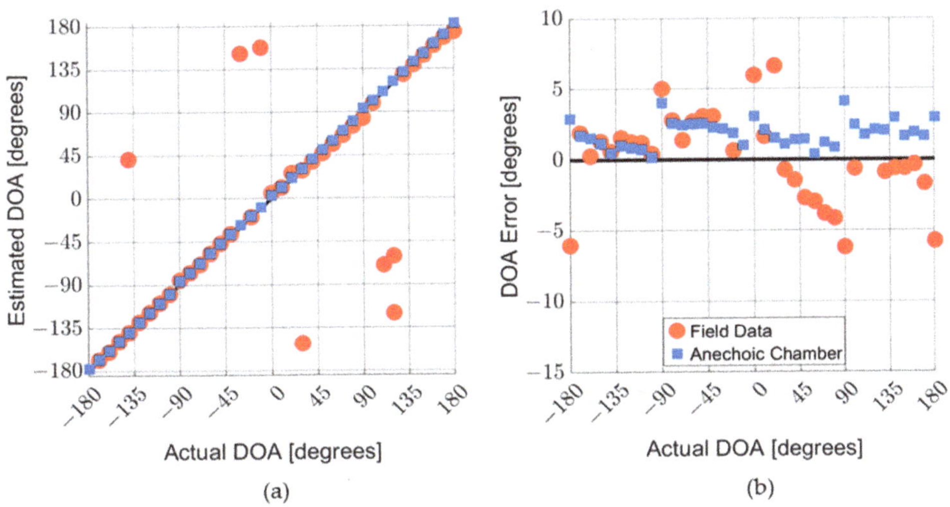

Figure 13. AVS characterization in air with gunshots. AVS was exposed to gunshot audio recording in an anechoic chamber and actual gun fire, from same time of weapon, in the field: (**a**) comparison of the actual DOA with the estimated DOA; and (**b**) detailed graph of DOA errors. Outlier data points are not shown on the DOA error graph.

The anechoic chamber gunshot data and drone data show a systematic error offset to the positive side. This indicates a misalignment of the turntable used to rotate the AVS when setting the DOA. Field data typically have larger errors due to reverberations, acoustic reflections, and background noise. In field experiments, the AVS occasionally collects data on gunshot sounds that reflect off nearby structures, which is the case with the outlier points in Figure 13a.

The DOA errors for both drones and gunshots (for data taken in an anechoic chamber) fall within a 0-to-5-degree range (which corresponds to a ±2.5-degree range when misalignment is corrected for). It should be noted that the drone sound used to characterize the AVS (as shown in Figure 14) aligned well the MEMS sensors. Different drone sounds produce DOA errors of up to ±5 degrees. The tonal nature of the acoustic signatures of drones tends to lead to larger DOA errors than gunshots.

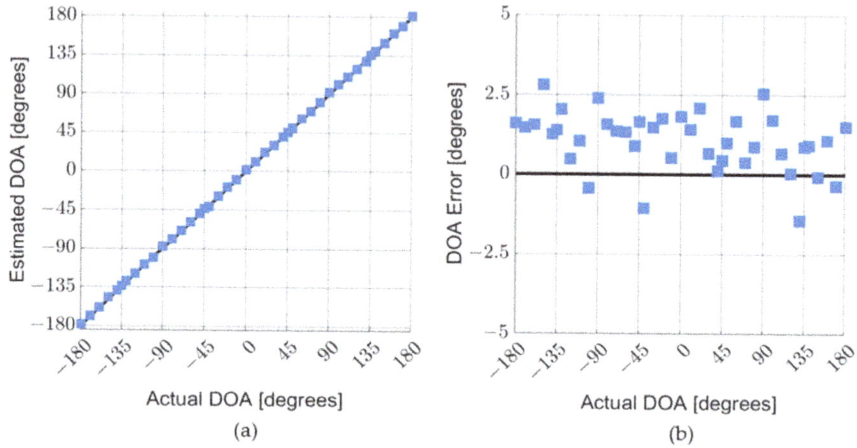

Figure 14. AVS characterization in air with drone sounds. The AVS was exposed to audio recordings of a four-rotor drone: (**a**) comparison of actual DOA with estimated DOA; and (**b**) detailed graph of DOA errors.

Figure 15a,b shows the typical acoustic signatures of pistol gunfire and a small four-rotor drone, respectively, measured with a broadband microphone (Piezotronics Model 378A21). The acoustic signatures are normalized to their maximum values and plotted against the normalized acoustic sensitivity of a typical MEMS sensor. Gunshot sounds, while bursts, are typically broadband in the frequency range interrogated by this sensor. Drone sounds exhibit high peaks at the harmonics of the blade passing frequency (BPF). These tones differ from drone to drone. Note that in Figure 15b, the MEMS sensor peak sensitivity is misaligned with the drone BPF harmonic at around 700 Hz. If this signature is known, a MEMS sensor can be designed to align with a specific harmonic. This will maximize drone detection while filtering out other acoustic sources. More details on these types of acoustic sources can be found in [49–53].

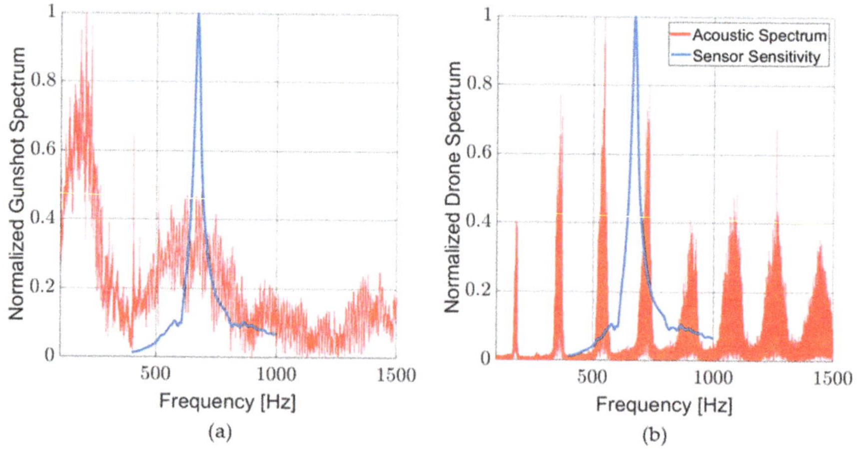

Figure 15. Acoustic spectrum of common sound sources: (**a**) typical gunshot acoustic spectrum with MEMS sensor sensitivity overlay; and (**b**) typical drone acoustic spectrum with MEMS sensor sensitivity overlay.

4.2. Operations Underwater

Data collection for underwater sensor operations was conducted, for the most part, in a vertical, water-filled, standing wave tube (SWT). A diagram of the experimental setup is shown in Figure 16a. A detailed description of the experiment can be found in Section 3. The acoustic properties of the SWT were characterized to verify that a flat acoustic wave front is produced in the section of the tube where the sensor is operated. Figure 16b shows the acoustic pressure value at a depth of 15.2 cm (6 in) below the surface for a constant frequency and speaker voltage. This depth corresponds to the depth of the sensor during data collection. The data show that the wave front is essentially flat and mimics a plane wave. The average pressure is 9.9 Pa with a standard deviation of 0.1 Pa. The most significant wavefront distortion occurs near the tube wall, with a maximum deviation of 3% from the average. During data collection, the sensor was placed in the center of the tube where the wavefront is the flattest.

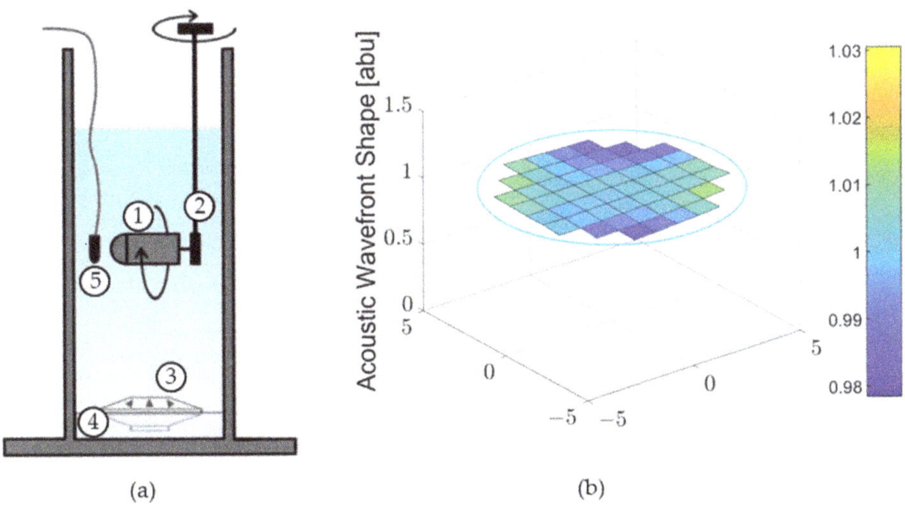

Figure 16. (**a**) Standing wave tube experimental setup: (1) sensor in housing, (2) rotating mechanism, (3) underwater speaker, (4) sound damping material, and (5) reference hydrophone; and (**b**) acoustic pressure wave front measured at 640 Hz, 6 in depth. The blue ring represents the 10-inch inner diameter of the tube.

The sensitivity and DOA response of the underwater sensor were measured in a similar manner to the in-air measurements. Figure 17 shows typical MEMS sensor characteristics when measured in the SWT. While there is no significant difference in the frequency response as compared to in-air frequency sweeps, there is a noticeable difference in lobe size between the front and back of the sensor. This is likely caused by the influence of the sensor housing. This lobe mismatch was observed for all underwater DOA measurements in the SWT.

Underwater AVS measurements were taken at the Transducer Evaluation Center (TRANSDEC) which is a six-million-gallon Navy facility designed to test underwater acoustic devices. The MEMS sensors were individually evaluated and the AVSs were calibrated at TRANSEC prior to conducting the DOA measurements. Figure 18a shows the directionality of each sensor in the AVS. A reduced back lobe was observed similar to the data obtained in the SWT. Figure 18b shows a diagram of the underwater AVS setup. Figure 19a,b shows the results of a DOA characterization for an AVS stimulated by a 670 Hz single tone. The average of the DOA error magnitude is approximately 6.7 degrees.

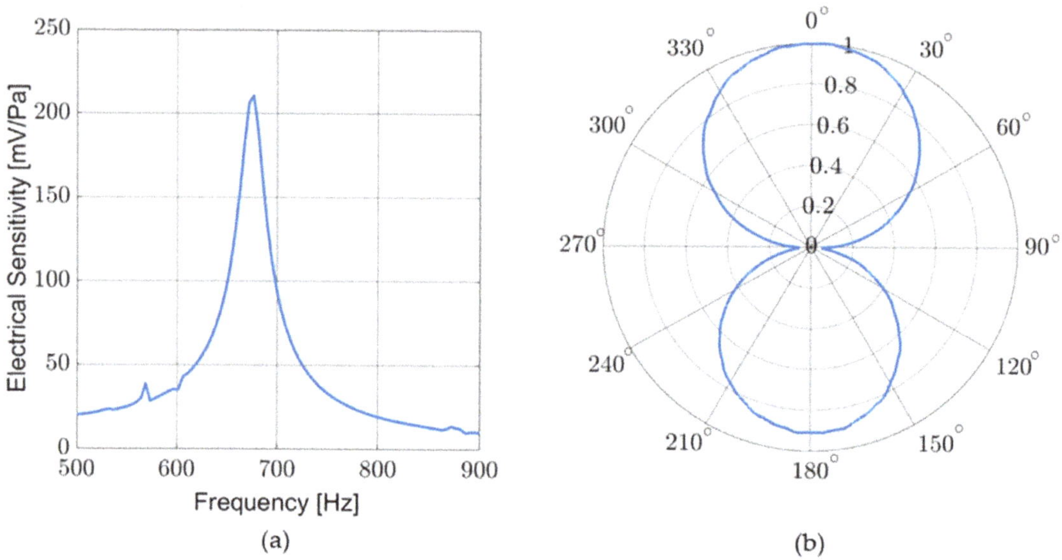

Figure 17. Sensor response in standing wave tube: (**a**) electrical sensitivity measured during a frequency sweep; and (**b**) normalized sensor response to a single tone (near resonance) while rotating sensor to sweep DOA.

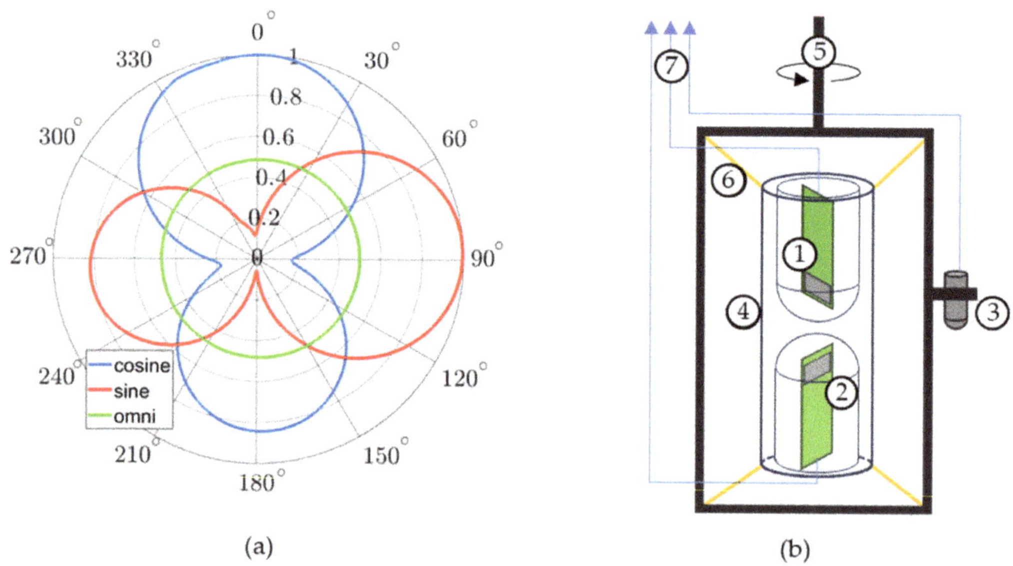

Figure 18. (**a**) Measured directionality response of AVS to 670 Hz tone at TRANSEC. Sine and cosine sensor responses are normalized to the maximum cosine response. The omni response is significantly less than the MEMS sensor but is enlarged to show the shape of its directionality; and (**b**) diagram of underwater AVS setup: (1) sine sensor in underwater housing, (2) cosine sensor in underwater housing, (3) omnidirectional hydrophone, (4) sensor alignment tube, (5) rotating mounting frame, (6) elastic bands connecting sensor housing to rotating frame, and (7) output signals to microprocessor.

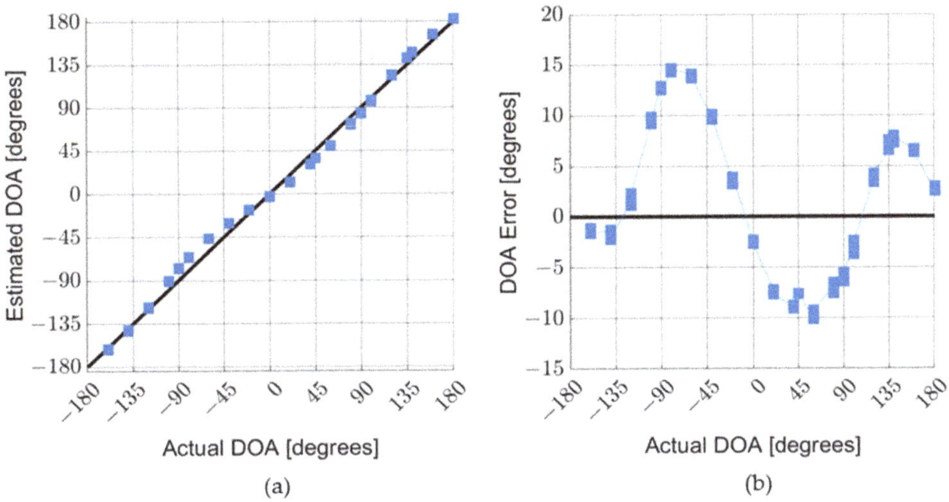

Figure 19. AVS underwater characterization. AVS was exposed to a 670 Hz tone in an anechoic pool: (**a**) comparison of actual DOA with estimated DOA; and (**b**) detailed graph of DOA errors.

It is noticeable that, even though the estimator used underwater was the same as that used for in-air operation, the error is higher. Possible sources of this error included surface reflections and effects from the AVS mounting apparatus. Surface reflections were observed for the frequency ranges investigated with this AVS. As noted above, the back lobes of the MEMS sensors are smaller than the front lobes when operated underwater. This phenomenon was seen both in the SWT and the anechoic pool using different mounting schemes. Therefore, the underwater housing is likely affecting the sensor response.

A sinusoid-like shape to the DOA error is observable for the underwater sensor. There are two primary causes of a sinusoid-like DOA error: amplitude and phase mismatches between the two MEMS sensors. The minimum error amplitudes are seen at cardinal angles (i.e., 0, ±90, and 180 degrees), where either vs. or V_C is near zero and the algorithm is less sensitive to fluctuations. The maximum error magnitudes are near the center of each quadrant (i.e., ±45 and ±135 degrees), where the amplitude mismatch is most significant. Phase mismatches (for small phase angles) lead to a different sinusoid-like DOA error centered at approximately half of the mismatch in the phase. The correction factor, M_S, compensates for the differences between vs. and V_C. However, vs. and V_C are determined by applying a Fourier transform to the sine and cosine sensor signals, respectively. Consequently, there are limitations on the frequency bin size of vs. and V_C. Broadband signals (e.g., white noise, gunshots) are naturally integrated across these frequency bins and the error is minimized. Single-tone sound sources do not benefit from this phenomenon and consequently show larger errors.

5. Discussion/Conclusions

This paper describes the design and experimental characterization of directional acoustic MEMS sensors and an AVS comprised of two collocated and orthogonally aligned sensors in combination with a commercial omnidirectional microphone (or hydrophone). The AVS can determine in-plane acoustic DOA over a non-ambiguous 360 degrees. This AVS is meant to operate near resonance and to be used in the air and underwater. The configuration of the MEMS sensors, AVS design, and algorithms used make this a novel approach to using *Ormia*-inspired MEMS acoustic sensors to determine acoustic DOA.

The small size, light weight, and low power requirements of the MEMS acoustic sensor and associated AVS show potential for their use as a man-portable sensor for detecting and locating acoustic contacts of interest. The MEMS acoustic sensor demonstrates very high

SNR near resonance. This makes the sensor ideal for detecting quiet and distant acoustic targets of interest. The resonant frequency of the MEMS sensor is based on the physical characteristics of the sensor (e.g., bridge and wing size) allowing for bespoke sensors to be designed for acoustic targets that emit specific acoustic tones.

5.1. Sensor Performance

The maximum SNR for this MEMS sensor was determined to be 88 dB with a corresponding maximum sensitivity of −84.6 dB re 1 V/μPa (59 V/Pa). The sensor demonstrates a cosine-like acoustic directionality for the bending eigenmode. This sensitivity is significantly larger than comparable MEMS acoustic sensors, as seen in Table 2. Note that the SNR reported in [27] is over the frequency band of resonance of the sensor (120 Hz bandwidth). The SNR reported for this sensor was calculated over the frequency range of its associated circuit (0 to 3 kHz). When calculating noise only about this sensor's resonant peak (120 Hz bandwidth), it would have an SNR of approximately 102 dB.

Table 2. Comparison of MEMS sensor performance.

	This Sensor(7-1)	Double-Wing MEMS [27]	Double-Wing Design [16]	16 Cantilever Beam Design [54]	Circular Membrane Design [12]	8 Cantilever Beam Design [55]
Sensitivity	59 V/Pa	13 V/Pa	110.5 mV/Pa	70.8 mV/Pa	4.36 mV/Pa	1.67 mV/Pa
SNR	88 dB [102 dB]	91 dB	71.3 dB	51 dB	66.77 dB	Not Discussed

5.2. AVS Performance

The accuracy of the AVS was measured both in the lab and in the field for use in both the air and underwater. While calculating the DOA of gunshots in the field, the average best case AVS accuracy was approximately 3.5 degrees. AVS accuracy, measured in the lab, was determined to be less than an average of 2 degrees over a 360-degree arc. This performance is an improvement on previous laboratory measurements reported by this group of 3.4-degree accuracy over a ±60-degree arc [23]. The performance details of comparable AVS designs are presented in Table 3. Note that the performance data for the AVS presented in this paper were collected in the field with actual gunfire. The tabulated data for all other AVS designs were taken in an anechoic chamber.

Table 3. Comparison of AVS performance.

	This Sensor (7-1)	Three-Sensor Array [19]	Double Diaphragm Design [15]	2 Sensor Array [13]	Canted Double-Wing AVS [23]
DOA Arc	360°	360° [1]	180°	90°	±60°
Average DOA Accuracy	3.5°	2°	2.6°	Not Specified	3.4°

[1] DOA determined with a priori knowledge.

These results indicate the great potential of this type of MEMS sensor for DOA determination in multiple domains. The specific characteristics and figures of performance can be modified by design, according to the application demands. Future work includes the development of multi-resonance MEMS sensors for use in detecting very quiet broadband acoustic sources.

Author Contributions: Conceptualization, F.A. and G.K.; methodology, F.A., G.K. and J.I.; formal analysis, J.I.; investigation, J.I., F.A. and G.K.; data curation, J.I. and F.A.; writing—original draft preparation, J.I.; writing—review and editing, F.A. and G.K.; supervision, F.A.; project administration, F.A.; funding acquisition, F.A. and J.I. All authors have read and agreed to the published version of the manuscript.

Funding: This research was funded by the Office of Naval Research, grant number: N0001420WX00976 and by a student fellowship from the Naval Information Warfare Center Pacific, grant number N6600123WX00492.

Institutional Review Board Statement: Not applicable.

Informed Consent Statement: Not applicable.

Data Availability Statement: The data that support the findings of this study are available from the corresponding author upon reasonable request.

Acknowledgments: The authors would like to thank Reed Herzog for his support in characterizing the standing wave tube, Jeffrey Catterlin for his support with design and rapid prototyping of the sensor housing, and Peter Crooker for his technical support and mentorship.

Conflicts of Interest: The authors declare no conflict of interest.

References

1. Rahaman, A.; Kim, B. Microscale Devices for Biomimetic Sound Source Localization: A Review. *J. Microelectromech. Syst.* **2022**, *31*, 9–18. [CrossRef]
2. Miles, R.N.; Robert, D.; Hoy, R.R. Mechanically Coupled Ears for Directional Hearing in the Parasitoid Fly *Ormia ochracea*. *J. Acoust. Soc. Am.* **1995**, *98*, 3059–3070. [CrossRef] [PubMed]
3. Arthur, B.J.; Hoy, R.R. The Ability of the Parasitoid Fly *Ormia ochracea* to Distinguish Sounds in the Vertical Plane. *J. Acoust. Soc. Am.* **2006**, *120*, 1546–1549. [CrossRef]
4. Akcakaya, M.; Nehorai, A. Performance Analysis of the *Ormia ochracea*'s Coupled Ears. *J. Acoust. Soc. Am.* **2008**, *124*, 2100–2105. [CrossRef] [PubMed]
5. Miles, R.N.; Su, Q.; Cui, W.; Shetye, M.; Degertekin, F.L.; Bicen, B.; Garcia, C.; Jones, S.; Hall, N. A Low-Noise Differential Microphone Inspired by the Ears of the Parasitoid Fly *Ormia ochracea*. *J. Acoust. Soc. Am.* **2009**, *125*, 2013–2026. [CrossRef] [PubMed]
6. Lisiewski, A.P.; Liu, H.J.; Yu, M.; Currano, L.; Gee, D. Fly-Ear Inspired Micro-Sensor for Sound Source Localization in Two Dimensions. *J. Acoust. Soc. Am.* **2011**, *129*, EL166–EL171. [CrossRef] [PubMed]
7. Miles, R.N.; Cui, W.; Su, Q.T.; Homentcovschi, D. A MEMS Low-Noise Sound Pressure Gradient Microphone with Capacitive Sensing. *J. Microelectromech. Syst.* **2015**, *24*, 241–248. [CrossRef]
8. Reid, A.; Windmill, J.F.C.; Uttamchandani, D. Bio-Inspired Sound Localization Sensor with High Directional Sensitivity. *Procedia Eng.* **2015**, *120*, 289–293. [CrossRef]
9. Zhang, Y.; Bauer, R.; Jackson, J.C.; Whitmer, W.M.; Windmill, J.F.C.; Uttamchandani, D. A Low-Frequency Dual-Band Operational Microphone Mimicking the Hearing Property of *Ormia ochracea*. *J. Microelectromech. Syst.* **2018**, *27*, 667–676. [CrossRef]
10. Zhang, Y.; Bauer, R.; Whitmer, W.M.; Jackson, J.C.; Windmill, J.F.C.; Uttamchandani, D. A MEMS Microphone Inspired by Ormia for Spatial Sound Detection. In Proceedings of the 2018 IEEE Micro Electro Mechanical Systems (MEMS), Belfast, UK, 21–25 January 2018; pp. 253–256.
11. Zhang, Y.; Reid, A.; Windmill, J.F.C. Insect-Inspired Acoustic Micro-Sensors. *Curr. Opin. Insect Sci.* **2018**, *30*, 33–38. [CrossRef]
12. Ishfaque, A.; Rahaman, A.; Kim, B. Bioinspired Low Noise Circular-Shaped MEMS Directional Microphone. *J. Micro/Nanolithogr. MEMS MOEMS* **2019**, *18*, 010501. [CrossRef]
13. Rahaman, A.; Kim, B. Fly-Inspired MEMS Directional Acoustic Sensor for Sound Source Direction. In Proceedings of the 2019 20th International Conference on Solid-State Sensors, Actuators and Microsystems & Eurosensors XXXIII (TRANSDUCERS & EUROSENSORS XXXIII), Berlin, Germany, 23–27 June 2019; pp. 905–908.
14. Rahaman, A.; Kim, B. AI Speaker: A Scope of Utilizing Sub–Wavelength Directional Sensing of Bio–Inspired MEMS Directional Microphone. In Proceedings of the 2020 IEEE SENSORS, Rotterdam, The Netherlands, 25–28 October 2020; pp. 1–4.
15. Rahaman, A.; Kim, B. Sound Source Localization by *Ormia ochracea* Inspired Low–Noise Piezoelectric MEMS Directional Microphone. *Sci. Rep.* **2020**, *10*, 9545. [CrossRef] [PubMed]
16. Rahaman, A.; Jung, H.; Kim, B. Coupled D33 Mode-Based High Performing Bio-Inspired Piezoelectric MEMS Directional Microphone. *Appl. Sci.* **2021**, *11*, 1305. [CrossRef]
17. Ren, D.; Qi, Z.-M. An Optical Beam Deflection Based MEMS Biomimetic Microphone for Wide-Range Sound Source Localization. *J. Phys. Appl. Phys.* **2021**, *54*, 505403. [CrossRef]
18. Shen, X.; Zhao, L.; Xu, J.; Yao, X. Mathematical Analysis and Micro-Spacing Implementation of Acoustic Sensor Based on Bio-Inspired Intermembrane Bridge Structure. *Sensors* **2021**, *21*, 3168. [CrossRef] [PubMed]
19. Rahaman, A.; Kim, B. An Mm-Sized Biomimetic Directional Microphone Array for Sound Source Localization in Three Dimensions. *Microsyst. Nanoeng.* **2022**, *8*, 66. [CrossRef] [PubMed]
20. Touse, M.; Sinibaldi, J.; Karunasiri, G. MEMS Directional Sound Sensor with Simultaneous Detection of Two Frequency Bands. In Proceedings of the 2010 IEEE SENSORS, Waikoloa, HI, USA, 1–4 November 2010; pp. 2422–2425.

21. Touse, M.; Sinibaldi, J.; Simsek, K.; Catterlin, J.; Harrison, S.; Karunasiri, G. Fabrication of a Microelectromechanical Directional Sound Sensor with Electronic Readout Using Comb Fingers. *Appl. Phys. Lett.* **2010**, *96*, 173701. [CrossRef]
22. Downey, R.H.; Karunasiri, G. Reduced Residual Stress Curvature and Branched Comb Fingers Increase Sensitivity of MEMS Acoustic Sensor. *J. Microelectromech. Syst.* **2014**, *23*, 417–423. [CrossRef]
23. Wilmott, D.; Alves, F.; Karunasiri, G. Bio-Inspired Miniature Direction Finding Acoustic Sensor. *Sci. Rep.* **2016**, *6*, 29957. [CrossRef]
24. Espinoza, A.; Alves, F.; Rabelo, R.; Da Re, G.; Karunasiri, G. Fabrication of MEMS Directional Acoustic Sensors for Underwater Operation. *Sensors* **2020**, *20*, 1245. [CrossRef]
25. Rabelo, R.C.; Alves, F.D.; Karunasiri, G. Electronic Phase Shift Measurement for the Determination of Acoustic Wave DOA Using Single MEMS Biomimetic Sensor. *Sci. Rep.* **2020**, *10*, 12714. [CrossRef]
26. Alves, F.; Park, J.; McCarty, L.; Rabelo, R.; Karunasiri, G. MEMS Underwater Directional Acoustic Sensor in Near Neutral Buoyancy Configuration. *Sensors* **2022**, *22*, 1337. [CrossRef]
27. Alves, F.; Rabelo, R.; Karunasiri, G. Dual Band MEMS Directional Acoustic Sensor for Near Resonance Operation. *Sensors* **2022**, *22*, 5635. [CrossRef] [PubMed]
28. Crooker, P.; Soule, I.; Ivancic, J.; Karunasiri, G.; Alves, F. Direction of Arrival Algorithm for Acoustic Sensors Operating Near Resonance. *IEEE Sens. Lett.* **2023**, *7*, 7002404. [CrossRef]
29. Rayburn, R. *Eargle's the Microphone Book: From Mono to Stereo to Surround—A Guide to Microphone Design and Application*, 3rd ed.; Routledge: New York, NY, USA, 2011; ISBN 978-0-240-82078-1.
30. Schoess, J.N.; Zook, J.D.; Burns, D.W. Resonant Integrated Micromachined (RIMS) Acoustic Sensor Development. In Proceedings of the 1994 North American Conference on Smart Structures and Materials, Orlando, FL, USA, 13–18 February 1994; SPIE: Bellingham, WA, USA, 1994; Volume 2191, pp. 276–281.
31. Schoess, J.N.; Zook, J.D. Test Results of a Resonant Integrated Microbeam Sensor (RIMS) for Acoustic Emission Monitoring. In Proceedings of the 5th Annual International Symposium on Smart Structures and Materials, San Diego, CA, USA, 1–5 March 1998; SPIE: Bellingham, WA, USA, 1998; Volume 3328, pp. 326–332.
32. Baumgartel, L.; Vafanejad, A.; Chen, S.-J.; Kim, E.S. Resonance-Enhanced Piezoelectric Microphone Array for Broadband or Prefiltered Acoustic Sensing. *J. Microelectromech. Syst.* **2013**, *22*, 107–114. [CrossRef]
33. Shkel, A.A.; Baumgartel, L.; Kim, E.S. A Resonant Piezoelectric Microphone Array for Detection of Acoustic Signatures in Noisy Environments. In Proceedings of the 2015 28th IEEE International Conference on Micro Electro Mechanical Systems (MEMS), Estoril, Portugal, 18–22 January 2015; pp. 917–920.
34. Liu, H.; Liu, S.; Shkel, A.A.; Kim, E.S. Active Noise Cancellation with MEMS Resonant Microphone Array. *J. Microelectromech. Syst.* **2020**, *29*, 839–845. [CrossRef] [PubMed]
35. Lee, T.; Nomura, T.; Su, X.; Iizuka, H. Fano-Like Acoustic Resonance for Subwavelength Directional Sensing: 0–360 Degree Measurement. *Adv. Sci.* **2020**, *7*, 1903101. [CrossRef] [PubMed]
36. Li, Y.; Omori, T.; Watabe, K.; Toshiyoshi, H. Bandwidth and Sensitivity Enhancement of Piezoelectric MEMS Acoustic Emission Sensor Using Multi-Cantilevers. In Proceedings of the 2022 IEEE 35th International Conference on Micro Electro Mechanical Systems Conference (MEMS), Tokyo, Japan, 9–13 January 2022; pp. 868–871.
37. Kusano, Y.; Segovia-Fernandez, J.; Sonmezoglu, S.; Amirtharajah, R.; Horsley, D.A. Frequency Selective MEMS Microphone Based on a Bioinspired Spiral-Shaped Acoustic Resonator. In Proceedings of the 2017 19th International Conference on Solid-State Sensors, Actuators and Microsystems (TRANSDUCERS), Kaohsiung, Taiwan, 18–22 June 2017; pp. 71–74.
38. MEMSCAP | SOIMUMPs and MEMS Multi Project Wafer Service. Available online: http://www.memscap.com/products/mumps/soimumps/ (accessed on 24 January 2023).
39. PolyJet Materials | StratasysTM Support Center. Available online: https://support.stratasys.com/en/materials/polyjet (accessed on 24 January 2023).
40. Leslie, C.B.; Kendall, J.M.; Jones, J.L. Hydrophone for Measuring Particle Velocity. *J. Acoust. Soc. Am.* **1956**, *28*, 711–715. [CrossRef]
41. Liu, C. *Foundations of MEMS*; Pearson Education India: Chennai, India, 2012.
42. Gologanu, M.; Bostan, C.G.; Avramescu, V.; Buiu, O. Damping Effects in MEMS Resonators. In Proceedings of the CAS 2012 (International Semiconductor Conference), Sinaia, Romania, 15–17 October 2012; Volume 1, pp. 67–76.
43. Sader, J.E. Frequency Response of Cantilever Beams Immersed in Viscous Fluids with Applications to the Atomic Force Microscope. *J. Appl. Phys.* **1998**, *84*, 64–76. [CrossRef]
44. Kinsler, L.E.; Frey, A.R.; Coppens, A.B.; Sanders, J.V. *Fundamentals of Acoustics*; John Wiley & Sons: Hoboken, NJ, USA, 2000; ISBN 978-0-471-84789-2.
45. Landau, L.D.; Lifshitz, E.M. *Fluid Mechanics: Landau and Lifshitz: Course of Theoretical Physics, Volume 6*; Elsevier: Amsterdam, The Netherlands, 2013; ISBN 978-1-4831-6104-4.
46. Hopcroft, M.A.; Nix, W.D.; Kenny, T.W. What Is the Young's Modulus of Silicon? *J. Microelectromech. Syst.* **2010**, *19*, 229–238. [CrossRef]
47. Knowles High SNR, High AOP Analog Bottom Port SISONIC Microphone. Available online: https://media.digikey.com/pdf/Data%20Sheets/Knowles%20Acoustics%20PDFs/SPM0687LR5H-1_DS.pdf (accessed on 21 September 2023).
48. Miniature Hydrophone | Type 8103 | Brüel & Kjær. Available online: https://www.bksv.com/en/transducers/acoustic/microphones/hydrophones/8103 (accessed on 24 January 2023).

49. Donzier, A.; Millet, J. Gunshot Acoustic Signature Specific Features and False Alarms Reduction. In Proceedings of the Sensors, and Command, Control, Communications, and Intelligence (C3I) Technologies for Homeland Security and Homeland Defense IV, Orlando, FL, USA, 28 March–1 April 2005; SPIE: Bellingham, WA, USA, 2005; Volume 5778, pp. 254–263.
50. Maher, R.C. Acoustical Characterization of Gunshots. In Proceedings of the 2007 IEEE Workshop on Signal Processing Applications for Public Security and Forensics, Washington, DC, USA, 11–13 April 2007; pp. 1–5.
51. Luzi, L.; Gonzalez, E.; Bruillard, P.; Prowant, M.; Skorpik, J.; Hughes, M.; Child, S.; Kist, D.; McCarthy, J.E. Acoustic Firearm Discharge Detection and Classification in an Enclosed Environment. *J. Acoust. Soc. Am.* **2016**, *139*, 2723–2731. [CrossRef]
52. Kolamunna, H.; Dahanayaka, T.; Li, J.; Seneviratne, S.; Thilakaratne, K.; Zomaya, A.Y.; Seneviratne, A. DronePrint: Acoustic Signatures for Open-Set Drone Detection and Identification with Online Data. *Proc. ACM Interact. Mob. Wearable Ubiquitous Technol.* **2021**, *5*, 20. [CrossRef]
53. Bernardini, A.; Mangiatordi, F.; Pallotti, E.; Capodiferro, L. Drone Detection by Acoustic Signature Identification. *Electron. Imaging* **2017**, *2017*, 60–64. [CrossRef]
54. Kang, S.; Hong, H.-K.; Rhee, C.-H.; Yoon, Y.; Kim, C.-H. Directional Sound Sensor with Consistent Directivity and Sensitivity in the Audible Range. *J. Microelectromech. Syst.* **2021**, *30*, 471–479. [CrossRef]
55. Jang, J.; Lee, J.; Woo, S.; Sly, D.J.; Campbell, L.J.; Cho, J.-H.; O'Leary, S.J.; Park, M.-H.; Han, S.; Choi, J.-W.; et al. A Microelectromechanical System Artificial Basilar Membrane Based on a Piezoelectric Cantilever Array and Its Characterization Using an Animal Model. *Sci. Rep.* **2015**, *5*, 12447. [CrossRef]

sensors

MDPI

Article

Directional Multi-Resonant Micro-Electromechanical System Acoustic Sensor for Low Frequency Detection

Justin Ivancic and Fabio Alves *

Department of Physics, Naval Postgraduate School, Monterey, CA 93943, USA
* Correspondence: fdalves@nps.edu

Abstract: This paper reports on the design, modeling, and characterization of a multi-resonant, directional, MEMS acoustic sensor. The design builds on previous resonant MEMS sensor designs to broaden the sensor's usable bandwidth while maintaining a high signal-to-noise ratio (SNR). These improvements make the sensor more attractive for detecting and tracking sound sources with acoustic signatures that are broader than discrete tones. In-air sensor characterization was conducted in an anechoic chamber. The sensor was characterized underwater in a semi-anechoic pool and in a standing wave tube. The sensor demonstrated a cosine-like directionality, a maximum acoustic sensitivity of 47.6 V/Pa, and a maximum SNR of 88.6 dB, for 1 Pa sound pressure, over the bandwidth of the sensor circuitry (100 Hz–3 kHz). The presented design represents a significant improvement in sensor performance compared to similar resonant MEMS sensor designs. Increasing the sensitivity of a single-resonator design is typically associated with a decrease in bandwidth. This multi-resonant design overcomes that limitation.

Keywords: MEMS acoustic sensor; multi-resonant acoustic sensor; directional acoustic sensor; underwater acoustic sensor

Citation: Ivancic, J.; Alves, F. Directional Multi-Resonant Micro-Electromechanical System Acoustic Sensor for Low Frequency Detection. *Sensors* **2024**, *24*, 2908. https://doi.org/10.3390/s24092908

Academic Editor: Hector Eduardo Roman

Received: 20 March 2024
Revised: 29 April 2024
Accepted: 30 April 2024
Published: 2 May 2024

1. Introduction

The design, modeling, and analysis of a multi-resonant, directional, micro-electromechanical system (MEMS) acoustic sensor is presented. Decades of research and development have been dedicated to better understanding microscale, directional acoustic devices. These small device designs are useful for creating small acoustic vector sensors (AVS), which are capable of determining the direction of arrival (DOA) of incoming sound [1,2]. MEMS devices are popular for use in acoustics because they allow detectors to be small, lightweight, and have low power consumption requirements. They are ideal for creating manually portable AVS systems. The motivation of this research is to improve upon existing MEMS resonant acoustic sensors with a multi-resonant design that increases the frequency bandwidth of the sensor while maintaining a high signal-to-noise ratio (SNR) and preserving directionality characteristics.

1.1. Subwavelength-Sized Directional Sensors

Maintaining the sensor's cosine-like directionality is a key factor in this research. Directional sensors have an acoustic sensitivity that varies with the sound's DOA. Frequently, although not necessarily, microscale acoustic sensors have a dipole (or cosine-like) directionality, where the maximum sensitivity is exhibited when the acoustic wave travels normally in relation to the face of the sensor. The sensitivity decreases, like a cosine, to zero when the wave direction is rotated 90 degrees so that it propagates parallel to the sensor face. This effect is due to the gradient that is formed by the incident sound pressure on the front and back of the sensor. The presented sensor has just such a cosine-like directionality. Understanding the directionality of an acoustic sensor is necessary to determine the acoustic DOA. Some other microscale sensor designs display different directionality patterns and are discussed in this section.

In 2018, Zhou and Miles [3] demonstrated an acoustic flow detector using nanofibers. The nanofibers were driven by viscous forces, created by the particle motion of the surrounding medium of the sensor when subjected to an acoustic wave. This design demonstrated a dipole directionality and a flat sensitivity curve of zero dB over a wide frequency range (100 Hz to 10 kHz).

Research presented by Lee et al. [4] in 2020 demonstrated how a sensor consisting of coupled Helmholtz resonators could be used for DOA determination. Two designs were presented, a dual resonator and a triple resonator, each with their own directionalities. By comparing the pressure response of the resonator chambers, the DOA of an acoustic source could be determined. The triple resonator design demonstrated a 360-degree DOA coverage.

In 2022, Chen et al. [5] presented an acoustic detector consisting of a four-sided Helmholtz resonator placed in the center of an array of phononic crystal cylinders. The sensor design demonstrated a cross-shaped directionality, with the maximum sensitivities being 90 degrees apart from each other. The design showed a 280:1 gain in acoustic pressure at resonance.

Also in 2022, Chen et al. [6] presented a gradient acoustic metamaterial coupled with a space-coiling structure acoustic device consisting of an array of metamaterial plates that incrementally increased in size. The design exploited wave compression effects to amplify the sound signal. The sensor has a unique directionality, with one large lobe at the front of the array and a small back lobe. Two of these sensors were aligned in a canted configuration to determine the acoustic DOA. Despite being small compared to acoustic wavelengths, many of these directional acoustic sensors are significantly larger than MEMS sensors.

1.2. Resonant MEMS Sensors

This research is interested in acoustic sensors that are capable of detecting quiet or distant acoustic sources. Operating the sensor at resonance helps achieve this goal. Typical microphones are designed to operate at frequencies that are far from their resonances so that they maintain a constant sensitivity over a large frequency range [7]. However, to maximize the acoustic sensitivity, achieve a high SNR, or mechanically filter unwanted acoustic noise, it is advantageous to operate acoustic detectors at or near resonance. One common MEMS acoustic sensor design consists of a vibrating cantilever beam or paddle connected to a substrate. Research has been steadily conducted on this kind of MEMS acoustic sensor.

In 2019, Rahaman and Kim [8] presented a disc-shaped double-wing MEMS acoustic sensor with a dipole directionality. The sensor utilized a piezoelectric sensing system. An AVS was constructed using two of these sensors. This AVS demonstrated an ability to calculate the acoustic DOA over a 90-degree arc. In 2020, Rahaman and Kim [9,10] presented a different double-wing sensor with rectangular wings. The sensor demonstrated a cosine-like directionality. The reported sensitivity was 3.45 mV/Pa (−49.2 dB re 1 V/Pa) at 1 kHz, with an SNR of approximately 68.5 dB. An array of three of these sensors was used to localize a sound source.

In 2020, Espinoza et al. [11] demonstrated two MEMS acoustic sensors: a double-wing design and a cantilever paddle design. These sensors were intended for use underwater by placing them in a silicone oil-filled housing. The housing was then submerged in water. The paddle and double-wing sensors demonstrated a peak sensitivity of approximately 5.5 mV/Pa and 6 mV/Pa (−45.2 dB and −44.4 dB re 1 V/Pa), respectively, at resonance. When operating in air, both sensors demonstrated a cosine-like directionality pattern; however, in water, the directionality pattern was distorted with unequal lobe sizes.

In 2020, Rabelo et al. [12] presented a double-wing design with a closed cavity behind the sensor. This configuration allowed for comparable rocking and bending modes. The acoustic DOA was demonstrated to be proportional to the phase shift between these two modes. This allowed for DOA determination, using a single sensor, over a 180-degree arc with an accuracy of 3 degrees. The sensitivity of the sensor was determined to be on the order of 1 V/Pa (0 dB re 1 V/Pa).

In 2021, Li et al. [13] presented methods to optimize the dimensions of a piezoelectric MEMS cantilever beam acoustic sensor. The peak sensitivity of the sensor at resonance (30 kHz) was 148 V/m/s.

In 2022, Li et al. [14] followed up their work of improving the piezoelectric MEMS acoustic sensor's bandwidth and sensitivity performance. They created an array of identical cantilever beams and optimized the layer thickness of the devices. A single cantilever sensor demonstrated a peak sensitivity of approximately 1 V/m/s with a narrow resonance peak at 48.7 kHz. An array of 210 cantilevers with identical designs improved the sensitivity to 2 V/m/s. The bandwidth also increased in frequency range. The sensitivity was essentially constant from 44.9 to 48.9 kHz. This work demonstrated how an array of beams can improve performance by broadening the response through multiple resonances.

In 2022, Rahaman and Kim [15] presented an AVS made from an array of three double-wing, resonant, MEMS acoustic sensors. The maximum sensitivity of the sensors was approximately 100 mV/Pa (−20 dB re 1 V/Pa) at the bending resonant mode (11.9 kHz). The AVS demonstrated 360 degrees of coverage in azimuth and elevation, but one required a priori information of the other.

In 2023, Ivancic et al. [16] demonstrated a symmetric double-wing design that emphasized the bending mode. The sensor demonstrated a sensitivity of 59 V/Pa (35.4 dB re 1 V/Pa) and an SNR of 88 dB at 1 Pa over the bandwidth of the sensor circuity. The sensor demonstrated a cosine-like directionality in air and a distorted cosine directionality in water (similar to [11]). An AVS was assembled, which consisted of two of these sensors and a commercial omnidirectional acoustic sensor. The AVS demonstrated a 360-degree DOA coverage with a 3.5-degree accuracy.

1.3. Multi-Resonant MEMS Sensors

A limitation with many resonant sensor designs is that they operate in a narrow frequency band, which makes the sensors less effective for detecting broadband acoustic sources. This research is interested in broadening that frequency band. As suggested by [14], combining multiple vibrating wings into a single sensor can be an effective way to broaden the frequency band of the sensor.

Multi-resonant MEMS acoustic sensors employ multiple resonators with differing resonant frequencies. This increases the overall bandwidth of the sensor. In 2013, Baumgartel et al. [17] presented a multi-resonant MEMS acoustic sensor that consisted of thirteen cantilevered paddles with a piezoelectric vibration sensing scheme. The resonant frequencies of each paddle varied from 860 Hz to 6.2 kHz, with a maximum sensitivity of 202.6 mV/Pa (−13.9 dB re 1 V/Pa). The sensitivity of the sensor remained above 2.5 mV/Pa (−52.0 dB re 1 V/Pa) over the designed frequency range of the sensor (240 Hz to 6.5 kHz). In 2015, Shkel et al. [18] followed up this research with thirteen cantilevered paddle designs, using the resonant frequencies of each paddle to mechanically isolate sound (human speech) from noisy background environments. They demonstrated that the sensor could improve automated speech recognition by 62.7% from a signal with a 15 dB SNR.

In 2020, Liu et al. [19] presented two piezoelectric MEMS cantilevered paddle arrays, one with ten paddles and the other with nine. The resonant frequencies of the ten-paddle and nine-paddle arrays ranged from 856 to 4889 Hz and 5380 to 8820 Hz, respectively. Using these arrays in conjunction demonstrated an improvement in SNR for typical human speech frequencies. The maximum acoustic sensitivity was 202.1 mV/Pa (−13.9 dB re 1 V/Pa) at 856 Hz.

In 2021, Kang et al. [20] demonstrated an MEMS acoustic device, inspired by the human cochlea, consisting of sixteen cantilever beams. The cantilevers were of different sizes and operated over multiple bending modes of each beam. The beams were designed so that the entire frequency band of the sensor was covered by one or more of these modes.

The sensor demonstrated a sensitivity of approximately 71 mV/Pa (−23 dB re 1 V/Pa) over a frequency range of 300 Hz to 8 kHz. The sensor demonstrated a cosine-like directionality.

In 2022, Alves et al. [21] presented a double-wing design where the torsional legs were offset from the center of the bridge that connected the wings. This configuration created two separate bending mode resonances (one for each wing). This allowed for a wider resonance bandwidth when the responses of each wing were combined. The sensor demonstrated a 13 V/Pa (22.3 dB re 1 V/Pa) maximum sensitivity with a 91 dB SNR. The sensor demonstrated a cosine-like directionality in air.

The resonant sensors discussed above provide high sensitivity and directionality but are limited in effective bandwidth. Most of the multi-resonant sensors demonstrated broader bandwidths but lacked high acoustic sensitivities. The sensor presented in this paper combines a high sensitivity with a broader bandwidth. It utilizes a wing design inspired by those described in [16]. However, instead of consisting of two mechanically coupled identical wings, this design consists of six independent wings. Each wing has a different resonant frequency so that the sensor has increased bandwidth while maintaining a high sensitivity and SNR across that bandwidth.

1.4. Environmental Sensing

The multi-resonant MEMS acoustic sensor presented in this paper is ideal for use in AVS designs. The acoustic sources of interest to this research are gunshots, drones, and underwater vehicles. However, this sensor design can be modified to detect and monitor a variety of sound sources (e.g., road vehicle noise, airborne noise, environmental noise) in a wide range of acoustic environments. While a single AVS can provide a bearing to a sound source, a distribution of these AVSs (alone or as part of a larger suite of sensors) can provide the ability to determine a sound source's location.

2. Design and Modeling

2.1. Design Requirements

The MEMS sensor was microfabricated out of a 400 μm thick silicon-on-insulator (SOI) wafer with a 25 μm device layer. The vibrating wings were etched into the device layer. Likewise, the substrate below the wing was etched all the way through to allow the wing to vibrate freely. Gold pads were deposited onto the device layer to provide ohmic contact. Insulating trenches were etched onto the device layer to electrically separate the vibrating wing from the fixed substrate. The sensor was fabricated by the MEMSCAP [22] commercial foundry.

Individual resonators in the array consist of a vibrating wing connected to a substrate via a bridge and torsional legs, as shown in Figure 1. When exposed to sound waves, the wings vibrate normally to the plane of the substrate. At the end of each wing, fishbone-style comb fingers are interlaced with corresponding comb fingers on the substrate. When the wing vibrates, the capacitance between the wing and substrate varies with the deflection of the wing. The sensor is cemented into an open cavity in a printed circuit board (PCB) and wire-bonded to a circuit that converts the sensor capacitance to an output voltage. Similar capacitive sensing schemes are described in more detail in [12,23].

The resonance frequency of a wing (or paddle)-shaped MEMS acoustic sensor is, in part, a function of the physical parameters of the wing and bridge (e.g., wing size, bridge length, layer thickness, material). The sensor parameters were selected to align each wing to different desired resonant frequencies.

While a sensor with a high quality factor is good for detecting a specific tone, it can limit the detection of broader acoustic sources or tones outside the passband [16]. One promising way to overcome these limitations is to use multiple resonators with near resonances to broaden the response [14]. To explore this idea, two similar multi-resonant sensors (versions V11 and V12) were produced. These designs operated nearly identically, except for slightly shifted resonant frequencies.

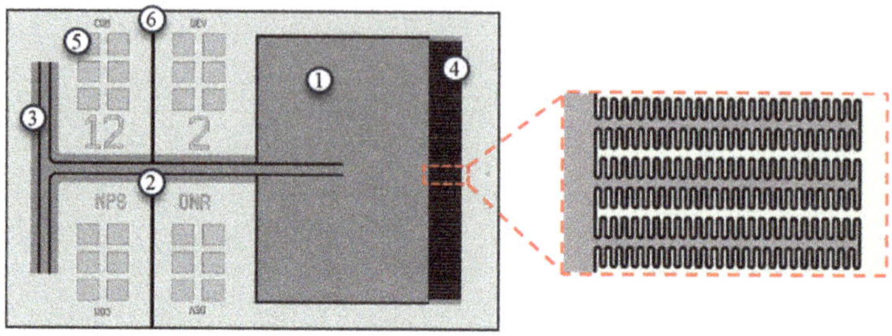

(A) (B)

Figure 1. Layout of single wing of the sensor design. (**A**) Layout of entire wing: (1) wing, (2) bridge, (3) torsional leg, (4) comb fingers, (5) gold wire bonding pad, and (6) a groove in the device layer electrically separating the wing from the substrate. (**B**) Zoomed-in view of fishbone shape of comb fingers. The dark grey areas under the wing and surrounding the bridge and torsional legs represent a trench that passes through the base layer of the sensor.

The design criteria was to support detection of sound from 300 to 500 Hz while maintaining a high sensitivity and SNR across the entire sensor bandwidth. The target SNR of the overall sensor should be comparable to the SNRs of the individual resonators. Additionally, the sensor should demonstrate a cosine-like directionality. The design was also constrained by manufacturing limitations (foundry design rules) [22].

To meet these criteria, a sensor consisting of six individual wings was conceptualized. Each wing was designed with a different resonant frequency to cover the target bandwidth. The response of the wings to the incoming sound was transduced in capacitance and linearly correlated with the vibration. The output of each wing was wire-bonded to the same port on the capacitive readout circuitry. This configuration places the capacitors of each wing in parallel, creating a single sensor output that can be modeled as the complex sum of the outputs of the individual wings. Figure 2A shows a finite element (FE) simulation of the frequency response for individual wing displacements and the complex addition of all the wings for sensor design V11. The graph is normalized to the maximum displacement of wing number 1. Figure 2B shows the phase response of each wing (with respect to a driving acoustic signal) during a frequency sweep. At resonance, each wing behaves like a harmonic oscillator. However, the phase response of the whole sensor is more complex than that of a single wing.

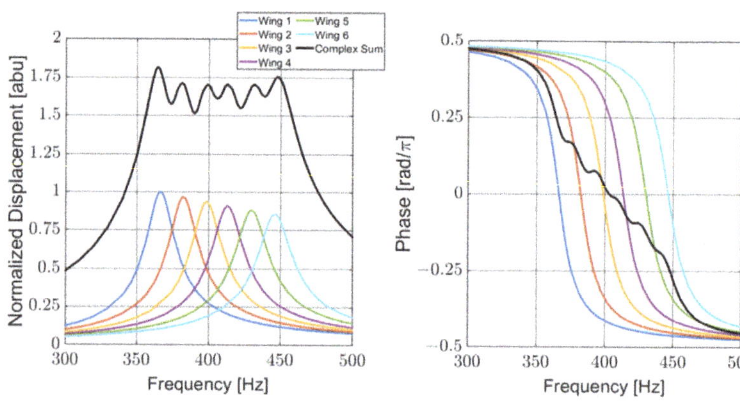

(A) (B)

Figure 2. Computer-modeled behavior of sensor: (**A**) normalized wing displacement for individual wings and the complex sum of the all the wings; (**B**) individual wing phases and complex sum. The phases are offset so that the phase equals zero at resonance.

2.2. Design Parameters

The sensor wings used in this design differ from similar paddle designs. For the purposes of this paper, a paddle design consists of a vibrating paddle that is directly connected to a substrate via a bridge. The bridge acts as a fixed cantilever. In the wing design, the bridge connects to torsional legs. The torsional legs then connect to the substrate and twist while the bridge bends. There are two primary reasons for including the torsional legs in this design. First, the torsional legs allow the resonance frequency of the wing to be lowered while still meeting size and manufacturing limitations. Second, our previous investigations into paddle designs revealed that the cantilever connection between the beam and substrate was structurally weak and prone to failure. Designs that include torsional legs reduce the stress on the pivot points and are less prone to failure.

Figure 3A shows a top-down picture of the sensor (version V12). A picture of the sensor mounted into a PCB is shown in Figure 3B. A scanning electron microscope (SEM) image of the fishbone-style comb fingers is shown in Figure 3C. Table 1 shows some of the key dimensional parameters of the sensor. The resonant frequency of each wing was set via the bridge length. The wing dimensions and torsional leg dimensions were maintained wing to wing.

(A) (B) (C)

Figure 3. Multi-resonant acoustic sensor. (**A**) Microscope image of V12 sensor; (**B**) image of MEMS sensor mounted in PCB; (1) multi-resonant MEMS sensor, (2) capacitive readout circuitry, and (3) wire connections for power and readout; (**C**) SEM image of fishbone-patterned comb fingers. Note that residual stress on wings from fabrication causes wings to bend slightly, lifting wing comb fingers while at rest, approximately halfway up from substrate comb fingers.

Table 1. Key sensor design dimensions.

Wing Width	Wing Length	Wing Thickness
2500 μm	1600 μm	25 μm
Design Freq (V11 Wing 1)	**Bridge Length (V11 Wing 1)**	**Torsional Leg Length**
366 Hz	2900 μm	1000 μm

2.3. Analytical Modeling

At resonance, each wing acts like a driven, damped harmonic oscillator. The sensor was limited to frequencies where only the first mode was excited. In this mode, we can think of a single wing as a mass-loaded spring system with two stiffnesses to consider: the bending of the beam and the twisting of the torsional legs. The analytical model discussed here follows from the analytical model presented in [16], with modifications to account for the twisting of the torsional legs.

The wing is modeled as an undamped, simple harmonic oscillator with three springs. Two springs are in parallel (each torsional leg). Those springs are in series with the third spring (the bending of the bridge). The overall stiffness of the wing is given by

$$k_{wing} = \left(\frac{1}{2k_{\mathrm{leg}}} + \frac{1}{k_{\mathrm{bridge}}}\right)^{-1}, \tag{1}$$

where k_{leg} is the stiffness of a torsional leg, and k_{bridge} is the stiffness of the bridge, which can be determined by the standard equation for a flexural beam [24]:

$$k_{brige} = \frac{Ewt^3}{4L^3}, \tag{2}$$

where E is Young's modulus of silicon. The parameters w, t, and L are the width, thickness, and length of the bridge, respectively. To determine the stiffness of the legs, first, the torsional stiffness, K_t, must be established based on the physical properties of the torsional legs [24], which is determined by

$$J_{leg} = \mathrm{G} * \frac{w_{leg}t^3}{16}\left[\frac{16}{3} - 3.36\frac{t}{w_{leg}}\left(1 - \frac{t^4}{12{w_{leg}}^4}\right)\right], \tag{3}$$

$$K_t = \frac{\mathrm{J}}{l} = \frac{\mathrm{T}}{\theta}, \tag{4}$$

where J_{leg} is the torsional rigidity of a single torsional leg, G is the shear modulus of the silicon, and the parameters w_{leg}, t, and l are the width, thickness, and length of the torsional leg, respectively. Note that for this wing design, the thickness is consistent across the entire wing. T is the applied torque to the beam and θ is the twist angle at the end of the leg. Figure 4 shows a diagram representing how the torsional stiffness of the torsional legs relates to flexural stiffness. This allows the effects of the torsional legs and bridge to be combined as shown in (1).

K_t can be related to k_{leg} based on the twisting angle and applied torque from the force applied to the wing by the acoustic wave as follows:

$$F = k_{leg}d = k_{leg}L * \tan(\theta), \tag{5}$$

$$T = F * L = K_t\theta. \tag{6}$$

Combining (5) and (6) yields

$$k_{leg} = \frac{K_t\theta}{L^2\tan(\theta)}. \tag{7}$$

However, for a small θ, tan $(\theta) \approx \theta$. Therefore,

$$k_{leg} = \frac{K_t}{L^2}. \tag{8}$$

The mass of the wing is approximated by an effective point mass, m_{eff}, located at the end of the bridge. The moment of inertia of the point mass is equivalent to the moment of inertia of the wing. This technique is discussed in more detail in [16]. Neglecting damping effects, the resonant frequency, f_0, of the wing can be modeled as follows:

$$f_0 = \frac{1}{2\pi}\sqrt{\frac{k_{wing}}{m_{eff}}}. \quad (9)$$

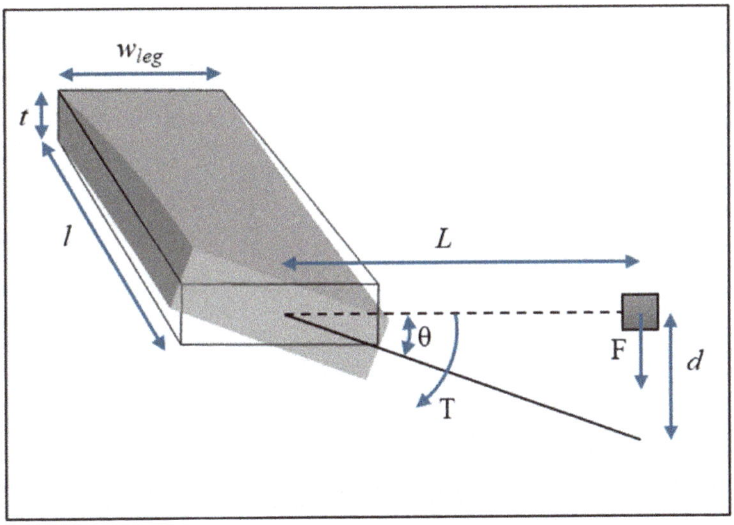

Figure 4. Relating torsional stiffness (K_t) to flexural stiffness (k_{leg}).

To include damping, this analytic model modifies the Sader [25] method to determine the resonant frequency and quality factor of a cantilever beam vibrating in a surrounding fluid. The Sader method is agnostic to the cross-sectional shape of the beam, but it assumes that the cross-section is constant across the length of the beam. The effective width of the beam, b, is determined based on its cross-sectional shape. The presented wing design does not meet this assumption. Therefore, the performance of previous wing designs was used to modify the method to determine b. The value of b is determined based on the widths of bridge and wing using the following formula:

$$b = w_{wing} - 0.016 * \left(\frac{{w_{wing}}^2}{w}\right). \quad (10)$$

The quality factor, Q, can be computed using

$$Q = \frac{\frac{4\mu}{\pi\rho b^2} + \Gamma_r}{\Gamma_i}, \quad (11)$$

where μ is the dynamic viscosity, and ρ is the density of the fluid. Γ_r and Γ_i are the real and imaginary parts of the hydrodynamic function detailed in [25].

The modification of Sader's method is discussed in more detail in [16]. The analytical model slightly underestimates the measured resonant frequency. The average modeled resonant frequency is 2% lower than the average measured resonant frequency for all wings. However, the analytical model underdamps the system with respect to the quality factor. The average modeled quality factor is approximately 1.8 times larger than the average measured quality factor. A detailed comparison of the analytical modeling, computer simulations, and measured sensor responses is provided in Section 4.

For an MEMS acoustic sensor of this type operating in air, only including damping from the modified Sader method is sufficient. However, if the sensor is operating in a more viscous fluid (e.g., water, silicone oil), additional damping effects such as Couette flow in the gaps between the wing and substrate and the capacitive comb fingers must be considered.

2.4. Finite Element Modeling

FE modeling of the sensor was conducted using COMSOL Multiphysics version 6.1 modeling software. The FE models were based on similar models to those described in [16]. Each wing in the sensor was modeled independently to determine its resonance frequency, response to driving frequency (i.e., wing displacement and phase), and directionality.

The device layer was modeled as anisotropic silicon, with the elasticity matrix aligned to the crystalline structure of the silicon. The sensor was enclosed within a sphere of air with standard properties from the COMSOL material library. A shell of air with perfectly matched layer properties was included around the sphere to prevent acoustic reflections. To reduce computational time, the FE model was bisected along the centerline of the sensor, and symmetry boundary conditions were applied along the bisection. Previously, similar FE models were used in the development of single-resonant MEMS sensors [16]. Based on the measured performance of those sensors, this FE model was updated to include an additional damping (drag) force so that the modeled behavior better matched the measured sensor performance.

The FE model used a free tetrahedral meshing for the sensor and surrounding sphere of air. The perfectly matched layer shell of air surrounding the sphere was meshed using a swept mesh method. Figure 5A shows a depiction of the device suspended in the sphere of air and surrounding a shell of air. Figure 5B shows a zoomed-in view of a single half-wing in the bending vibration mode. The solid mechanics module was used to set fixed constraints, boundary loads on the wings, and symmetry conditions. The pressure acoustic module was used to apply a plane wave pressure field to the sensor. The plane wave direction of propagation was adjusted with a parametric sweep to model the acoustic source rotating around the sensor to obtain the directionality pattern, as seen in Figure 6A. The simulation shows a cosine-like response, as expected. The frequency of the acoustic wave was adjusted with a separate parametric sweep to measure the displacement and phase response of the sensor, as seen in Figure 6B. The results show a harmonic oscillator behavior near resonance. An arbitrary phase offset was applied so that the phase equals zero at resonance. This offset was applied to match the algorithms used for DOA estimation.

The bending mode is the lowest-frequency resonant mode of the sensor design. An eigenfrequency analysis was conducted to determine the frequency of the second major resonant mode of the sensor. That mode consists primarily of the wing rocking back and forth laterally, pivoted at the point where the bridge meets the wing. The second mode for wing 1 is approximately 3018 Hz. Its deflection magnitude is approximately 15% that of the first resonant mode. This is outside of the range of interest and was filtered out by the electronics readout.

The FE model's quality factor was determined by calculating the magnitude of wing displacement with respect to the frequency:

$$Q = \frac{f_0}{f_h - f_l} \tag{12}$$

where f_0 is the resonant frequency, and f_h and f_l are the upper and lower bounds of the frequencies, where the displacement magnitude is 70.7% of the maximum. The FE modeling results are compared with measured results in more detail in Section 4. However, the average modeled quality factor agrees with the measured quality factors within 0.6%.

The average modeled resonant frequencies agree with the measured values within 0.9%. This demonstrates that FE modeling is an effective tool for sensor design.

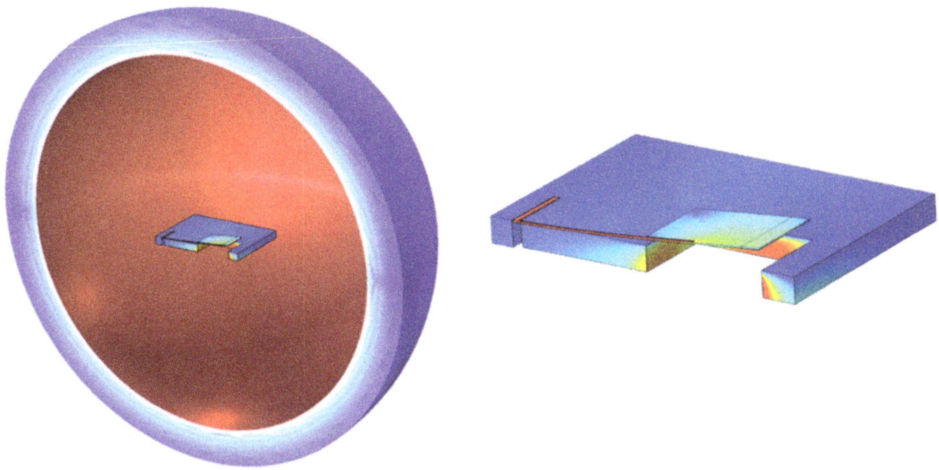

(A) (B)

Figure 5. FE model images of single wing. (**A**) Depiction of the FE model. Wing located in center of sphere of air. (**B**) Single wing in bending mode. The FE model consists of only half the model and sphere or air. Symmetry conditions are applied along the bisection to account for the entire sensor.

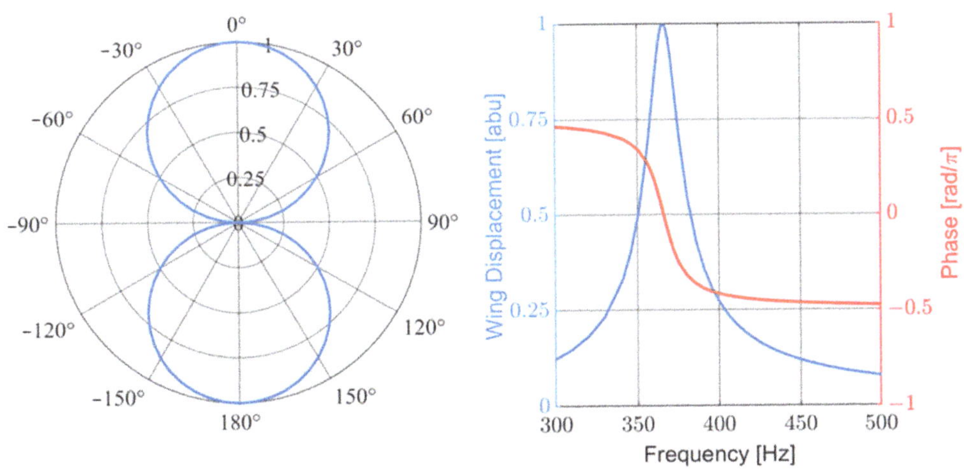

(A) (B)

Figure 6. An FE model of wing behavior. (**A**) The modeled wing directionality matches an ideal cosine-like shape. (**B**) The Frequency response of a single wing showing normalized displacement (blue) and phase (red). A single wing behaves like a harmonic oscillator in terms of resonance.

3. Experimental Methods

3.1. Mechanical Sensitivity

Prior to cementing the sensor into the PCB and wire-bonding it to the capacitive readout circuit, the mechanical sensitivity was measured via laser vibrometry utilizing a Polytech data management system (DMS) computer, OFV-5000 controller, and an OFV-534 laser unit. Data collection was conducted in an anechoic chamber. The sensor was held in place in the path of a laser beam so that the beam terminated at the wing center line, near the far edge of a wing, just before the comb fingers. The DMS generated an audio signal (250 to 510 Hz frequency sweep) that was sent through a Techron 5507 amplifier to a JBL 7-inch speaker, which faced the sensor. The DMS measured the deflection of the sensor wing. The acoustic pressure was measured with a Piezotronics Model 378A21 reference microphone. The microphone signal was sent through a Piezotronics Model 482C signal conditioner to the DMS. The DMS would calculate the average deflection amplitude per acoustic pressure (mechanical sensitivity) of the wing, as a function of frequency, over the course of five frequency sweeps. Once the mechanical sensitivity of a given wing was measured, the sensor would be repositioned so that a different wing was moved into the path of the laser, and the process was repeated for each wing in the sensor. Figure 7 shows the experimental setup of the laser vibrometry.

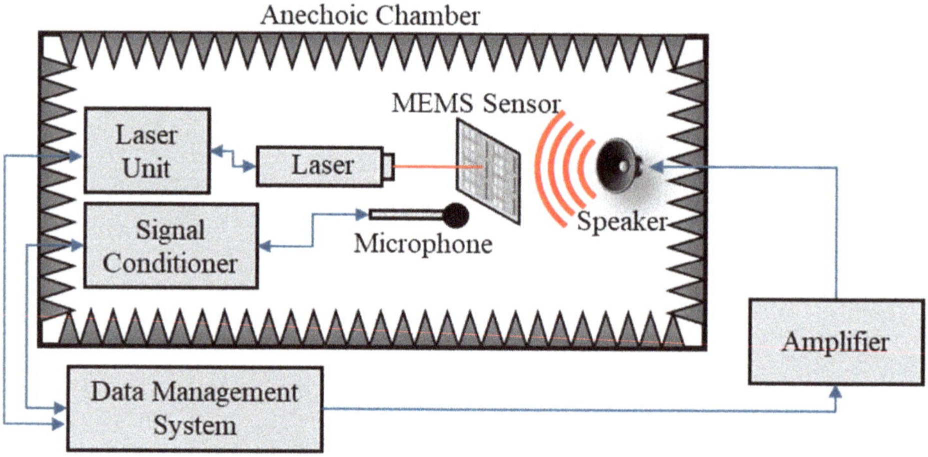

Figure 7. Laser vibrometry experimental setup.

3.2. Electrical Characterization in Air

After laser vibrometry measurements were taken, the sensor was cemented into the host PCB and wire-bonded to the capacitive readout circuit for directionality and frequency response measurements. The sensor was mounted on a precision turntable (B&K Model 5960) in an anechoic chamber, with a stationary speaker (7-inch JBL cone speaker) pointed at the sensor. Rotating the sensor changed the DOA at which the acoustic wave was incident upon the sensor. The MEMS sensor was connected to a control box, which provided power to the sensor and distributed the output to other devices. A calibrated reference microphone (Piezotronics Model 378A21) was mounted near the MEMS sensor. The signal from the microphone was sent to a signal conditioner (Piezotronics Model 482C). The outputs of the microphone and MEMS sensor were read by separate Zurich Instruments multifunction lock-in amplifiers (MFLIs).

The MFLIs and a signal generator (Agilent 33220A) were used to produce various sounds (e.g., steady tones, white noise, frequency sweeps) to characterize the MEMS sensor. Signals from the MFLI and signal generator were sent to an amplifier (Techron 5507) and then to the speaker in the anechoic chamber. Figure 8 shows the experimental layout to determine the frequency response, directionality, and SNR of the MEMS sensor.

To determine the SNR, the MEMS sensor was mounted in an anechoic chamber. All electrical and acoustic equipment and noise sources were secured in the chamber, except for the MEMS sensor. The output of the sensor was read by an MFLI, which measured the noise spectral density over a bandwidth of 0 to approximately 12 kHz. To distinguish the electronic noise of the sensor circuitry from the mechanical noise of the MEMS sensor chip, two sets of noise spectral density measurements were taken. One set was for an unmodified sensor. The second set of data was with the wings glued in place to prevent their vibration, which removed the mechanical noise from the system.

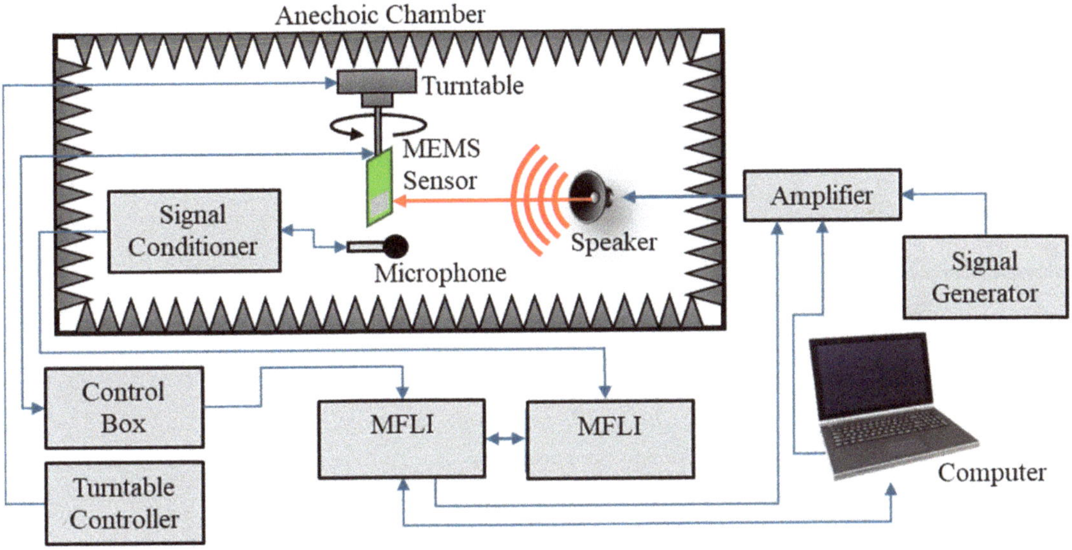

Figure 8. Experimental setup for sensor characterization (frequency response, directionality).

3.3. *Underwater Electrical Characterization*

Underwater sensor characterization was conducted at the Naval Transducer Evaluation Center (TRANSDEC), a six-million-gallon, anechoic pool operated by the US Navy, to perform a wide range of underwater sensor characterizations.

The MEMS sensor was enclosed in an air-filled, water-tight housing, as shown in Figure 9A. The sensor and housing were nearly neutrally buoyant. In this condition, the acoustic wave causes the sensor housing to vibrate, and the MEMS sensor acts as an inertial sensor, detecting the vibration of the housing rather than the acoustic wave directly. A similar experiment was described in [16].

The MEMS sensor and an omnidirectional reference hydrophone (B&K Type 8103) were mounted 6 feet deep on a pole with a motorized rotation mechanism. An underwater speaker (Electro Voice UW30) was suspended 6 m deep and 2 m away from the sensor. The output of the MEMS sensor was sent to a similar control box, discussed in Section 3.2. The output of the reference hydrophone was sent to a preamplifier (Stanford Research Systems SR560) and then to the control box. The control box directed the MEMS sensor and hydrophone outputs to the MFLIs for data collection. Acoustic signals were produced by either an MFLI, signal generator (Keysight 33500B), or computer. These signals were sent through an amplifier to the underwater speaker.

The characterization consisted of frequency response and directionality measurements. These measurements were conducted in a similar manner to those performed for the sensor in air, as discussed in Section 3.2. Figure 9B shows the experimental setup for data collected at TRANSDEC.

Additional frequency response measurements were taken in a water-filled standing wave tube (SWT). The sensor was mounted on the end of a pole, facing an underwater speaker (Electro Voice UW30) on the bottom of the SWT. The SWT produces a flat standing wave front at the sensor location. The SWT experimental setup was similar to the TRANSDEC setup, with an MFLI supplying a frequency sweep signal through an amplifier to the underwater speaker. The sensor output was then directed to the control box and then to the MFLI. A similar experimental setup using an SWT was discussed in [16].

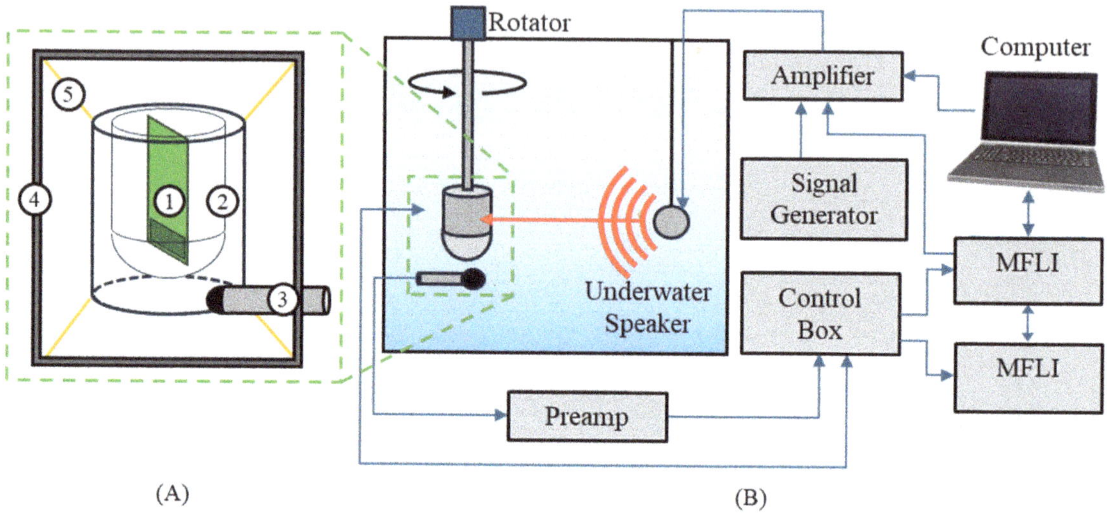

Figure 9. Underwater experimental setup. (**A**) Detailed diagram of the underwater sensor: (1) MEMS sensor, (2) air-tight sensor housing, (3) omnidirectional hydrophone, (4) mounting frame, and (5) elastic bands connecting housing to mounting frame. (**B**) Diagram of underwater experimental setup.

4. Experimental Results

4.1. Frequency Response

The mechanical sensitivities of the individual wings on the MEMS sensor were measured via laser vibrometry by measuring the deflection of the wing with respect to the applied acoustic pressure. Figure 10A shows the mechanical sensitivity of each wing individually. The complex sum of the wing sensitivities was calculated and plotted to predict an effective mechanical sensitivity of the entire sensor. The measured resonant frequencies and quality factors of the measured mechanical sensitivity are consistent with the FE models.

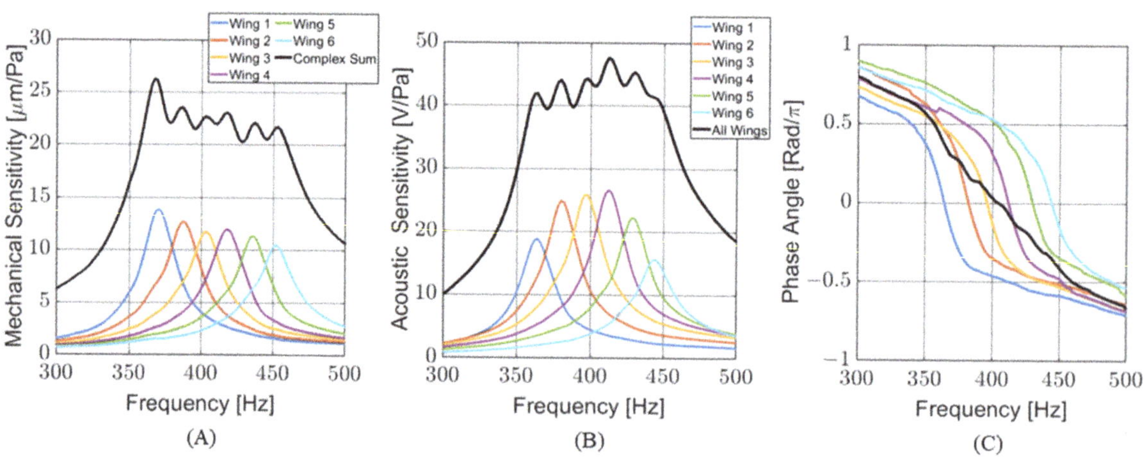

Figure 10. Measured sensor response: (**A**) mechanical sensitivity measured via laser vibrometry. Overall sensor response is calculated. (**B**) Acoustic sensitivity measured from output of capacitive sensing circuit. (**C**) Phase associated with acoustic sensitivity. Overall sensor response is measured for (**B**,**C**).

The acoustic sensitivity (output voltage per applied acoustic pressure) of the MEMS sensor was measured with the sensor cemented into the PCB with the capacitive readout circuit. The acoustic sensitivity is comparable to the mechanical sensitivity of the MEMS sensor. Figure 10B shows the sensitivity for each wing individually, as well as the entire sensor when all the wings, connected in parallel, were being read by the capacitive sensing circuit. The maximum sensitivity of the MEMS sensor was measured at 47.6 V/Pa (33.6 dB re 1 V/Pa). Figure 10C shows the phase response of the individual wings and their combination. Each wing behaves like its own harmonic oscillator. As predicted by the FE models, when the outputs of all the wings are combined, the phase response becomes more complex than those of single wings.

Table 2 shows the modeled and measured resonant frequencies and quality factors of individual wings for each sensor design, while Figure 11 presents these data graphically. The average percent difference in resonant frequencies between the FE model and measured electrical output is 0.43%. The quality factors agree within 0.59%. This shows that the FE model is an excellent predictor of the sensor's frequency response. When comparing the laser vibrometry and electrical sensor performance, the average resonance frequencies agree to within 0.36%, and the average quality factors agree to within 4%. This suggests that any electrical damping effects created when applying a voltage across the MEMS sensor are not significant.

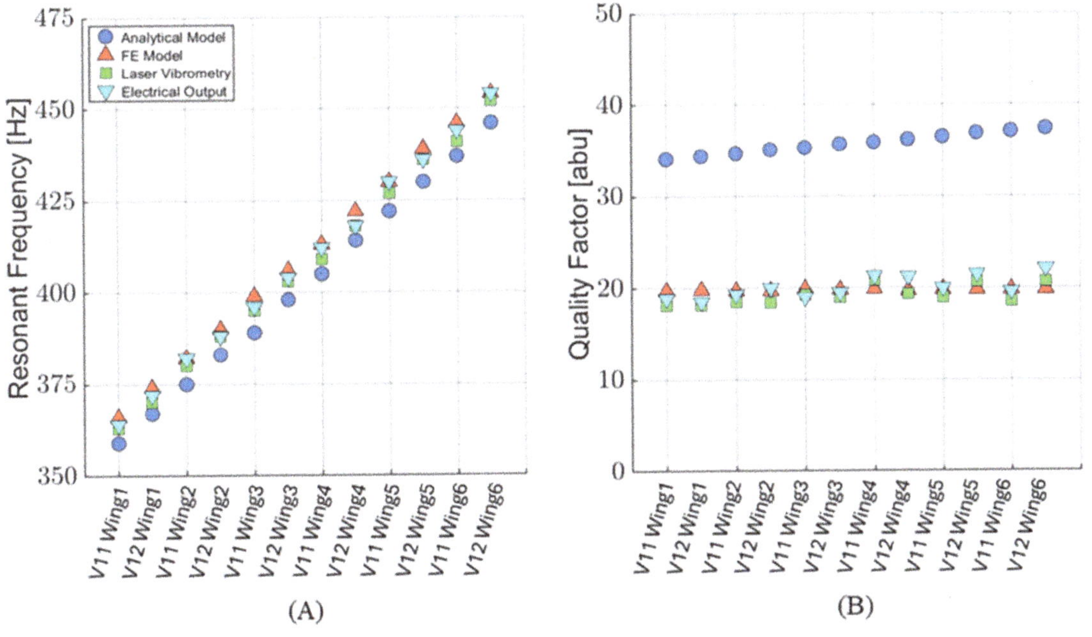

Figure 11. Comparison of analytical and FE models with measured results for sensor versions V11 and V12. (**A**) Modeled and measured resonant frequency. (**B**) Modeled and measured quality factor. The analytical model overestimates the quality factor of the sensor.

The acoustic sensitivity was measured underwater, with the sensor mounted in an air-filled, water-tight housing mounted in an SWT. Figure 12A shows the sensitivity response of the sensor with respect to frequency. Figure 12B shows the phase response of the sensor, measured at the TRANSDEC facility. The wavy nature of the phase response is due, in part, to acoustic reflections and interference patterns generated in the pool during the frequency sweep. As expected, the frequency response of the sensor in an air-filled underwater housing is comparable to its response in air.

Table 2. Resonant frequency and quality factor comparison of modeled and measured values.

Version V11 Resonant Frequency [Hz]/Quality Factor												
Data Source	**Wing 1**		**Wing 2**		**Wing 3**		**Wing 4**		**Wing 5**		**Wing 6**	
Analytical Model	359	34.1	375	34.7	389	35.3	405	35.9	422	36.5	437	37.1
Finite Element Model	366	19.7	382	19.7	399	19.9	413	19.9	430	19.8	446	19.8
Laser Vibrometry	363	18.1	380	18.5	395	19.3	409	20.8	427	19.0	441	18.6
Electrical Output	364	18.9	382	19.3	396	19.0	412	21.3	430	20.0	444	19.6
Version V12 Resonant Frequency [Hz]/Quality Factor												
Data Source	**Wing 1**		**Wing 2**		**Wing 3**		**Wing 4**		**Wing 5**		**Wing 6**	
Analytical Model	367	34.4	383	35.1	398	35.7	414	36.2	430	36.9	446	37.4
Finite Element Model	374	19.8	390	19.7	406	19.8	422	19.7	439	19.8	454	19.8
Laser Vibrometry	370	18.1	388	18.4	403	19.0	418	19.4	436	20.7	452	20.7
Electrical Output	372	18.5	388	20.0	404	19.5	418	21.2	436	21.5	454	22.1

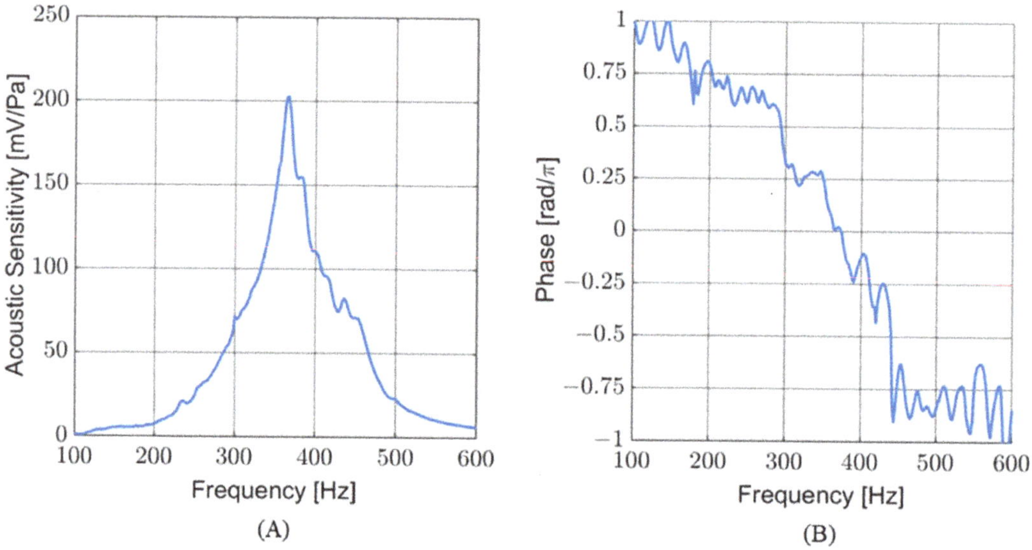

Figure 12. Underwater frequency response of the sensor. (**A**) Sensitivity of the sensor in SWT. (**B**) Phase response of the sensor at TRANSDEC.

4.2. Directionality

Ideally, the sensors produce a cosine-like directionality pattern. Figure 13A shows the directionality of the sensor operating in air with a 429 Hz acoustic stimulus. The directionality very closely matches the ideal cosine-like shape. This directionality was consistent for all frequencies within the target bandwidth of the sensor (300 to 500 Hz). However, this is not the case when the sensor is operating under water.

Figure 13B shows the directionality patterns for the sensor while stimulated at two different frequencies. The solid blue line shows the directionality at 367 Hz and the red line at 432 Hz. A dotted blue line shows the ideal cosine-like directionality for comparison. While only two patterns are shown, they represent the varying directionality patterns measured over the bandwidth of the sensor. All patterns are pseudo-cosine-like (opposing lobes pointing towards 0 degrees and 180 degrees) with significant deviations from the ideal pattern: lobe size, lobe angle (lobe does not point directly at 0 degrees), and failure to go to zero at +/− 90 degrees. This inconsistent directionality is likely due to both the underwater acoustic environment where the data were collected and the sensor housing. Further investigation is needed to positively identify the causes.

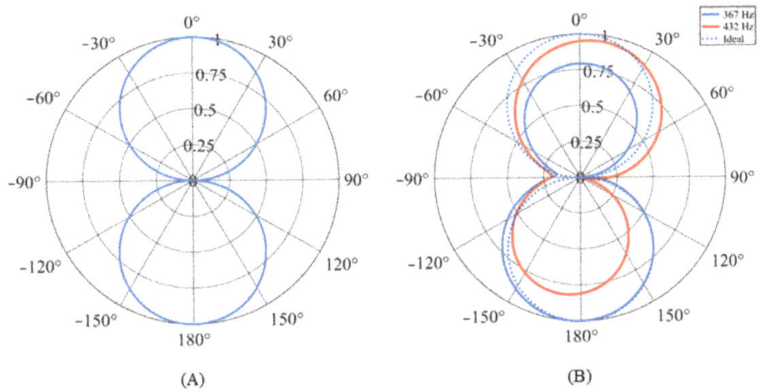

Figure 13. Directionality pattern of MEMS sensor: (**A**) in air and (**B**) under water.

4.3. Signal-to-Noise Ratio

Determining the SNR is critical to understanding the capabilities of this sensor. The noise spectral density (NSD) of the sensor (both individual wings and the entire sensor) was measured in an anechoic chamber, with all possible acoustic and electrical noise sources secured. The NSD data represent the mechanical and electronic noise of the MEMS sensor and associated readout circuitry. Figure 14A shows the NSD of each individual wing and the entire sensor (with all wings bonded to the readout circuit). The readout circuit has a bandwidth of 100 Hz to 3 kHz. The peaks in the NSD of individual wings correspond to their resonant frequencies. To isolate the mechanical and electrical portions of the NSD of a single wing, measurements were taken with the wing free to vibrate and again with the wing fixed (glued in place). The NSD curves with fixed wings closely match the curves with free wings, except for these resonant peaks. Figure 14B shows the NSD of a single wing, focusing on its resonance. The fixed wing's NSD curve closely matches that of the free wing except for the resonant peak.

The NSD data were used to calculate the noise level of the sensor over the bandwidth of the sensor circuitry (100 Hz to 3 kHz) and over the design bandwidth of the sensor (300 to 500 Hz). The acoustic sensitivity data were used to determine the signal level, at 1 Pa, with respect to frequency. Figure 15 shows the SNR for individual wings and the entire sensor, with the noise level based on the bandwidth of the sensor circuitry. The maximum SNR of the sensor over the circuit bandwidth is 88.6 dB, and over the design bandwidth, it is 97.4 dB. The sensor maintains a high SNR over the design bandwidth of the sensor.

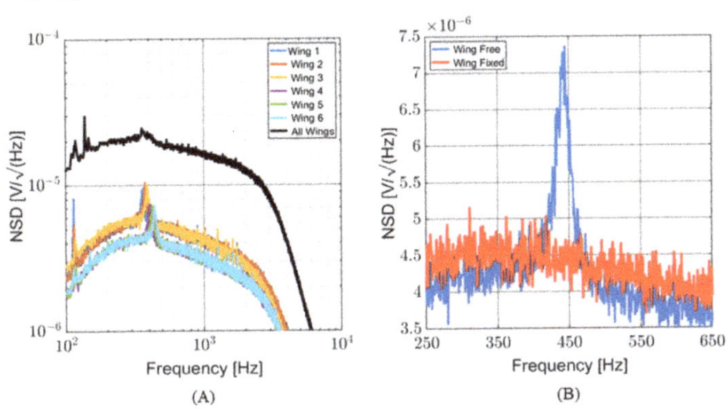

Figure 14. Noise spectral density of the MEMS sensor. (**A**) NSD of individual wings and the entire sensor. (**B**) Comparison of single wing while free to vibrate (blue line) and while fixed (red line). This demonstrates the mechanical contribution to the overall NSD.

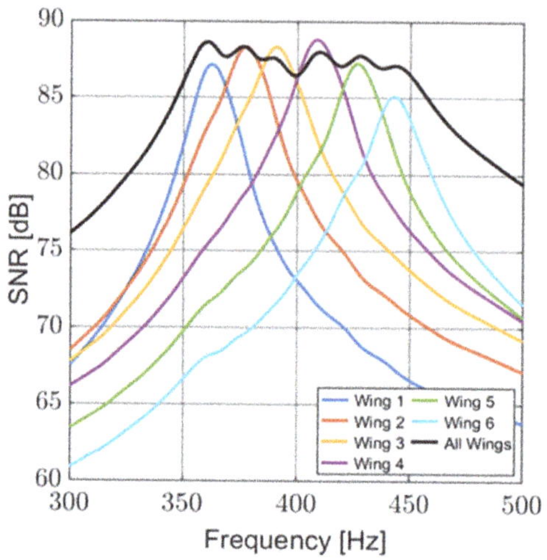

Figure 15. Signal-to-noise ratio of the sensor (individual wings and the entire sensor). The sensor demonstrates a high SNR over its design bandwidth.

5. Discussion and Conclusions

A multi-resonant, MEMS acoustic sensor was designed using analytical and FE modeling techniques. The sensor was characterized in air both mechanically (using laser vibrometry) and electrically to determine the directionality, frequency response, and SNR of the sensor. Additionally, the sensor was characterized underwater while contained in an air-filled housing. The sensor improves upon previous designs by broadening the effective bandwidth while maintaining a high SNR and cosine-like directionality.

5.1. Sensor Characterization

As seen in Figure 15, this sensor provides a very high SNR over a 200 Hz bandwidth. This is a significant improvement when compared to other resonant acoustic sensors. It was demonstrated in [21] for vibrating wing sensors operating at resonance that the SNR is proportional to the square root of the quality factor, tying a high SNR to narrow bandwidths. This multi-resonant design overcomes that limitation by maintaining a comparable SNR with nearly 4.5 times the bandwidth.

The sensor displays the same cosine-like directionality as other vibrating wing sensors. Deviations from the ideal directionality, which were observed for the underwater configuration of the sensor, were also seen in underwater version of previous designs [16].

5.2. Comparison with Similar Sensors

Table 3 compares the multi-resonant sensor's SNR and effective bandwidth (based on the full width and half max of the resonant peak) performance. It shows the improved performance of the multi-resonant sensor when compared to similar resonant MEMS acoustic sensors.

Table 3. Sensor performance comparison.

Sensor	Sensitivity	SNR [dB]	Bandwidth
Multi-Resonant	48 V/Pa	88.6 (97.4) [1]	300 Hz–500 Hz
Double-Wing Design [16]	59 V/Pa	88 (102) [1]	658 Hz–684 Hz
Dual-Band Design [21]	13 V/Pa	91 [1]	650 Hz–725 Hz [2]
Double-Wing [9]	3.45 mV/Pa	68.5	Not Discussed

[1] Noise based on sensor resonance bandwidth instead of bandpass of circuit. [2] Approximate values.

5.3. Improving Sensor Designs

This versatile multi-resonant sensor can be scaled to different frequency ranges. The pass band can be expanded, and the size of the device can be reduced. Several techniques, commonly used in MEMS devices, can be applied to change the stiffness of the bridges and torsional beams, as well as the mass of the paddles, allowing for adjusting the spectral response of the sensor as desired while preserving a small size. The detection and localization of quiet sources with specific acoustic signatures such as sniper fire, multi-rotor small UAVs (drones), single- or multi-tone communication, or sonar signals when used underwater, etc., can be achieved. A combination of such sensors can be used to make acoustic vector sensors to provide full 3D coverage (azimuth and elevation). These applications are of particular interest to Defense and law enforcement.

Another interesting aspect of this approach is that if the readout mechanism is changed from capacitive comb fingers to piezoelectric films, which can be achieved without adding complexity to the sensor, this sensor can easily become a mechanical energy harvester [26,27]. Moreover, by broadening the resonant response, as demonstrated in this manuscript, or tuning the response to desired bands, a very efficient and flexible harvester can be designed.

Author Contributions: Conceptualization, J.I. and F.A.; methodology, J.I. and F.A.; formal analysis, J.I.; investigation, J.I. and F.A.; data curation, J.I. and F.A.; writing—original draft preparation, J.I.; writing—review and editing, F.A.; supervision, F.A.; project administration, F.A.; funding acquisition, F.A. and J.I. All authors have read and agreed to the published version of the manuscript.

Funding: This research was funded by the Office of Naval Research, grant number: N0001420WX00976 and by a student fellowship from the Naval Information Warfare Center Pacific, grant number N6600123WX00492.

Institutional Review Board Statement: Not applicable.

Informed Consent Statement: Not applicable.

Data Availability Statement: The data that support the findings of this study are available from the corresponding author upon reasonable request.

Acknowledgments: The authors would like to thank Jeffrey Catterlin for his rapid prototyping support and Peter Crooker for his technical support and mentorship.

Conflicts of Interest: The authors declare no conflicts of interest.

References

1. Schoess, J.N.; Zook, J.D. Test results of a resonant integrated microbeam sensor (RIMS) for acoustic emission monitoring. *J. Intell. Mater. Syst. Struct.* **1998**, *9*, 947–951. [CrossRef]
2. Rahaman, A.; Kim, B. Microscale Devices for Biomimetic Sound Source Localization: A Review. *J. Microelectromech. Syst.* **2022**, *31*, 9–18. [CrossRef]
3. Zhou, J.; Miles, R.N. Directional Sound Detection by Sensing Acoustic Flow. *IEEE Sens. Lett.* **2018**, *2*, 1501204. [CrossRef]
4. Lee, T.; Nomura, T.; Su, X.; Iizuka, H. Fano-Like Acoustic Resonance for Subwavelength Directional Sensing: 0–360 Degree Measurement. *Adv. Sci.* **2020**, *7*, 1903101. [CrossRef]
5. Chen, T.; Jiao, J.; Yu, D. Strongly coupled phononic crystals resonator with high energy density for acoustic enhancement and directional sensing. *J. Sound Vib.* **2022**, *529*, 116911. [CrossRef]
6. Chen, T.; Wang, C.; Yu, D. Pressure amplification and directional acoustic sensing based on a gradient metamaterial coupled with space-coiling structure. *Mech. Syst. Signal Process.* **2022**, *181*, 109499. [CrossRef]
7. Rayburn, R. *Eargle's The Microphone Book: From Mono to Stereo to Surround—A Guide to Microphone Design and Application*, 3rd ed.; Routledge: New York, NY, USA, 2011; ISBN 978-0-240-82078-1.
8. Rahaman, A.; Kim, B. Fly-Inspired MEMS Directional Acoustic Sensor for Sound Source Direction. In Proceedings of the 2019 20th International Conference on Solid-State Sensors, Actuators and Microsystems & Eurosensors XXXIII (TRANSDUCERS & EUROSENSORS XXXIII), Berlin, Germany, 23–27 June 2019; pp. 905–908. [CrossRef]
9. Rahaman, A.; Kim, B. Sound source localization by Ormia ochracea inspired low–noise piezoelectric MEMS directional microphone. *Sci. Rep.* **2020**, *10*, 9545. [CrossRef] [PubMed]
10. Rahaman, A.; Kim, B. AI speaker: A scope of utilizing sub–wavelength directional sensing of bio–inspired MEMS directional microphone. In Proceedings of the 2020 IEEE SENSORS, Rotterdam, The Netherlands, 25–28 October 2020; pp. 1–4.

11. Espinoza, A.; Alves, F.; Rabelo, R.; Da Re, G.; Karunasiri, G. Fabrication of MEMS Directional Acoustic Sensors for Underwater Operation. *Sensors* **2020**, *20*, 1245. [CrossRef] [PubMed]
12. Rabelo, R.C.; Alves, F.D.; Karunasiri, G. Electronic phase shift measurement for the determination of acoustic wave DOA using single MEMS biomimetic sensor. *Sci. Rep.* **2020**, *10*, 12714. [CrossRef] [PubMed]
13. Li, Y.; Omori, T.; Watabe, K.; Toshiyoshi, H. Improved Piezoelectric MEMS Acoustic Emission Sensors. In Proceedings of the 2021 21st International Conference on Solid-State Sensors, Actuators and Microsystems (Transducers), Orlando, FL, USA, 20–24 June 2021; pp. 238–241.
14. Li, Y.; Omori, T.; Watabe, K.; Toshiyoshi, H. Bandwidth and Sensitivity Enhancement of Piezoelectric MEMS Acoustic Emission Sensor Using Multi-Cantilevers. In Proceedings of the 2022 IEEE 35th International Conference on Micro Electro Mechanical Systems Conference (MEMS), Tokyo, Japan, 9–13 January 2022; pp. 868–871. [CrossRef]
15. Rahaman, A.; Kim, B. An mm-sized biomimetic directional microphone array for sound source localization in three dimensions. *Microsyst. Nanoeng.* **2022**, *8*, 66. [CrossRef] [PubMed]
16. Ivancic, J.; Karunasiri, G.; Alves, F. Directional Resonant MEMS Acoustic Sensor and Associated Acoustic Vector Sensor. *Sensors* **2023**, *23*, 8217. [CrossRef] [PubMed]
17. Baumgartel, L.; Vafanejad, A.; Chen, S.-J.; Kim, E.S. Resonance-Enhanced Piezoelectric Microphone Array for Broadband or Prefiltered Acoustic Sensing. *J. Microelectromech. Syst.* **2013**, *22*, 107–114. [CrossRef]
18. Shkel, A.A.; Baumgartel, L.; Kim, E.S. A resonant piezoelectric microphone array for detection of acoustic signatures in noisy environments. In Proceedings of the 2015 28th IEEE International Conference on Micro Electro Mechanical Systems (MEMS), Estoril, Portugal, 18–22 January 2015; pp. 917–920. [CrossRef]
19. Liu, H.; Liu, S.; Shkel, A.A.; Kim, E.S. Active Noise Cancellation With MEMS Resonant Microphone Array. *J. Microelectromech. Syst.* **2020**, *29*, 839–845. [CrossRef] [PubMed]
20. Kang, S.; Hong, H.-K.; Rhee, C.-H.; Yoon, Y.; Kim, C.-H. Directional Sound Sensor With Consistent Directivity and Sensitivity in the Audible Range. *J. Microelectromech. Syst.* **2021**, *30*, 471–479. [CrossRef]
21. Alves, F.; Rabelo, R.; Karunasiri, G. Dual Band MEMS Directional Acoustic Sensor for Near Resonance Operation. *Sensors* **2022**, *22*, 5635. [CrossRef] [PubMed]
22. MEMSCAP | SOIMUMPs and MEMS Multi Project Wafer Service. Available online: http://www.memscap.com/products/mumps/soimumps/ (accessed on 24 January 2023).
23. Touse, M.; Sinibaldi, J.; Simsek, K.; Catterlin, J.; Harrison, S.; Karunasiri, G. Fabrication of a microelectromechanical directional sound sensor with electronic readout using comb fingers. *Appl. Phys. Lett.* **2010**, *96*, 173701. [CrossRef]
24. Garrett, S.L. *Understanding Acoustics: An Experimentalist's View of Sound and Vibration*; Springer Nature: Berlin/Heidelberg, Germany, 2020; ISBN 978-3-030-44787-8.
25. Sader, J.E. Frequency response of cantilever beams immersed in viscous fluids with applications to the atomic force microscope. *J. Appl. Phys.* **1998**, *84*, 64–76. [CrossRef]
26. Zhao, L.-C.; Zou, H.-X.; Wei, K.-X.; Zhou, S.-X.; Meng, G.; Zhang, W.-M. Mechanical Intelligent Energy Harvesting: From Methodology to Applications. *Adv. Energy Mater.* **2023**, *13*, 2300557. [CrossRef]
27. Zhao, L.-C.; Zhou, T.; Chang, S.-D.; Zou, H.-X.; Gao, Q.-H.; Wu, Z.-Y.; Yan, G.; Wei, K.-X.; Yeatman, E.M.; Meng, G.; et al. A disposable cup inspired smart floor for trajectory recognition and human-interactive sensing. *Appl. Energy* **2024**, *357*, 122524. [CrossRef]

MDPI AG
Grosspeteranlage 5
4052 Basel
Switzerland
Tel.: +41 61 683 77 34

Sensors Editorial Office
E-mail: sensors@mdpi.com
www.mdpi.com/journal/sensors

www.ingramcontent.com/pod-product-compliance
Lightning Source LLC
Chambersburg PA
CBHW040801250926
51850CB00007BA/1053
* 9 7 8 3 7 2 5 8 2 2 7 7 5 *